The Cable and Telecommunications Professionals' Reference

The Cable and Telecommunications Professionals' Reference

Transport Networks

Edited by Goff Hill

Focal Press
Taylor & Francis Group

NEW YORK AND LONDON

First published 2008

This edition published 2013
by Focal Press
70 Blanchard Road, Suite 402, Burlington, MA 01803

Simultaneously published in the UK
by Focal Press
2 Park Square, Milton Park, Abingdon, Oxon OX14 4RN

Focal Press is an imprint of the Taylor & Francis Group, an informa business m

Library of Congress Cataloging-in-Publication Data
The cable and telecommunications professionals' reference / edited by Goff Hill. — 3rd ed.
    p. cm.
Rev. ed. of: Telecommunications engineer's reference book / edited by Fraidoon Mazda.
    2nd ed. 1998.
Includes bibliographical references and index.
ISBN-13: 978-0-240-80748-5 (alk. paper)
1. Telecommunication: transport networks. I. Hill, Goff. II. Telecommunications
    engineer's reference book.
TK5101.M37 2007
621.382—dc22

                                                2007003442

British Library Cataloguing-in-Publication Data
A catalogue record for this book is available from the British Library.

ISBN 13: 978-0-240-80748-5 (pbk)

To my family

Contents

Preface

The past decade has seen massive change in the telecomunications industry. Industry structure, regulation, technology, and services have all seen or caused substantial development. New technologies and services are displacing long established ones, and the convergence of different networks, technologies, and services compound the change. The predecessor to this reference was called the *Telecommunication Engineer's Reference Book*. Recent trends toward convergence of voice, data, and video technologies and services have suggested that the title, scope, and content of the book should reflect these changes. For this reason the new reference is entitled *The Cable and Telecommunications Professionals' Reference*. Some of the content in the earlier edition has become dated, and a substantial amount of new material is now included so that it remains an authoritative reference book. However, some of the earlier content remains fundamentally relevant and is retained in this volume. Inevitably this means that some of the earlier material relating to the displaced technologies should, in some cases, be downsized and in other cases removed.

Even so, the huge amount of new material entering the scene means that the full scope would lead to a significant increase in the length of the book. The current edition is therefore divided into three volumes, and I have attempted to associate related material in each. Because of the many relationships between different topics, however, it is not possible to split them into three completely independent volumes.

The first volume dealt with the telecommunications industry and with voice and data and their convergence in both landline and mobile cellular networks. This seemed a logical grouping because of the historical influence of telephone networks on today's communications systems. This volume deals with core transport network technologies and the changes that are enabling them to deal with converged voice, data, and video traffic. The third volume will deal with the broadband access network technologies and the way in which they are used to support new services.

Volume 2 of *The Cable and Telecommunications Professionals' Reference* is arranged into three parts. The parts deal firstly with the digitization and coding of source information, secondly with the properties of the transmission paths across which the signals are carried, and thirdly with the networking techniques that are used to ensure that the signals reach their destination.

Because of the growth in demand for data services and the way in which voice, data, and video services have converged, core transport networks have needed to adapt to become more "data friendly." Transport networks and the services they carry are therefore adopting packet principles to a far greater extent than they did 10 years ago.

Part 1 examines the way voice, data, and video sources are digitized and coded in preparation for transmission over telecommunications networks. This provides a solid introduction to the way signals are carried over transport networks.

Part 2 deals with the physical media commonly used in core transport networks—namely optical fiber and radio. Optical fiber carries the vast majority of telecommunications traffic at regional, national, and international levels. The basic properties of optical fiber itself are discussed, and the principle characteristics and limiting factors of high-capacity, long-range optical systems are explained. Radio systems are used to a lesser extent in core networks. Very good discussions of these were included in the earlier *Telecommunication Engineer's Reference Book*, and selected chapters are reproduced here, with only minor additions, on the grounds that the principles of radio propagation have not changed.

Part 3 presents the main network technologies used today in core transport networks. While the trend is toward data-networking technologies, such as IP/MPLS and carrier-grade Ethernet, the concept of "next-generation networks" is not specific about which technologies should predominate, other than that a packet-networking technology should act as a convergence layer. Indeed, a key principle is that converged traffic can be carried over any underlying transport technology. For example, while SDH is generally considered to be a legacy technology now, its installed base is so huge that it will continue to be a key transport technology for many years to come. Moreover, its life is effectively extended with the application of Next Generation SDH. Other "legacy" technologies are also described, including ISDN, leased lines, Frame Relay, and ATM. In conclusion, IP/MPLS, which is now favored by many operators as a core transport network for data traffic, and the emerging carrier-grade Ethernet, are described.

Where appropriate, chapters that were included in the Second Edition of *Telecommunications Engineer's Reference Book* are reproduced in this volume with changes and updates as needed. I was unfortunately unable to contact four of the previous contributors for their updates. In Part 2 this applies to Chapters 7, 8, and 9 and in Part 3 to Chapter 10. An assessment was that these chapters contained a lot of information that is still very important, relevant, and clearly presented. In most cases the changes to the original text are minor. As for Chapter 10, the previous edition's chapter focused on SDH; for this book, I have rearranged the original text slightly and made some additions to highlight the similarities and differences between SONET and SDH.

I am greatly indebted to the authors who have contributed to this volume of the *Cable and Telecommunications Professionals' Reference*. They are all experts in their fields and working at the forefront of technology. I should like to express my thanks to them and to their respective organizations for all the support they have given.

Goff Hill
GTel Consultancy Ltd.

Contributors

Tahmina Ajmal, B.Eng., M.Eng., Ph.D.
University of Essex

Tahmina Ajmal received her Ph.D. from Essex University in 2007, having earlier earned her B.Eng. in 1990 and her M.Eng. in 1993 from Aligarh Muslim University in India. Before starting her doctoral studies at the University of Essex, she was a lecturer at Aligarh Muslim University and was upgraded to the post of reader there in August 2000. Since June 2007, Tahmina has worked as a part-time research officer at the University of Essex on the MUSE (Multi-Service Access Everywhere) project for the development of a low-cost multi-access network. Her research interests include fiber sensors, physical-layer cryptography, and different modulation formats for transmission.

J. H. Causebrook, B.Sc., Ph.D., C.Eng., MIEE
Independent Consultant

John Causebrook graduated in physics from the University of London in 1970 and obtained his doctorate for research into radio wave propagation in 1974. He worked for the BBC research department on broadcast coverage and frequency planning and later for the IBA, directing the work of several teams of propagation and broadcast-planning engineers. Dr. Causebrook participated in engineering the mobile phone network at Vodafone, the main product of which was a computer system for showing the coverage of networks that is used in the broadcasting-planning process. He also worked on the parameters for base stations and radio health of the mobile phone industry, was chairman of an international group that studied the subject, and lectured to University M.Sc. students in this area. Dr. Causebrook has published many technical papers and books and has presented several papers at international conferences. In 1978 he received the Oliver Lodge Premium Award from the IEE for his paper on propagation.

Ken Guild, M.Eng, Ph.D.
University of Essex

Ken Guild is currently a senior lecturer within the Department of Computing and Electronic Systems Engineering at the University of Essex, U.K. He received an M.Eng. degree in electronic and electrical engineering from the University of Surrey in 1993. Between 1993 and 1997, he was with Alcatel Submarine Networks, and worked on a variety of projects related to advanced optical transport networks and systems. In 1997, Ken joined the University of

Essex as a research assistant and concurrently undertook obtaining a Ph.D. degree in optical packet-switched networks. After cofounding a startup that developed optical cross-connects in 2000, he moved to Marconi (now Ericsson) in Germany and was responsible for the modeling and design of next-generation networks. His current research interests include dynamic multilayer optical networks, bandwidth provisioning mechanisms for fixed/mobile services, and future optical packet-switched networks. He has published more than 50 papers and holds 12 patents.

Anagnostis Hadjifotiou, B.Sc., M.Sc., Ph.D.
Visiting Professor

Anagnostis Hadjifotiou graduated from Southampton University in 1969 with a B.Sc. degree in electronic science with final-year options on communications, semiconductor physics, electronic devices, microwaves, and quantum electronics. He received his M.Sc. in 1970 from Manchester University in control systems, specializing in optimal control, stochastic processes, stochastic control, and space control systems. In 1974 he was awarded a Ph.D. from Southampton University for work on digital-signal processing for target extraction in radars. Anagnostis joined the Nortel Research Laboratories in Harlow, England (then Standard Telecommunications Laboratories) in 1984 and worked on nonlinear electronic circuits, computer-aided circuit design, optical systems (submarine, terrestrial, and free space), optical technologies, device and system simulation, and nonlinear optics. From 2000 until his retirement from Nortel in 2004, he was responsible for research on optical communications and optical technologies. He has also been a visiting professor at Essex University and University College London and a lecturer at a number of U.K. universities. Anagnostis is a fellow of the Royal Academy of Engineering, a fellow of the IET, a chartered engineer, and a member of the IEEE and the OSA. He has 104 publications to his credit.

Goff Hill, B.Sc., MIET
GTel Consultancy Ltd.

Goff Hill is the managing director of GTel Consultancy, which provides advice on broadband and core transport networks and is active in supporting Europe's Framework Program research. Globally recognized as a pioneer of "optical network" technology, Goff has more than 40 years' experience in telecommunications, including 17 years in optical telecommunications as a technical group leader and project manager with BT and 2 years as a chief network architect with ilotron and later with Altamar Networks. He has been a special editor for both *IEEE Communications Magazine* and *IEEE Journal of Lightwave Technology* and has published more than 70 technical papers. He also delivers lectures for university and professional short courses. Goff obtained an honors degree in electrical engineering from the University of Newcastle upon Tyne in 1969 and is a member of the IET.

David K. Hunter, B.Eng., Ph.D., MIET, SMIEEE
University of Essex

David Hunter, now a reader in the Department of Computing and Electronic Systems at the University of Essex, received a first-class honors B.Eng. in electronics and microprocessor engineering from the University of Strathclyde in 1987, and a Ph.D. from the same university in 1991 for research on optical TDM switch architectures. He then researched optical networking and optical packet switching at Strathclyde. In August 2002, he began teaching at the University of Essex, concentrating on TCP/IP, network performance modeling, and computer networks. David has authored or co-authored more than 110 publications. From 1999 until 2003 he was associate editor of *IEEE Transactions on Communications* and was associate editor of *IEEE/OSA Journal of Lightwave Technology* from 2001 until 2006. He is a chartered engineer, a member of the IET, a senior member of the IEEE, and a professional member of the ACM.

Malcolm D. Macleod, B.A., Ph.D.
QinetiQ Ltd

Malcolm Macleod graduated from the University of Cambridge with a B.A. in electronic engineering and a Ph.D. for his work on the discrete optimization of DSP systems. From 1978 to 1988, he worked for Cambridge Consultants Ltd. on a wide range of research and development projects. In 1988 he joined the engineering department of Cambridge University as a lecturer in signal processing and communications and was subsequently the department's director of research. Malcolm joined QinetiQ Ltd in Malvern as a senior research scientist and technical manager in November 2002. In 2007 he was appointed a visiting professor at the University of Strathclyde. A fellow of the IEE, Malcolm is a chartered engineer and an associate editor of *IEEE Signal Processing Letters*. He has published nearly 100 papers on nonlinear and adaptive filtering, efficient implementation of DSP systems, detection, estimation, beamforming, and wireless communications and image processing and their applications.

Alan McGuire, M.Sc.
British Telecommunications Plc

Alan McGuire is a principal engineer at BT, leading a multidisciplinary team working on next-generation transport networks. He is also active in the standards arena, where he has made numerous contributions and has acted as editor on a variety of ITU-T Recommendations concerning network architecture, control and management, optical networking, and Ethernet. Alan graduated from the University of St. Andrews in 1987 with a first in physics and earned his M.Sc. in medical physics one year later from the University of Aberdeen. He is a chartered engineer and chartered physicist, as well as a member of the Institute of Physics, a member of the IET, and a senior member of the IEEE.

Rouzbeh Razavi, M.Sc.
University of Essex

Rouzbeh Razavi received his master's degree with distinction in telecommunications and information systems from the University of Essex in 2005. He is currently pursuing his research toward a Ph.D. at that university and since June 2006 has been a research officer on the IST European Project, MUSE. Rouzbeh has been a reviewer for various journals and has published more than 30 papers in scholarly journals and conference proceedings. He is the recipient of both a best-paper award and an outstanding-paper award from two international conferences.

Martin Reed, B.Eng., Ph.D., MIET, MIEE
University of Essex

Martin Reed is a lecturer at the University of Essex. He is also an active researcher in the fields of transmitting media over packet networks and traffic engineering in IP networks and has published more than 25 papers in the fields of network control and audio signal processing. Martin was an early experimenter in spatialized (multichannel) voice for multi-party conferencing transmitted over ATM and IP networks, having worked on a 1998 project funded in the United Kingdom by BT. This earlier work has inspired projects investigating a number of issues with transmission of the next generation of voice applications over packet networks ranging from network-level traffic control to signal-processing requirements such as echo cancelation.

Emilio Hugues Salas, M.Sc.
University of Essex

Emilio Salas received his master's degree in telecommunications and information systems from the University of Essex in 2002. He is currently pursuing his Ph.D. degree at that university. Since June 2006, Emilio has been a research officer on the IST European Project, MUSE, and has worked for ALCATEL Mexico as a product and network design engineer in the transmission networks area. Emilio has been a reviewer for various journals and has published more than 15 papers in scholarly journals and conference proceedings.

Matthew R. Thomas, B.Eng.
University of Essex

Matthew Thomas studied electronic engineering at the University of Hertfordshire, graduating in 1994 with a B.Eng. He began configuring routers for the IBM Global Network in 1995 and gained CCIE status in 1996. From 1996 Matthew has specialized in running CCIE boot camps and OSPF training courses. As a contractor, he has worked in the Far East and South America and has spent extended periods in Switzerland, Brussels, and the United Kingdom. Today Matthew is working toward his Ph.D. in the Department of Computing and Electronic Systems at the University of Essex. Along with carrying out research in Internet routing protocols and carrier-class Ethernet, he teaches mathematics courses in the department.

Eur. Ing. Paul Urquhart, B.Sc., M.Sc., DIC, Ph.D., C.Eng., MIET
Universidad Pública de Navarra

Paul Urquhart is a senior fellow at the Universidad Pública de Navarra, Pamplona, Spain, and a visiting professor at the University of Essex. He was educated at the University of Edinburgh, Imperial College (London), and the University of Glasgow, where he obtained his bachelor's, master's, and doctoral degrees, respectively. For 15 years Paul worked at BT Laboratories in research and development on a wide range of optical component technologies and optical networks. He was also a professor in the optics department at ENST Bretagne, Brest, France, for three years and has taught master's-level and doctoral students in Germany, Poland, Mexico, France, Spain, and the U.K. In this capacity, he makes regular visits to the Hochschule Niederrhein, Krefeld, Germany. A member of the IET, Paul is a chartered engineer as well as an FEANI-registered European engineer (Eur. Ing.).

Paul A. Veitch, M.Eng., Ph.D.
British Telcom

Paul Veitch received his M.Eng. and Ph.D. degrees in electrical and electronic engineering from the University of Strathclyde in 1993 and 1996, respectively. He joined BT in September 1996, working on IP, ATM, SDH, and 3G network architectural design. In 2000, he joined UUNET (now Verizon Business) and led a number of projects on IP backbone network design. Paul returned to BT in 2003 and is currently the infrastructure solution design authority for BT's UK IPVPN platform, based at Adastral Park, Suffolk.

Stuart D. Walker, B.Sc., M.Sc., Ph.D.
University of Essex

Stuart Walker received a B.Sc. (Hons.) in physics from Manchester University in 1973, following that with an M.Sc. in telecommunications systems in 1975 and a Ph.D. in electronics from Essex University in 1981. In 1982 he joined BT Research Laboratories in Martlesham Heath, U.K., where he helped pioneer unrepeatered submarine optical system technologies. Stuart became a senior lecturer at Essex University in 1988 and then, in 2003 and 2004, reader and full professor, respectively. His group is extensively involved in U.K. government and European Union projects. Stuart has published more than 200 journal and refereed conference papers and has 10 patents granted.

John Charles Woods, Ph.D., MIEE, MIEEE
University of Essex

Dr John Woods is a senior lecturer in the Department of Computing and Electronic Systems at the University of Essex. His extensive research interests include telecommunications, renewable energy, and domestic appliance control, although his principal area of interest is video and image coding, where he is widely published. Having gained a Ph.D. in model-based image coding in 2000, John is one of the leaders in the field of object-based image coding and segmentation.

Part 1

For most of the past century, the main purpose of telecommunications networks has been to enable voice communications between people at a distance. To a lesser extent it has included communications between people and voice recording or playback devices, such as automatic voice response systems. In more recent decades, there has also been a steady growth in the use of telecommunications networks to provide data connections for applications such as remote sensing, transfering data between data centers, or carrying TV signals in digital format. However, more recently, the emphasis has shifted very much toward the provision of connections between information systems or between people and information systems capable of carrying voice, video, or data signals.

Voice and video signals are analogue at their source and in early systems were transported over telecommunications networks in analogue form. With the relentless development of digital technology, however, analogue systems have progressively given way to digital transmission, with signals remaining in analogue format for only a short distance before being converted to digital. Data signals, on the other hand, often start out in digital format—for example, from a file or data storage device.

Digital signals generally need more bandwidth for their transmission than analogue ones, but they offer the advantage (with careful systems design) of being highly resistant to noise and other impairments and of being highly extendable. In recent years considerable emphasis has been placed on ways to reduce the bitrate (and therefore bandwidth) needed to carry analogue signals that have been digitized without significant loss of quality.

Part 1 of Volume 2 of *The Cable and Telecommunications Professionals' Reference* looks at the process of digitizing analogue signals and at the properties of the resulting voice, data, and video signals. Chapter 1 describes the properties of speech signals and the principles involved in converting voice signals into digital format. Pulse code modulation, which has been the basic format for digital voice signals in the Public Switched Telephone Network since the first digital systems appeared in the 1960s, is explained along with basic digital multiplexing and transmission performance. Chapter 2 describes the basic principles of data transmission, including modulation schemes and synchronization, and summarizes recent trends in data transmission.

Chapter 3 discusses the principles of digital video transmission, including an overview of compression methods and a description of the ways in which spatial, temporal, and statistical redundancy can be exploited to minimize the information content to be transmitted. The particular requirements of video signals for transport over the Internet are a topic of particular importance, and this is also described.

When digital information is transmitted across a telecommunications channel, a very small proportion of the digits transmitted might be incorrectly detected at the receiver. While Chapter 3 is concerned with removing redundancy from the source signal ("source coding"), Chapter 4 is concerned with adding a small amount of redundancy to the transmitted signal so that errors arising during transmission can be detected and corrected at the receiver. This is known as "error control coding."

1 Voice Transmission

Stuart D. Walker
Rouzbeh Razavi
University of Essex

Speech-Coding Principles

Speech coding is the process of obtaining a compact representation of voice signals for efficient transmission over band-limited channels or for storage. Today, speech coders are essential components in telecommunications and in the multimedia infrastructure. This section describes the speech signal properties with regard to the human auditory system. It summarizes current capabilities in speech coding and describes how the field has evolved to reach these capabilities.

Speech Signal Properties

There are three basic types of speech sound—voiced, unvoiced, and plosive—which produce different spectra because of their sound formation. Voiced speech requires air pressure from the lungs to force open the normally closed vocal cords, which then vibrate. The pitch or vibration frequency varies from about 50 to 400 Hz (as determined by the speaker's age and gender); the vocal tract then resonates at odd harmonics. These resonant peaks are known as formants. All vowels (and some of the consonants) are considered voiced sound.

On the other hand, unvoiced sounds, or fricatives (derived from the word *friction*; e.g., *s*, *f*, *sh*), are formed by forcing air through an aperture without causing the vocal cords to vibrate. This lack of resonant characteristics produces a largely aperiodic structure (as might be expected). Figure 1.1(a) shows the time domain presentation of a sample voiced sound, while Figure 1.1(b) shows the autocorrelation. In general, autocorrelation is considered as a robust indicator of periodicity. Similarly, Figure 1.1(c) and Figure 1.1(d) show the time domain and autocorrelation of a sample unvoiced sound.

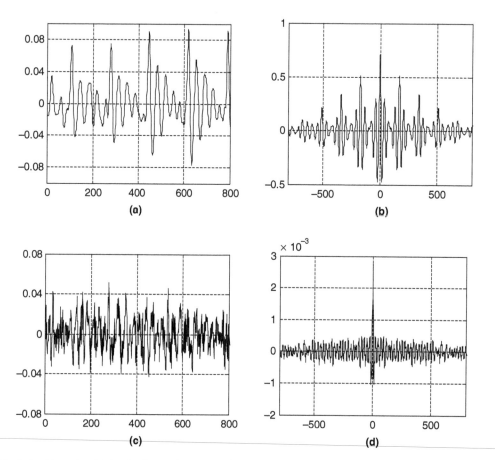

Fig. 1.1   Example of (a) voiced sound segment, (b) autocorrelation of voiced sound segment, (c) an unvoiced sound segment, and (d) autocorrelation of unvoiced sound segment.

The third category of sounds is plosive sound. Plosive sounds start with a closure of the mouth and/or nose and therefore contain a high degree of low-frequency energy. The terms "plosive" and "stop" are usually used interchangeably, but they are not exact synonyms. Notice that some sounds still cannot be categorized as in any of the three classes above but are a mixture.

Speech occupies a total bandwidth of approximately 7 kHz, but the spectrum has a typical average value of 3 kHz (combined voiced and unvoiced sounds and dependent on age and gender). As far as hearing is concerned, the auditory canal acts as a resonant cavity at this 3-kHz average frequency. Figure 1.2 shows the spectrum of a sample speech segment.

As can be seen from the spectra in the figure, there is little energy above 3.5 kHz. Overall, speech-encoding algorithms can be made less complex than general encoding by concentrating (through filters) on this region. Additionally, because traditional voice telecommunications allow a passband of less than 4 kHz, fricative-dependent high frequencies are removed. This can cause occasional issues with fricative names, such as "Frank."

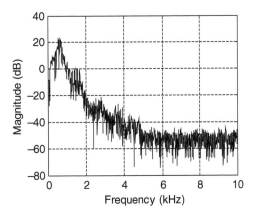

Fig. 1.2 Spectrum of a sample speech segment.

The subject area of speech production has a vast literature, and this introduction has served to consider only the basic issues.

Classification of Speech Coders

In most telecommunications applications, the original speech signal is compressed for improved storage density and more effective use of the transmission system. On this basis, the encoding process can be designated *lossy* or *lossless*. In the latter case, the original waveform is reproduced exactly in end-to-end network propagation. In fact, lossless coding is most suitable for source material, which generally fits a certain statistical behavior. In the case of a speech signal, this means a source in which low-frequency content dominates. Examples of lossless coding include Variable Length Coding (VLC) schemes, such as Huffman and arithmetic coding. Although lossy systems can imply sacrifices in perceived quality, they are widely accepted. Quantization, for example, in which the speech samples are allocated prescribed values, is the simplest and most fundamental lossy technique used by all digital speech-coding schemes.

Speech coding techniques can be classified in two main categories: speech *waveform-based* coding and Analysis-by-Synthesis (AbS) methods. The waveform-based method attempts to reproduce the time domain speech waveform as accurately as possible, while in the AbS method, the goal is to minimize the perceptual distortion by using linear prediction techniques. In other words, the coder parameters are evaluated by locally decoding the signal and comparing it to the original speech signal.

Pulse Code Modulation (PCM), considered to be the most common voice-coding technique, merely involves sampling and quantizing the input waveform. PCM is described in more detail in the next section. As shown in Figure 1.2, assuming standard, voice-only applications, the spectral extent of the input signal is taken as 300 Hz to 3.4 kHz (as evidenced by the formant/fricative sound generation mechanisms). With strict band limiting at 4 kHz (allowing 600 Hz for filter roll-off), standard sampling theory requires 8 ksamples/s. In contrast with linear PCM, where the sample amplitudes are presented on a linear scale, logarithmic PCM (log-PCM) uses a nonlinear quantizer to reproduce low-amplitude signals. There are two major variants of log-PCM: A-law, which is used in North America and Japan; and A-law, which is used mostly in Europe. These companding laws are explained with more details in the next section.

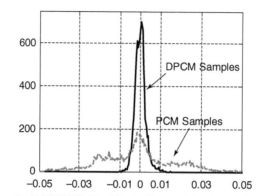

Fig. 1.3 Comparison of the histogram of the PCM and DPCM samples.

Adjacent speech samples are highly correlated. This means that the compression efficiency can be improved by removing this short-term redundancy. Differential PCM (DPCM) predicts the current sample from previous samples and, instead of the original speech sample, the prediction error is quantized and coded. With an accurate prediction, for most of the time, the prediction error is concentrated near zeros and can be coded with fewer bits compared to the original signal. Figure 1.3 compares the histogram of the PCM samples and DPCM samples for a segment of a speech signal where prediction is achieved by using a third-order Linear Predictive Filter (LPF). The accuracy of the predicted signal depends on the order of the predictor in use, which implies a trade-off between the compression efficiency and the coder complexity.

The short-term correlation in a speech signal is time varying. Therefore, to reduce the prediction error, the predictor should be adapted from frame to frame, where a frame is a group of speech samples (typically 20 ms long). DPCM systems with adaptive prediction and quantization are referred to as Adaptive Differential PCM (ADPCM). Basically, there are two types of adaptation: backward and forward. In backward adaptation, the predictor and quantizer are adapted based on the previous data frame. In this case it is not necessary to send the predictor and quantizer information along with the coded signal.

By contrast, in forward adaptation, before coding a data frame, the optimal predictor is found and the resulting predictor and quantizer information have to be transmitted with the coded signal as side information; however, compared to the backward scheme the prediction error is less. ADPCM is considered to be the most efficient variation of PCM and is used frequently at 32 kbit/s for landline telephony (G.726 standard). It is also widely employed for voice over IP and voice over Frame Relay as well as to store sound along with text and images on a CD-ROM.

Analysis-by-synthesis schemes are entirely different from waveform-based techniques. The most common AbS technique is referred to as Code Excited Linear Predictive (CELP). In CELP speech coders, a segment of speech (say 5 ms) is synthesized, using a linear prediction model for all possible excitations, in what is called a *codebook*. The excitation that results in minimumly perceptual coding error is selected. CELP coders operate at datarates between 4.8 kbit/s and 16 kbit/s. The U.S. Department of Defense standardized a version of CELP that operates at 4.8 kbit/s (FS 1016). Unfortunately, this method introduces an end-to-end delay up to 100 ms (due to block processing and searching the codebook) that is not acceptable for voice conversational applications. In 1992, AT&T Bell Laboratories standardized a

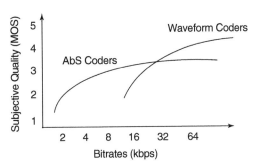

Fig. 1.4 Quality comparison of speech-coding schemes. *Source:* Reconstructed from Kondoz (2005); used with permission.

version of CELP (G.728 Recommendation), which uses a backward prediction and satisfies the ITU G.721 requirements. G.728-compatible coders are referred to as low-delay CELP coders, with the approximate end-to-end delay of 2.5 ms. AbS techniques can generally achieve better compression ratios than waveform coders; however, they are known for providing a more artificial speech quality.

In multimedia applications, the Mean Opinion Score (MOS) is the most common subjective quality measure, which provides a numerical indication of the perceived quality of received media after compression or transmission. The MOS is expressed as a single number in the range 1 to 5, where 1 is lowest perceived quality and 5 refers to highest perceived quality. Examples of other subjective quality measures are Diagnostic Rhyme Test (DRT) and Paired Comparison Test (PCT). Mean Squared Error (MSE) and signal-to-noise ratio (SNR), on the other side, are examples of widely used objective measures. Figure 1.4 compares the quality of the waveform-based and AbS coders in terms of MOS.

Standards for Voice Communications

This section describes the relevant details of current standardized voice codecs in the context of the network for which they were designed. Most speech coding systems in use today are based on telephone bandwidth narrowband speech, nominally limited to about 300 to 400 Hz and sampled at a rate of 8 kHz. Public Switched Telephone Network (PSTN) is the most familiar network for conversational, circuit-switched, Time Division Multiplexed (TDM) voice communications. The primary voice codec used for PSTN is a 64-kbit/s, log-PCM codec standardized by ITU-T as G.711 in 1988. Because of the very low bit error rate performance of PSTN circuits, no error concealment scheme is provided by G.711 when it is used for PSTN. Examples of other voice codecs standardized over the years by ITU-T for PSTN include G.726 and G.727, both of which were standardized in 1990 and use ADPCM technique. G.726 covers the transmission of voice at rates of 16, 24, 32, and 40 kbit/s, but the most commonly used mode is 32 kbit/s on international trunks in phone networks.

Digital cellular networks technologies have experienced enormous growth over the last decade. Digital cellular networks provide wireless voice connectivity by employing high-quality voice codecs as well as error detection and concealment techniques. The future performance can be improved by unequal protection of the bits according to their importance. In addition, because errors mainly appear as bursts in wireless channels, interleaving can be used to spread the errors, which can effectively simplify the concealment process.

The adaptive multirate wideband (AMR-WB) speech coder, standardized by the 3GPP, is one of the major codecs used for digital cellular networks. This coder operates at rates of

6.6, 8.85, 12.65, 14.25, 15.85, 18.25, 19.85, 23.05, and 23.85 kbit/s and is based on an algebraic CELP (ACELP) analysis-by-synthesis codec. The coder is also standardized by ITU-T as G.722-2. Mention should also be made of cdma2000 Variable-Rate Multimode Wideband (VRM-WB) speech-coding standard, which was standardized by the 3GPP2. The codec has five different modes and can be used to provide different Quality-of-Service (QoS) levels.

Voice over Internet Protocol (VoIP) is a revolutionary technology that enables the use of packet-switched networks to deliver voice traffic. Examples of considerations in designing good VoIP systems include voice quality, latency, jitter, packet loss performance, and packetization. Today's VoIP product offerings typically include G.711, G.729, and G.722, in addition to G.723.1, where the latter is often favored for videophone applications. G.711 coders are mainly known for their low complexity and delay while resulting in toll-quality voice. Appendix I in G.711 defines a Packet Loss Concealment (PLC) algorithm to help hide transmission losses in a packetized network. G.729, on the other hand, provides a toll-quality voice with acceptable delay level but at the lower rate. G.729 operates at 8 kbit/s, but there are extensions that provide 6.4 kbit/s and 11.8 kbit/s rates. G.722 is an ITU standard codec that provides 7-kHz wideband audio at datarates from 48 to 64 kbit/s. This is especially useful in fixed-network VoIP applications, where the required bandwidth is typically not prohibitive. G.722 provides a significant improvement in speech quality compared to other narrowband codecs such as G.711. Finally, G.723.1 can operate at 6.3 kbit/s (using 24-byte frames) or at 5.3 kbit/s by using 20-byte frames.

Human Auditory Perception and Subband Coding

In another method of speech coding, the speech signal is divided into separate frequency bands called subbands. Each subband can be coded by using either a waveform-based or an AbS technique (or a combination of both). Because not all the speech signal is relevant to the human ear, Subband Coding (SBC) enables the human auditory perception model to be considered as well. As an example, the Absolute Threshold of Hearing (ATH), which is the minimum sound level that an average ear with normal hearing can perceive, varies from one frequency to another. The ATH can be approximated using the following equation (Painter and Spanias, 1997):

$$T(f) = 3.64\left(\frac{f}{1000}\right)^{-0.8} - 6.5e^{-0.6\left(\frac{f}{1000}\right)^2} + 10^{-3}\left(\frac{f}{1000}\right)^4 \qquad (1.1)$$

where $T(f)$ is the minimum sound pressure level and f represents the frequency (see Figure 1.5).

Notice that the value of ATH depends on the listener's age and gender as well. In addition, human hearing performs space selection of different frequency bands called critical bands. In other words, the human inner ear behaves as a bank of bandpass filters with nonuniform bandwidth. Table 1.1 shows the critical bands of the human inner ear.

This means the human ear cannot distinguish between two sounds of the same frequency band. This property is referred to as the *frequency masking* effect. Now, by using a subband coding technique, the coder can dynamically allocate the bits to different subbands. In this case the coder analyzes the relevance of different bands and allocates the bits accordingly (i.e., more important subbands are quantized with more bits). One of the disadvantages of subband coding is the overlap between the adjacent subbands due to the imperfection of the bandpass filters. However, this can be compensated by using Quadrature Mirror Filter (QMF) banks.

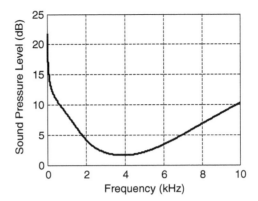

Fig. 1.5   Absolute threshold of hearing.

Table 1.1   Critical Band of the Human Ear

Critical Band	Center Frequency (Hz)	Bandwidth (Hz)	Critical Band	Center Frequency (Hz)	Bandwidth (Hz)
1	50	0–100	13	1850	1720–2000
2	150	100–200	14	2150	2000–2320
3	250	200–300	15	2500	2320–2700
4	350	300–400	16	2900	2700–3150
5	450	400–510	17	3400	3150–3700
6	570	510–630	18	4000	3700–4400
7	700	630–770	19	4800	4400–5300
8	840	770–920	20	5800	5300–6400
9	1000	920–1080	21	7000	6400–7700
10	1175	1080–1270	22	8500	7700–9500
11	1370	1270–1480	23	10500	9500–12000
12	1600	1480–1720	24	13500	12000–15500

Source: Adapted from Painter and Spanias (1997).

Because PCM is one of the most common speech coding techniques with many applications and variations, the next section is devoted to describing the PCM process in more detail, although most of the fundamental concepts such as sampling and quantization are common between all speech coding methods.

Pulse Code Modulation

PCM is mainly used for converting analogue speech signals into discrete digital bits. Sampling, quantizing, and coding are considered to be the foundation of any PCM system.

Sampling

Sampling is the first step in any pulse modulation system. In fact, by sampling, a signal is represented by a set of discrete samples. If the frequency of the sampling is high enough, the original signal can be recovered from the samples. This concept will be examined in more detail later.

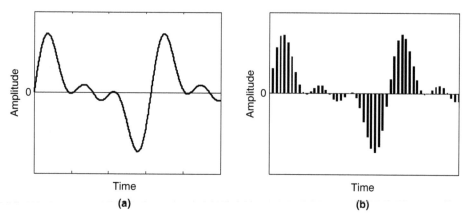

Fig. 1.6 Example of sampling process: (a) continuous analogue signal; (b) discrete analogue PAM signal.

Pulse Amplitude Modulation (PAM) is a direct result of sampling a signal. It is a pulse modulation technique in which the amplitude of the sample pulses is in direct proportion to the message signal amplitude. Therefore, the result of the sampling process sometimes is called a PAM-sampled signal. Figure 1.6 shows the sampling process.

From a mathematical point of view, the sampling process can be considered to be the multiplication of the message signal by a train of Dirac impulses. A Dirac impulse is mathematically defined as a very narrow, very high pulse, represented by δ:

$$\delta(t) = \begin{cases} 1/\omega & -\omega < t < \omega \text{ and } \omega \to 0 \\ 0 & \text{otherwise} \end{cases} \tag{1.2}$$

If the input speech signal, $x(t)$, is sampled by a sampling function, $p(t)$, which is a train of pulses with period of T, the sampling result, $x_p(t)$, would be

$$x_p(t) = x(t) \cdot p(t) \quad \Rightarrow \quad x_p(t) = x(t) \sum_{n=-\infty}^{+\infty} \delta(t - nT) \tag{1.3}$$

On the other hand, because $\delta(t - t_0)$ is 0 for all values of t except t_0, we have

$$x(t) \cdot \delta(t - t_0) = x(t_0) \cdot \delta(t - t_0) \tag{1.4}$$

The preceding equation shows the sampling property of a Dirac impulse. Using this, equation 1.3 can be written as

$$x_p(t) = \sum_{n=-\infty}^{+\infty} x(nT) \cdot \delta(t - nT) \tag{1.5}$$

It is known that multiplication of two waveforms in the time domain is equivalent to the convolution of their frequency domain transforms:

$$x_p(t) = x(t) \cdot p(t) \Leftrightarrow X(j\omega) = \frac{1}{2\pi}[X(j\omega) * P(j\omega)] \tag{1.6}$$

Using the Fourier transform tables we have:

$$P(j\omega) = \frac{2\pi}{T} \sum_{n=-\infty}^{+\infty} \delta(\omega - \omega_s) \tag{1.7}$$

where ω_s is the sampling frequency ($\omega_s = \frac{2\pi}{T}$). It is also known that convolving any waveform with a shifted Dirac impulse will shift the waveform. This means that

$$X(j\omega) * \delta(\omega - \omega_0) = X(j(\omega - \omega_0)) \tag{1.8}$$

Now equation 1.6 can be written as:

$$X(j\omega) = \frac{1}{T} \sum_{k=-\infty}^{+\infty} X(j(\omega - k\omega_s)) \tag{1.9}$$

This equation is important because it clearly shows the effect of sampling in the frequency domain. According to this, the sampling causes a periodic repetition of the message spectrum in the frequency domain. The period of this repetition is the same as the sampling period. Figure 1.7 shows the effect of sampling on the power spectrum of a message signal sampled by a train of Dirac impulses with sampling frequency ω_s.

Nyquist Sampling Theorem and Aliasing Distortion

It can be understood from both equation 1.9 and Figure 1.7 that if there is no overlap between the main message signal spectrum and its replicas, the original message can be retrieved using a lowpass filter to eliminate the repeated replicas. However, if the sampling frequency used to sample a message signal with a maximum bandwidth of Δf is lower than $2 \times \Delta f$, the overlap will take place between the replicas, so the message signal cannot be

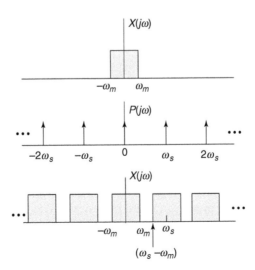

Fig. 1.7 Effect of sampling in the frequency domain.

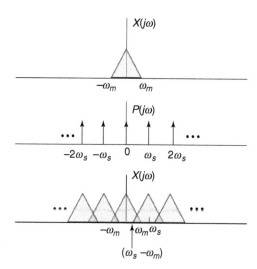

Fig. 1.8 Aliasing distortion due to slow sampling.

retrieved by filtering or any other means. The distortion that is caused by slow sampling is known as *aliasing* distortion. Figure 1.8 shows the aliasing distortion.

The fact that efficient sampling needs a sampling frequency at least twice the maximum frequency of the message signal was first proven by Nyquist in 1928 and is known as the Nyquist sampling theory. Obviously, for a message to be presented by sample pulses, the power spectrum should be band limited; otherwise, sampling cannot be used to represent the signal; mathematically speaking, there is no sampling frequency that satisfies the Nyquist criteria. Respectively, an anti-aliasing filter is a filter that is used before a signal is sampled. This is to restrict the bandwidth of the signal to satisfy the Nyquist sampling theorem.

Unfortunately, in contrast to digital filters, it is very difficult to implement sharp cut-off analogue anti-aliasing filters. Therefore, a technique called *oversampling* is used. A signal is called oversampled if it is sampled with a frequency higher than the Nyquist sampling frequency limit. The oversampling ratio, β, is defined as

$$\beta = \frac{\omega_s}{2 \cdot \omega_m} \tag{1.10}$$

where again ω_s represents the sampling frequency and ω_m is the highest frequency of the signal (signal bandwidth).

By using the oversampling technique, the transient range of the anti-aliasing filter can be extended. This implies that the filter can be implemented with a reduced complexity (i.e., more cheaply). For example, considering a 4-kHz speech signal, without oversampling, the anti-aliasing filter should sharply cut off all frequencies above 4 kHz. Figure 1.9(a) shows an eighth-order Chebyshev II lowpass filter. It can be seen that the filter roll-off is very sharp, and the corresponding phase response is nonlinear toward the higher frequencies.

These variations in phase can cause undesirable effects in audio signal. By contrast, if the oversampling technique is employed, with $\beta = 64$, a lowpass filter with a cutoff frequency of 256 kHz can be used. Figure 1.9(b) shows a very simple first-order Chebyshev II lowpass

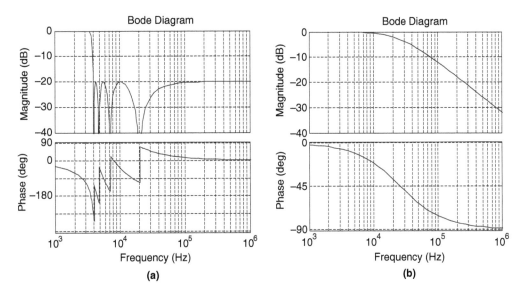

Fig. 1.9 Analogue filter at (a) Nyquist limit (no oversampling) and (b) 64× Nyquist limit.

filter. As it is shown, the roll-off is much gentler but the phase response is remarkably better. Needless to mention, the implementation of a first-order filter is much easier compared to an eighth-order filter.

Another advantage of oversampling is the noise reduction. In fact it is shown by Watkinson (1994) that by taking N samples from the same quantity with a random noise signal, the noise is reduced by factor $1/\sqrt{N}$ when the samples are averaged. Oversampling can also be used to reduce the quantization noise. A detailed description is given in the next section. After all, the signal will be again down-sampled to the target rate. In fact, oversampling enables digital processing of the signal.

Mention should also be made of the impact of the sampling pulses. In fact, in contrast with impulses, the practical sample pulses have limited amplitude and finite duration. If $x(t)$ represents the message signal, the sampled waveform, $x_s(t)$, can be written as

$$x_s(t) = \sum_{n=-\infty}^{\infty} x(nT_s)p(t-nT_s) = p(t) * \left[\sum_{n=-\infty}^{\infty} x(nT_s)\delta(t-nT_s) \right] \tag{1.11}$$

where $p(t)$ represents the sampling pulse. Comparing this with equation 1.3, it can be seen that the sampling waveform has appeared as a new term in equation 1.11. The convolution of this term in the time domain corresponds to a multiplication of its spectrum in the frequency domain.

In other words, the spectrum of the sampled pulse is shaped by the spectrum of the sampling waveform. In most cases the sampling waveform consists of narrow, square (flat-topped) pulses. This means that the power spectrum of the sampling waveform, $P(f)$, is a sinc function, which implies an attenuation of the high-frequency component of the message signal. This phenomenon is referred to as the *aperture effect* and can be compensated for by employing equalizers at the receiver.

Quantization

Quantization is a process in which the amplitude of the sampled pulses is approximated to discrete values. The quantization is referred to as uniform if all quantization steps have an equal length. The error of approximation in the quantization process is called quantization error. If the quantization noise level is donated by e, then the quantization noise power can be given by

$$E(e^2) = \int_{-\infty}^{+\infty} e^2 P(e) de \tag{1.12}$$

where $P(e)$ is the probability density function of e. Because the quantization error is distributed uniformly during each quantization level, the $P(e)$ can be written as

$$P(e) \begin{cases} 1/\delta & -\delta/2 < e < \delta/2 \\ 0 & \text{otherwise} \end{cases} \tag{1.13}$$

where δ is the quantization step size. Therefore, equation 1.12 can be written as

$$E(e^2) = \int_{-\frac{\delta}{2}}^{+\frac{\delta}{2}} e^2 \frac{1}{\delta} de = \frac{1}{\delta} \int_{-\frac{\delta}{2}}^{+\frac{\delta}{2}} e^2 de = \frac{\delta^2}{12} \tag{1.14}$$

Equation 1.14 shows that the quantization noise depends only on the quantization step size. Evidently, the smaller quantization step size results in a lower quantization noise value but requires more bits for coding. If quantization is realized by using N bits, the maximum signal-to-noise ratio (SNR) can be written as

$$\left(\frac{S}{N}\right)_{max} = \frac{(2^N \cdot \delta/2)^2}{\delta^2/12} = 3 \times 2^{2N} \tag{1.15}$$

Expressing SNR in decibels yields

$$\left(\frac{S}{N}\right)_{max} = 10 \cdot \log_{10}(3 \times 2^{2N}) = 4.8 + 6N \; dB \tag{1.16}$$

This shows that the SNR is improved by 6 dB per bit. PCM voice telephony usually uses 8 bits, which implies a maximum SNR of 52.8 dB.

The *crest factor* is defined as the peak amplitude of a waveform divided by its RMS value. Unfortunately for voice applications, the crest factor is usually very high, implying that the actual SNR is significantly less that the theoretical upper bound.

Figure 1.10 shows another aspect of quantization noise, the noise power spectrum. As illustrated, the quantization noise is uniformly distributed in the frequency range of $[-\omega_s/2, \omega_s/2]$. Figure 1.9(a) shows the case where oversampling is not used (i.e., $\omega_s = 2 \cdot \omega_2$). In this case, the noise is distributed in the same frequency range as the message signal. On the other hand, Figure 1.9(b) illustrates the spectrum of the noise when oversampling is in use (i.e., $\omega_s > 2 \cdot \omega_2$). The area under the shaded curves in both figures is equal because, as is shown by equation 1.13, the quantization noise power depends only on the quantization step size and not the sampling rate. When using oversampling, the lowpass filter rejects the higher-frequency noise components, which results in overall noise reduction. The noise can

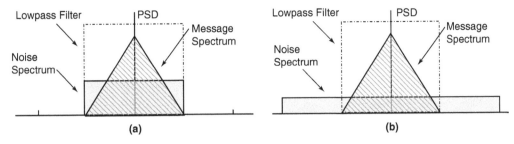

Fig. 1.10 Spectrum of quantization noise (a) without oversampling and (b) with oversampling.

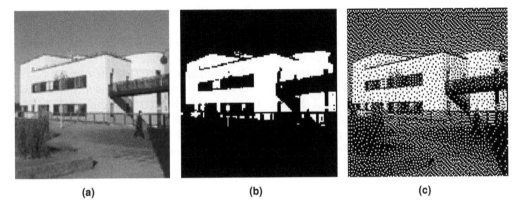

Fig. 1.11 Example of two-level quantization: (a) original image, (b) thresholded image, (c) dithered image.

be further reduced by employing noise-shaping filters. In fact, a noise-shaping filter contours the noise so that it is reduced in the audible regions and increased in the inaudible regions

Dithering is another technique that can be used to improve the perceptual quality of speech signals. The idea of dithering is to randomize quantization error when a high-resolution signal is converted to low resolution. Dithering is widely used in image-processing applications. For example, because most printing facilities use a binary (blacks and whites only) scheme to print a picture, given a gray-scale picture, a conversion is needed before printing. This is a special case of quantization with only two quantization levels. One solution is to use a simple fixed threshold level. However, a more efficient way is to use dithering to spread the quantization noise spatially across the image pixels.

In other words, when a pixel is quantized, the quantization error is calculated and is added to the neighboring pixels that have not been quantized yet. This is sometimes referred to as error diffusion as well. Figure 1.11 illustrates the level of quality improvement that can be achieved by spatially spreading the quantization noise over the image pixels. Notice that the root mean square (r.m.s.) of the quantization error is almost the same for the thresholding and dithering algorithm.

However, in audio engineering, dithering can be used to spread the quantization noise over the frequency range to minimize the perceptual distortion. Surprisingly, this is done

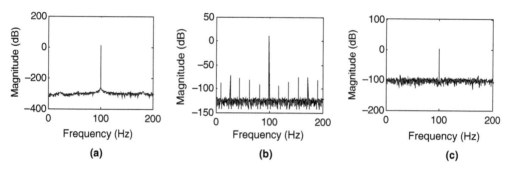

Fig. 1.12 Impact of dithering: (a) spectrum of the original sine wave, (b) spectrum of the quantized sine wave without dithering, (c) spectrum of the quantized sine wave with dithering.

by adding low-level random noise to the signal. In fact, the nonlinear effect of the quantization leads to audible distortion. This is mainly because of the "stairsteps" waveform of the quantized signal and is more significant for signals of sufficiently low level to be comparable with the quantization step size. The idea is to add random noise to the waveform to "break up" the statistical determinability of the stairstepped waves. In fact, by adding dither to the audio signal with a level that is approximately half the level of the least significant bit (LSB), the frequency selective noise (distortion) is eliminated at the cost of a very-low-level constant noise. Because the human brain is less sensitive to a constant background noise, the perceptual quality will be improved.

As an example, consider a sine wave at the frequency of 100 Hz that is quantized using 16 quantization levels. Figure 1.12(a) shows the spectrum of the sine wave before quantization. As expected, the power spectrum contains a single peak at 100 Hz. Figure 1.12(b) shows the power spectrum of the quantized signal without dithering. As can be seen, the signal contains many harmonic distortion components. Figure 1.12(c) shows the spectrum of the quantized signal when dither noise was added to the signal before quantization. It can be seen that, with dithering, the harmonic distortions are eliminated at the cost of overall shift in the noise floor.

The dither noise can be selected from a Rectangular Probability Density Function (RPDF), a Triangular Probability Density Function (TRDF), or a Gaussian PDF. RPDF dithering is the most basic method where a uniformly distributed noise is generated and added to the signal before it is quantized. The dither noise can be also reduced by employing an error feedback loop around the quantizer, which operates as a noise-shaping filter.

Peak clipping is another source of distortion in the quantization process. Peak clipping will take place if the signal amplitude is larger than maximum quantization level. In practice, signal limiters are used to avoid peak clipping.

A uniform quantizer is optimum when the input has uniform distribution. However, this might not be the case for speech signals with a high crest factor. In fact, the quantization process can be optimized (i.e., minimum quantization error) by considering the PDF of the input signal. Experimental results have shown that the probability distribution of a voice signal, $P_x(x)$, can be approximated by a Laplace *distribution* in the form (Carlson, Rutledge, and Crilly, 1968)

$$P_x(x) = \frac{1}{2 \cdot |\bar{x}|} \times \exp\left(-\frac{|x|}{|\bar{x}|}\right)$$

(1.17)

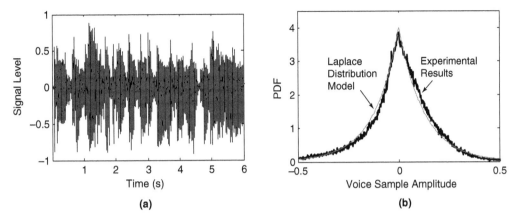

Fig. 1.13 (a) Six-second sample of a voice segment; (b) PDF for the voice signal in Figure 1.10(a).

As shown by Figure 1.13(b), this approximation is very accurate, especially when the number of samples is high (long-term approximation). The experimental results presented in Figure 1.13(b) belong to six seconds of a voice segment (Figure 1.13(a)). As is evident, the frequency of samples with smaller amplitudes is considerably high compared to high-amplitude samples. In other words, efficient speech quantizations require nonuniform quantization levels that concentrate more on small amplitudes.

Employing a logarithmic scale allows quantization intervals to increase with amplitude, and ensures that low-amplitude signals are digitized with smaller quantization error. In other words, fewer bits per sample will be necessary to provide a specified SNR for small signals and an adequate dynamic range for large signals. Companding laws are used to implement logarithmic quantization. A-law and μ-law are companding schemes commonly used in telephone networks. While μ-law is mainly used in North America and Japan, European countries mostly use A-law.

According to μ-law, if the normalized input signal is presented by V_i, the normalized compressed signal, V_c, is defined as

$$V_c = \frac{\log(1 + \mu \cdot |V_i|)}{\log(1 + \mu)} \cdot \text{sgn}(V_i) \qquad -1 < V_i < 1 \tag{1.18}$$

where μ is defined as the compression parameter (i.e., a higher value of μ results in more compression). The μ-law companding is commonly implemented with a value of μ equal to 255. The transfer function of A-law companding is also given by

$$V_c = \text{sgn}(V_i) \times \begin{cases} \dfrac{A \cdot |V_i|}{1 + \ln(A)} & |V_i| < \dfrac{1}{A} \\[3mm] \dfrac{1 + \ln(A \cdot |V_i|)}{1 + \ln(A)} & \dfrac{1}{A} \le |V_i| \le 1 \end{cases} \tag{1.19}$$

where A is the compression parameter. In Europe, the E1 lines use A-law companding with $A = 87.7$, although the value of $A = 87.6$ is also used. Figure 1.14 shows the transfer function of μ-law and A-law companding schemes for different values of μ and A.

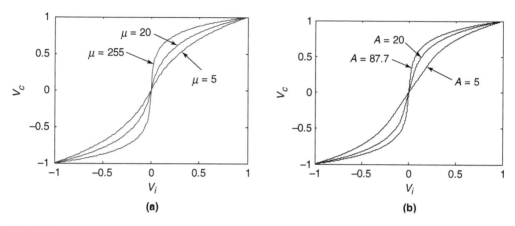

Fig. 1.14   Transfer function of (a) μ-law companding and (b) A-law companding.

There are many similarities between these two schemes. For example, both companding laws are implemented using 8-bit codewords (256 levels, one for each quantization interval). In addition, in both schemes, the whole dynamic range is divided into 16 segments (8 positive and 8 negative) where the length of each segment is twice the length of the proceeding segment. Considering the total 256 quantization levels, this means there are 16 quantization levels inside each segment. These steps are uniformly distributed during each segment. In an 8-bit word, the first 4 bits identify the segment, and the rest quantize the signal value within the segment.

Despite all similarities there are some differences between these two schemes. A-law provides a slightly larger dynamic range compared to μ-law, while μ-law is more efficient in terms of quantization noise reduction for small signal levels. A-law is also used for international connections. If any conversion between these two companding schemes is required, it should be accomplished by the μ-law country. Also notice that this conversion is a lossy process because it involves requantization of the signal.

Encoding

After sampling and quantization, the PCM signal is encoded into the bits of 0 and 1. As mentioned before, the coding will be more efficient if the number of quantization levels is a power of 2. Most PCM systems use 8 bits to represent each sample (256 quantization levels). There are different coding schemes that can be used to represent the data.

The *binary number code* is an example of a more general category known as weighting codes. In binary number coding, each binary digit (either 0 or 1) carries a certain weight according to its position in the binary number. The binary code can be found in different forms. For example, in 8421 Binary Coded Decimal (BCD) representation, each decimal digit is converted to its 4-bit pure binary equivalent. The 4221 BCD code is another example where each bit is weighted by 4, 2, 2, and 1, respectively. Mention should also be made of another important class of codes called unit-distance codes in which the distance between two code characters is the number of digits by which they differ. In other words, a code is considered to be unit-distance if it changes at only one digit position when going from one number to the next in a consecutive sequence of numbers.

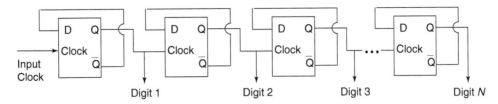

Fig. 1.15 *N*-digit binary counter using master–slave flip-flops.

Gray code (also referred to as reflected binary code) is an example of a unit-distance code that facilitates the error correction process and is widely used in many applications, including analogue–digital converters. Another category of codes consists of *symmetrical codes*. For many message signals (such as speech) that extend more or less above and below the quiescent level, it might beneficial to select the first codebit to represent the sign of the sample value. The resulting code is symmetrical above its center.

One common requirement in digital circuits is counting. Most of the counters are implemented using individual flip-flops. A flip-flop is an electronic device that that has two stable states (i.e., a device with a single-bit memory). A control signal(s) and/or a clock signal are usually used to change the state of a flip-flop from one state to another. The output often includes the complement as well as the normal output. Among all different types, D flip-flops are commonly used for implementing binary counters. The term is "D flip-flop" because the output is the delayed version of the input signal. In fact, a D flip-flop tries to follow the input but cannot make the required transitions unless it is enabled by the clock.

The associated delay is referred to as *propagation delay*. The propagation delay can be removed by using a master–slave D flip-flop in which two D flip-flops are connected in series and the clock of the second flip-flop (slave) is the inverted version of the first (master). Figure 1.15 shows the diagram of an *N*-digit binary counter using master–slave flip-flops.

Pulse code modulation can be realized by using a sawtooth reference signal and a binary counter. The overall diagram is shown in Figure 1.16. If the slope of the ramp signal, K^{-1}, is known, then each amplitude sample can be presented using Pulse Position Modulation (PPM) by a pulse displacement at $t_s = K \cdot a_s$. Pulse Width Modulation (PWM) can be derived from this by generating a pulse of width t_s for each sample. The pulse width–modulated signal and the reference clock signal are combined using a logical AND gate, and the resulting signal is sent to a digital counter. In this situation the digital count is proportional to the analogue input value.

Notice that the frequency of the clock corresponds to the quantization step size; that is, a higher clock frequency results in a smoother quantization. The digital counter is reset to its initial condition after the end of each sample.

After an analogue signal is converted into a digital format, it can be easily combined with other digital signals by interleaving in the time domain. This is referred to as *multiplexing* and is covered in the next section.

Multiplexing

Multiplexing is a term used to refer to a process where multiple analogue message signals or digital datastreams are combined into one signal over a shared medium. It is often practi-

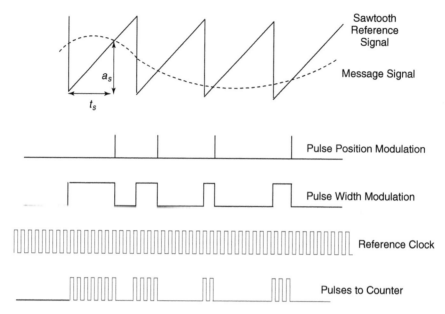

Fig. 1.16 Implementation of a coder using a sawtooth signal.

cal to combine a set of low-bitrate streams, each with a fixed and predefined bitrate, into a single high-speed bitstream that can be transmitted over a single channel. This technique is called Time Division Multiplexing (TDM).

Time Division Multiplexing Principles

Time division multiplexing is the earliest and simplest form of digital multiplexing. It was initially developed to provide multiaccess to the computing resources for groups of co-located users during the 1960s. Using a *serializer* the parallel input signals of different sources are combined into a single serial output. In fact, each channel (input) is allocated a timeslot, which defines the channel information within the serial output. Based on the channel time-slot duration we can have two types of multiplexing: If each timeslot is long enough to hold the full length of the input codeword, the multiplexed signal is referred to as *word interleaved*. Alternatively, the multiplexer may split each input code into binary bits and take one bit from each input, channel at a time; in this case, the multiplexed signal is called *bit interleaved*.

Bit interleaving the simplest form of multiplexing, was used by the first commercially available multiplexers. It is simple because it only requires the multiplexer to be able to detect the boundaries of the incoming input data. In addition, if the sequence of the databits is known at the receiver end, the de-multiplexing process can be easily done for any type of input data. However, because this means sending all databits for an asynchronous char-acter, including the stop, start, and parity bits, the adopted multiplexing technique quickly moved to byte interleaved. In fact, when choosing the size of the timeslots, there is a trade-off between efficiency and delay. With very small timeslots (such as in bit interleaving), the multiplexer should be very fast to switch between different input channels. On the other hand, if timeslots are longer than a bit length, then data from each input should be stored while the other input channels are served. In the case of very long timeslots, a significant delay will be introduced.

Notice that data interleaving can occur inside each input channel as well. This is usually achieved in the application layer before the data is sent to the lower layers to be multiplexed with other input channels. This is entirely different from the interleaving during the multiplexing process. Because most of the transmission errors occur as bursts, the aim of higher-layer interleaving is to spread the erroneous data to improve the perceived quality of the received data.

Frame Alignment

Needless to mention, the de-multiplexer needs to identify which timeslot is allocated to which input channel. Therefore, a predetermined binary sequence is periodically repeated during the multiplexed data. A *frame* is then defined as a set of successive timeslots that can be identified by reference to a frame alignment signal. The frame alignment signals commonly occupy a finite number of timeslots (usually one or two) during each frame. These timeslots are referred to as *Frame Alignment Words* (FAWs). Theoretically, the frame alignment words can be placed anywhere during the frame; however, in practice the FAWs are placed in a way that reduces the average time taken for the frame alignment process. Particularly, the framing strategy is called *bunched (grouped)* if all frame alignment timeslots are concentrated at the beginning of each frame. By contrast, *distributed framing* refers to the case where the frame alignment words are spread during the frame.

There is also a danger of simulating the framing pattern by information-carrying bits. One way to reduce this risk is to choose a framing pattern with a low correlation so that it is impossible to imitate by shifting and infringement on random neighboring bits. Alternatively, it is possible to reject all the channels of the frame when the framing is lost at the receiver end. However, this might not be a good practical solution because it requires a return path to the sender so that the receiver can send a Negative Acknowledgment (NACK) message. Especially if the size of the FAW is small compared to the frame size, the possibility of such patter imitation is high, and requesting a new alignment word for every single imitation case might not be an efficient solution.

Among all of the different alignment techniques, *serial frame alignment* is the most commonly used. In this technique, when the alignment is lost, the system tries to find a FAW in the next coming frames (say two or three frames); if a FAW is found, then the system can return to the full-alignment state. Otherwise, the system enters the search mode. In the search mode, the system scans the incoming frames until it finds a valid FAW. However, to improve stability, the system does not enter full-alignment mode when it finds the first FAW during the scanning state; instead, it waits until a number of successive FAWs (say two or three) are received correctly. This mode is referred to as the waiting state. Losing any FAW during the waiting state returns the system to the search state.

Figure 1.17 shows an example where the scanning starts from point *a*. The imitation of the FAW is represented by *I* in the figure. At the beginning, when the scanning process starts (point *a*), there is no previous alignment, and therefore the system starts in a search state. During this state the bit sequence is scanned to find the FAW pattern. In the example shown in the figure, the first FAW happens to be imitated by information bits (point *b*). Therefore, the system assumes that a FAW is found and by mistake enters a waiting state.

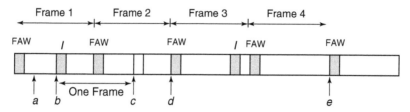

Fig. 1.17   Example of the frame alignment procedure.

However, because there is no FAW at one frame distance from point *b* (point *c*), the system returns back to the search state at point *c*. At point *d* the system again enters the waiting state. However, the state does not change until the third correct alignment word is recognized at point *e*, where the system starts operating in full-synchronized mode. Notice that the imitation between points *d* and *e* has no effect. In fact, when a partial synchronization is established, the system stops searching for the FAW pattern and only checks the frame alignment pattern every frame interval.

The serial alignment algorithm can be implemented in a very simple fashion; however, performance (in terms of average search time) can be improved by using *parallel searching*. Parallel searching can minimize the impact of frame alignment imitation by selecting all possible FAWs (the real or imitated) and rejecting the ones that are not repeated after a frame interval, though this advantage might not be very significant, especially when imitations of FAW occur rarely.

Another interesting property is the stability of the system against the transmission errors. In fact, when the system enters a full-synchronization state, the occasional bit corruptions of FAWs cannot return the system to the search state; this later event occurs only if a FAWs is corrupted during some number of consecutive frames. This feature is referred to as the system's *flywheel*. The minimum number of failed FAWs that can lose the alignment of the system is called *momentum* of the flywheel. The transmission errors appear as short bursts, and therefore the momentum of the flywheel should be long enough to accommodate the longest possible burst.

Higher-Order Multiplexing

The principles of PCM were discussed in the previous section. As mentioned there, the input voice data is fed into a bandpass filter with the range from 300 to 3400 Hz. The filter output is then sampled at 8 kHz, giving the sample duration of 125 μs. Then each sample is quantized using 8 bits. This means that the bitrate will be equal to (8/125) bits/μs or 64 kbit/s. This is also referred to as the basic PCM rate.

There are basically two worldwide transmission systems: one with 24 and the other with 32 channels. While the former is used in Japan and North America, the latter is employed by Europe and the rest of the world. The American system is based on the PCM system, originally developed by AT&T. The name T1 is derived from the identification number of the committee set up by the American National Standards Institute (ANSI). The European system, on the other hand, is referred to as E1, which is based on the CEPT (Conference of Post and Telecommunication) recommendation.

The T1 carrier framing was designed to carry 24 channels of speech in each frame. This was based on one of the earliest framing formats called D4 framing. A T1 frame consists of

193 bits, with 192 bits (8 × 24) dedicated to the voice samples and the very last bit to framing and signaling. Twelve frames are combined to form a *superframe*. This means that each superframe contains 12-bit words, composed of individual bits coming from each of the 12 frames. The framing bits are the odd-number framing/signaling bits while the signaling bits are the even-numbered bits. The framing bits alternate every other frame.

On the other hand, in an E1 connection, out of 32 slots, 30 are used to carry the speech data, and the other two are used for signaling and synchronization. Therefore, the framing format is also referred to as CEPT PCM-30. Timeslot 0 can be used for synchronization and alarm transport. In addition, in all frames the first bit of timeslot 0 is dedicated to international carrier use. The Frame Alignment Signal (FAS) is transmitted in bit position 2–8 of timeslot 0 every other frame. The FAS pattern recommended by CEPT PCM-30 is 0011011, which is impossible to imitate by shifting and infringement. If a frame does not contain the FAS, it can be used to carry 5 bits of national signaling and an alarm indication bit for loss of frame alignment. Furthermore, in each CEPT PCM-30 frame, timeslot 16 is used to transmit signaling data. Because a common channel is dedicated to the signaling data of all speech circuits, this method of signaling is referred to as Common Channel Signaling (CCS).

The signaling consists of 4 bits per channel, which are grouped in the two halves of timeslot 16. This means that 15 frames are required to carry the information of the 30 channels. Together with the first frame (frame 0), a multiframe consisting of 16 frames is formed. The 16th timeslot of the first frame contains a Multiframe Alignment Signal (MAS), which allows unambiguous numbering of the frames within a multiframe. Figure 1.18 shows the structure of a CEPT PCM-30 multiframe.

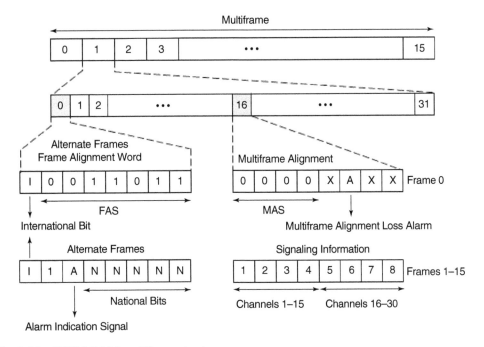

Fig. 1.18 CEPT PCM-30 multiframe structure.

Table 1.2   Comparison of American and European Digital Hierarchies

	USA			Europe	
Order	Number of Channels	Total Bitrate (Mbit/s)	Order	Number of Channels	Total Bitrate (Mbit/s)
DS0	1	0.064	E0	1	0.064
DS1	24	1.544	E1	30	2.048
DS1-C	48	3.152	E2	120	8.448
DS-2	96	6.312	E3	480	34.368
DS-3	672	44.736	E4	1920	139.264
DS-4	4032	274.176	E5	7680	565.148

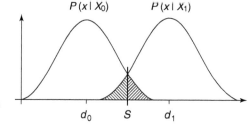

$P(x|X_0)$ $P(x|X_1)$

d_0 S d_1

Fig. 1.19   Conditional probability distribution of the received bit position.

Finally, notice that in normal applications there is no error-checking capability provided within CCITT recommendation G.704. However, if there is a need to provide additional protection against imitation of frame alignment signals or where there is a need for an enhanced error-monitoring capability, then the international bit can be used for the cyclic redundancy check (CRC).

Higher-order multiplexing is accomplished by multiplexing the standard TDM frames. In this case each of the PCM 64-kbit/s channels is referred to as DS0 in the United States and 0-order system (or E0) in Europe. Table 1.2 summarizes the American and European digital hierarchies. Because multiplexing of the tributaries needs very accurate synchronization, it is necessary to bring all the bits to exactly the same rate, which is achievable by employing retiming buffers.

Transmission Performance

After the speech data is converted into a multiplexed bitstream, it is transmitted over a communication channel. This section discusses the issues and challenges related to the transmission of digital speech signals over error-prone transmission channels.

Digital Data Transmission

After an analogue voice signal is converted to digital format, it is transmitted to the receiver through a communication channel where random noise is added to the PCM signal. The impact of this noise can appear as erroneous digits in the codeword, which can mislead the decoder. Figure 1.19 shows the conditional probability distribution of the received bit position where $P(x\,|\,X_0)$ represents the probability distribution when 0 bits are sent and $P(x\,|\,X_1)$ corresponds to the case where a marked bit is sent. Assuming an equal frequency of 0 and marked bits, the optimum performance is achieved when the decision threshold, S, is placed exactly in the middle of the two constellation points (i.e., $S = \dfrac{d_1 - d_0}{2}$). The shaded parts of the figure represent the area where an erroneous decision is made by the receiver.

Notice that in a PCM signal, not all the bit errors have the same impact, but the effect of the bit errors depends on the position of the erroneous bits. Considering a binary codeword of length $N(b_1, b_2, \ldots, b_N)$ in which the mth bit distinguishes between quantum levels spaced by 2^m times the step size, δ, an error in the mth bit shifts the decoded level by $\varepsilon_m = \pm\, 2^m \times \delta$. The average of ε_m^2 over the N bits equals the mean square error for random bit error location.

$$\overline{\varepsilon_m^2} = \frac{1}{N}\left(\sum_{m=1}^{N}(\delta \cdot 2^m)^2\right)^2 = \frac{4}{3N} \times \frac{4-\delta^2}{4} \approx \frac{4}{3N} \tag{1.20}$$

Assuming that the probability of bit error, P_e, is very small ($P_e \ll 1$), then the probability of a single error in N bits can be approximated by $N \cdot P_e$ and the probability of more than a single bit error can be considered to be 0. Therefore, the decoding channel noise power can be written as

$$\delta_d^2 = N \cdot P_e \cdot \overline{\varepsilon_m^2} \approx \frac{4}{3}P_e \tag{1.21}$$

The total noise power can be written as the summation of the quantization and the decoding channel noise. Therefore

$$N_T = \sigma_d^2 + \sigma_q^2 = \frac{\delta^2 + 16 \cdot P_e}{12} \tag{1.22}$$

and

$$\left(\frac{S}{N}\right)_T = \frac{12 \cdot S_x}{\delta^2 + 16 \cdot P_e} \tag{1.23}$$

Figure 1.20 shows the graph of the total SNR versus the bit error rate for different values of δ and for the case where S_x is set to 1. As the figure illustrates, the impact of the quantization noise is dominant when the bit error rate is very low. In this case the advantage of employing a smaller step size can be clearly observed. However, with an increase in the bit error rate, the impact of the channel noise becomes more pronounced and the curves start to converge.

Eye Diagram

Normally, digital transmission systems employ pulse-shaping techniques to limit the effective bandwidth of the transmitting signal. This is very useful for controlling the effect of Intersymbol Interference (ISI) caused by the channel. The impact of the noise and other transmission impairments can be clearly observed by superimposing the various transitions

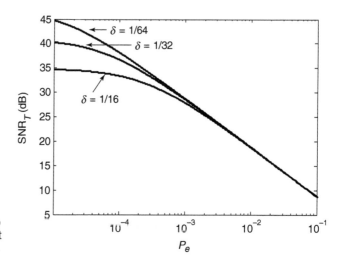

Fig. 1.20 Signal-to-noise ratio versus bit error rate for different values of the quantization step.

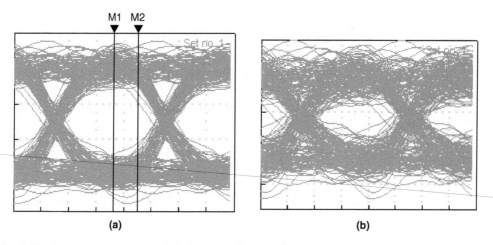

Fig. 1.21 Example eye diagrams: (a) wide-open diagram of a transmission system with a low bit error rate (2.45×10^{-9}); (b) diagram of a noisy system with a bit error rate of 1.79×10^{-4}.

of the signals on top of each other. The resulting pattern is known as the *eye diagram*. Respectively a wide-open eye diagram represents a system with high performance while noise, jitter, and other disturbances have the effect of closing the eye and increasing the probability of bit errors. Figure 1.21(a) shows an example of a wide-open eye that corresponds to a bit error rate of 2.45×10^{-9}. Similarly, Figure 1.21(b) represents the eye diagram of a noisy transmission system with a bit error rate of 1.79×10^{-4}.

Notice that the optimum bit error rate performance is achieved if both the horizontal and vertical decision thresholds are placed at their optimal place, which is, in fact, in the middle of the eye. The vertical threshold can change because of temperature effects or component inaccuracy. Inaccuracy in timing is also possible and is known as *jitter*. Jitter can considerably affect the bit error rate performance of a receiver. As an example, if the sampling takes place at the point shown by marker M2 in Figure 1.21(a), the bit error rate will be equal to 2.27×10^{-5}. Marker M1, however, shows the optimum sampling time.

Jitter and Time Extraction

As mentioned before, jitter is defined as the short-term variation of digital signals from their ideal position. Though this is mainly due to the mistuning of the digital regenerators and repeaters (e.g., random thermal noise from a crystal), it can also be affected by the pattern of the transmitted data as well. There are basically two main impairments associated with jitter. The first problem is the impact on the bit error rate performance of the receiver as previously discussed. Second, jitter can cause irregular spacing between the coded samples, which can result in low-frequency distortion component. Fortunately, this latter effect is not a serious problem in primary PCM voice systems.

The problem of jitter can become more significant with an increase in the number of repeaters and regenerators in the transmission system. In fact, the amplitude of the jitter is directly proportional to the number of these repeaters and regenerators.

There are different ways to compensate for the impact of jitter. The most common is to use a Phase Lock Loop (PLL) for clock recovery. In its simplest form, a PLL consists of a phase comparator, a lowpass filter and a Voltage Controlled Oscillator (VCO) in a closed feedback loop. By employing a PLL when a gap occurs between the reference timing signals, the extracted clock signal can be used to stimulate the phase comparator. The performance of a PLL is evaluated based on two main metrics: phase noise and lock time. For a carrier frequency at a given power level, the phase noise of a PLL is defined as the ratio of the carrier power to the power found in a 1-Hz bandwidth at a defined frequency offset. The lock time of a PLL is also defined as the time it takes to move from one specified frequency to another specified frequency within a given frequency tolerance. After all, dejitterizers can be employed to reduce the effect of jitter in digital signals. Dejitterizers consist of an elastic buffer in which the signal is temporarily stored and then retransmitted at a rate based on the average rate of the incoming signal.

Resources

Carlson, B., Rutledge, J. C., and Crilly, P. B. (2001) *Communication Systems: An Introduction to Signals and Noise in Electrical Communication*, Fourth Edition, McGraw-Hill.
Gibson, J. D. (2005) Speech Coding Methods, Standards, and Applications, *IEEE Circuits Syst. Mag.* 5(4):3049.
Glover, I., and Grant, P. (2003) *Digital Communications*, Second Edition, Prentice Hall.
Kondoz, A. M. (2005) *Digital Speech: Coding for Low Bit Rate Communication Systems*, John Wiley & Sons.
Owen, F. F. E. (1982) *PCM and Digital Transmission Systems*, McGraw-Hill.
Painter, T., and Spanias, T. A. (1997) A Review of Algorithms for Perceptual Coding of Digital Audio Signals, *Proceedings 13th Int. Conf. Digital Signal Process.*

2 Data Transmission

Stuart D. Walker
Emilio Hugues-Salas
Rouzbeh Razavi
Tahmina Ajmal
University of Essex

Introduction

Data transmission is the transfer of information from one entity (transmitter) to another (receiver) via a communication path. The data transfer can be local or remote. For example, it can be a simple file transfer within a computer or it can be Internet browsing where the information is exchanged with the Web over more than one communication medium. Other forms of data communication include e-mail, data storage, database queries and updates, and home shopping.

Three types of digital signal can be differentiated on the basis of bit error rate performance, bandwidth utilization, and delay requirements. They are data, voice, and video. For data transmission, the bit error rate requirements are more severe in comparison to voice and video applications, but there is a higher tolerance against delay constraints. Utilized bandwidth is an important parameter in the design of a transmission link. For a voice communication, typical bitrates extend from 4.8 kbit/s to 64 kbit/s (see Chapter 1). The bandwidth requirement for a video application depends on a number of factors, such as the type of video codec in use, picture size, frame rate, and so on (see Chapter 3). Given their natures, both voice and video communications can stand a certain level of bit error rate compared to data transmission, but they are more sensitive to delay. Another major difference between the data and voice/video traffic is that in the latter case, not all the digits in a byte have the same level of importance. Therefore, the end-to-end performance of video and voice applications can be considerably improved by unequal protection of bits.

Data transmission in its simplest form (binary digital transmission) is related to the transfer of information between terminals, computers, and other equipment for data processing. This binary data is converted to two-level pulses for transmission through the channel. The bandwidth of binary data will depend on the shape of the waveform and the order of the 1 and 0 binary bits.

This chapter starts with the basics of a typical data transmission system, including the classification and important characteristics of data. Issues relating to the loss of data and its retrieval are dealt with in subsequent subsections. Recent trends are finally covered in the last section.

Data Transmission Basics

A basic data transmission system consists of a transmitter unit, a transmission medium (also known as a transmission channel), and a receiver unit (see Figure 2.1). The channel or transmission medium is a physical path by which a message travels from the transmitter to the receiver.

Whenever an input message signal is to be transmitted, a processing subsystem is needed for adaptation of the data according to the transmission link characteristics. The input signal is known as the *baseband signal*, and the process of making the signal suitable for transmission is called *modulation*. As the signal travels through the transmission path, it is affected by transmission impairments such as the noise, interference, and distortion. Noise is undesired random and unpredictable signals added to the transmitted signal, interference is contamination by extraneous signals, and distortion is waveform perturbation caused by the imperfect channel response. Unlike noise and interference, distortion disappears when there is no signal. The message needs to be extracted (recovered) from the received signal. This process is referred to as *demodulation*.

Figure 2.1 illustrates transmission in one direction only, from one point to another point. This kind of transmission is known as *simplex*. If facilities are provided to transmit in the return direction also, then both stations can transmit, either one at a time or simultaneously. *Half-duplex* transmission refers to the case where only one transmitter can send data at a time. If the transmission is simultaneous from both blocks, it is known as *full-duplex*.

Another classification can be depicted according to the number of receivers addressed by a transmitter. A point-to-point transmission generally refers to a connection restricted to two endpoints, which are usually host computers. In their simplest form, point-to-point communications refer to a direct data transfer without any data or packet formatting. The

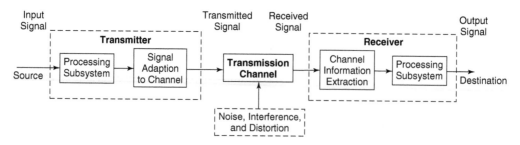

Fig. 2.1   Data transmission system.

end nodes are responsible for data formatting, synchronization, and so on. In such cases the connection is typically established through a serial protocol like RS-232, which cannot be extended over very long distances. However, for longer-range communications, a point-to-point connection is established through a public network, which is generally packet switched (for example, peer-to-peer file-sharing networks).

Point-to-multipoint communications (also known as broadcast), on the other hand, refer to the case where a transmitter sends data for multiple receivers, which can save a considerable amount of network resources.

The efficiency of the transmission is mainly affected by the quality of the media and the characteristics of the sources and receivers that generate and retrieve the signal. Therefore, an understanding of channel characteristics and different types of transmission signals plays a significant role in optimal transmission system design.

Analogue and Digital Signals

Any transmission signal can be represented as an electromagnetic signal that can vary in time as well as in frequency; therefore, two domains will exist to analyze the signal: time and frequency.

Within these two domains, two types of signal are possible: analogue and digital. Analogue signals take a level that can vary continuously in time without any disruption, whereas a digital signal is represented by a limited number of discrete values with sudden changes between varying signal levels.

An analogue signal can be represented as

$$s(t) = M(t)\sin(2\pi ft + \theta) \tag{2.1}$$

where $M(t)$ is the amplitude (varying in time), f is the frequency, and θ is the phase of the signal. Any signal can be represented using these fundamental parameters. In contrast to the analogue signal, which contains only one harmonic, a digital signal is represented by a square wave, which is composed of odd-integer harmonics. A square wave can be represented by the following infinite Fourier series:

$$\begin{aligned} s(t) &= \sum_{k=1}^{\infty} \frac{\sin((2k-1)2\pi ft)}{(2k-1)} \\ &= \sin(2\pi ft) + \frac{1}{3}\sin(6\pi ft) + \frac{1}{5}\sin(10\pi ft) + \ldots \end{aligned} \tag{2.2}$$

An ideal square wave requires that the signal changes from the high to the low state instantaneously. This requires infinite harmonics to be added and would need infinite bandwidth for transmission. In fact, transmitting a signal at a high modulation rate through a band-limited channel can create Intersymbol Interference. Therefore, a pulse is shaped before its transmission. The purpose of pulse shaping is to make the transmitted signal better suited to the communication channel by limiting the effective bandwidth of the transmission. By filtering the transmitted pulses this way, the intersymbol interference caused by the channel can be kept in control.

When the digital signal is passed through a real channel (of limited bandwidth), effects like ringing or ripple are exhibited. Figure 2.2 illustrates these effects on the square wave when the Fourier series is truncated to only 7, 11, and 21 terms, respectively.

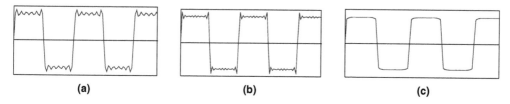

Fig. 2.2 Square wave composed of increasing numbers of odd-integer harmonics: (a) 7 harmonics, (b) 11 harmonics, (c) 21 harmonics.

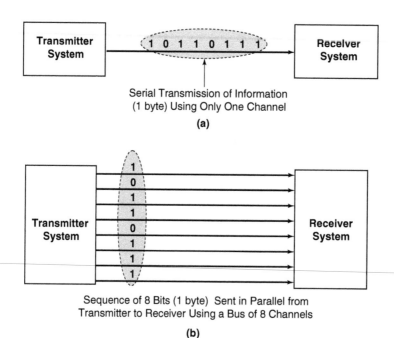

Fig. 2.3 (a) Serial and (b) parallel transmission.

Serial and Parallel Transmission

Data can be sent in either serial or parallel format (see Figure 2.3). In serial communication, the transmitter sends one bit at a time. In contrast, as the name implies, parallel data transmission involves several bits simultaneously.

The serial transmission mode uses only one communication channel (for example, one pair of wires) to send a binary datastream (Figure 2.3(a)). In this case bits are sent into the channel sequentially according to a clock instance. In the case of parallel transmission (Figure 2.3(b)), a group of bits is transferred simultaneously from transmitter to receiver via a set of channels (one bus of information). Generally, parallel transmission is faster compared to the serial scheme but requires a greater number of transmission links. Parallel

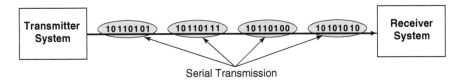

Fig. 2.4 Serial transmission of individual bytes in asynchronous mode.

transmission is often used within or between co-sited equipment. It can also be used in longer-distance communications such as virtual concatenation in Next Generation SDH (see Chapter 11) or in inverse multiplexing in ATM.

In addition, clock skew between different channels is not an issue in the case of serial communications while it can be problematic when data is transmitted in parallel channels. Finally, crosstalk is less of an issue in the case of serial communications because of the limited number of transmission links. However, serial transmission is more prone to dispersion.

Serial-to-parallel conversion can be carried out at the transmitter when the channel is set to parallel transmission. In the opposite way, parallel-to-serial transmission conversion can occur if only one channel is used for the transmission.

Serial transmission can be subdivided into asynchronous and synchronous data transfer. In asynchronous serial transmission (see Figure 2.4), no external clock is used for synchronization of the transmission process. In this case, timing information can be embedded in the signal. To achieve a data transmission without synchronization, an agreement (handshake) between the transmitter and the receiver should be made prior to start of the data transfer. As an example, a specific bit pattern can be used to prepare the receiving mechanism for the reception and registration of a symbol (start symbol), and a stop symbol can be employed to bring the receiving mechanism to rest in preparation for the reception of the next symbol.

The efficiency of an asynchronous transmission will depend on the number of additional bits that are sent together with the information. Synchronization is still present at the bit level in an asynchronous transmission. This is because the receiver needs to know the timing of each bit within each group of bytes. Besides, a timer is usually needed to detect the length of the transmitted group of information bytes. This kind of data transmission might need higher processing power at the receiver to accurately identify the beginning and end of each datastream. Because of this operation, the transmission might be slower compared to a synchronous transmission. However, this procedure might be cheaper to implement because synchronization is not needed. Asynchronous transmission applications can be observed in simple keyboard-to-computer communications, where each character typing is converted in a binary transmission.

In synchronous serial transmission (see Figure 2.5), the transmitter is in charge of assigning a specific time to the bytes to be sent through the channel. In this case no gaps exist in the transmission link and the channel is totally occupied. In serial transmission, the reason is that even if just a group of bits is required to be sent through the channel, rather than allowing the bytes to travel through the channel in an ad hoc manner, a longer "frame" is assigned to these bits, which corresponds to multiple bytes. An analogy is a transport railroad train carrying passengers as an alternative to several cars carrying different groups of persons. With this method a device is not required to detect the beginning of each group of

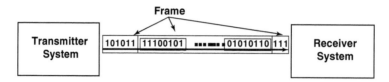

Fig. 2.5   Serial transmission in synchronous mode.

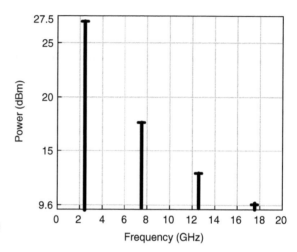

Fig. 2.6   Spectrum of a square wave.

bits. On the other hand, synchronization is a very important issue because of the requirement for accuracy in the bit timing and the count of the arriving bits.

Bandwidth and Datarate

A square-wave signal is composed of many frequencies. This square signal is analogous to a continuously constant level with abrupt changes in amplitude, which are the digital signals. A bit is the minimum unit of information of a digital signal. The bit can be represented by a 1, if the continuous level is high, or by a 0, if the continuous level is low. If an alternating sequence of 1s and 0s is considered as a digital signal, "1010 . . . ," to be transmitted through the system, a datarate will characterize the signal. In this case, the duration of each bit is half the period of repetition.

From the frequency analysis, the difference between the highest frequency and the lowest frequency will represent the transmission bandwidth of the signal. Figure 2.6 illustrates the spectrum of data signal for a square wave at 2.5 GHz. In this case, the frequencies f, $3f$, $5f$, and $7f$ were used, where f is 2.5 Gbit/s. Therefore, the bandwidth of the transmitted signal would be $7f - f = 6f = 15$ GHz.

In general terms, to transmit a higher datarate through a system, a higher transmission bandwidth is required. The next section gives a more detailed relationship between datarate, channel capacity, bandwidth, and signal-to-noise ratio.

Channel Capacity

Channel capacity is the upper bound on the amount of information that can be transmitted over a transmission channel. According to the Nyquist (1928) theory, a noise-free channel will be constrained only by the bandwidth of the signal. Therefore, the highest signal rate that can be transported is 2B for a given bandwidth of B. Signals at higher rates will deteriorate because of delay distortion and ISI. If multilevel signaling is used to represent the data, the Nyquist formula will be

$$C = 2B\log_2 M \qquad (2.3)$$

where M is the number of discrete signal or voltage levels.

The datarate can be increased for a given bandwidth by increasing M. However, the receiver becomes more complex because of the requirement of distinguishing one of M possible signal levels. Furthermore, the addition of noise will limit the practical value of M for the transmission. Shannon's capacity formula states that the capacity, in bits per second, of a channel is limited by its bandwidth and its *signal-to-noise ratio* (SNR). The SNR of a signal corresponds to the ratio of the signal power to the power contained in the noise at a specific point of the transmission system. Shannon's capacity formula can be expressed as

$$C = B\log_2(1 + \text{SNR}) \qquad (2.4)$$

where C is the capacity of the channel in bits per second, B is the bandwidth of the channel in Hertz, and SNR is the signal-to-noise ratio. This formula represents the theoretical maximum achievable datarate.

Baseband Data Transmission

Baseband data transmission refers to the transfer of data without modification. Figure 2.7(a) shows the waveform of a possible digital baseband signal. After transmission over the restricted bandwidth of a channel, the signal might look like Figure 2.7(b).

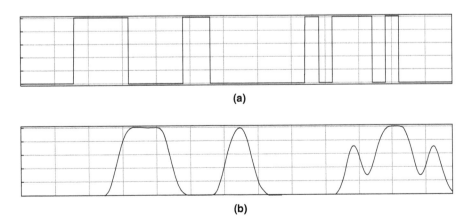

(a)

(b)

Fig. 2.7   Binary data transmission (a) before the channel and (b) after the channel.

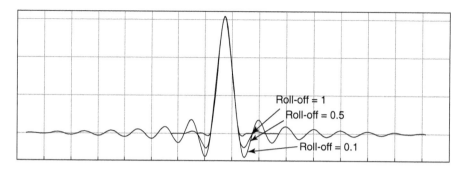

Fig. 2.8   Raised cosine filter response for different roll-offs.

Another effect known, as *Intersymbol Interference* (ISI), is also observed in Figure 2.7(b). This effect is due to the spreading in time of the pulses, provoking their overlapping. ISI can be reduced if the minimum transmission bandwidth allows all frequencies from zero to ½ T and rejects all the others (T is the time spacing between symbols). This was proved by Nyquist (1928). Then, if a brick-wall filter with a sharp cutoff is used before introducing the data signal to the channel, an ISI-free signal can be observed at the receiver side. However, this ideal case is not possible. As a consequence, errors will be observed at decision timing.

An alternative is to use a raised cosine function in the equalizer block. With this, the rectangular pulses are shaped to obtain ISI-free transmission. Figure 2.8 shows the raised cosine response in time of a filter that is fed by a rectangular pulse. The shape of the filter corresponds to different raised cosine roll-offs.

For a binary sequence, a single decision threshold is required. Each amplitude decision will have to be made in the presence of any distortion and interference superimposed on the data sequence during transmission. The decision-making process will be more reliable if it is timed by a clock that distinguishes when each received symbol has reached an optimum value. The accuracy with which the decision threshold and sample times must be placed will depend on the severity of distortion suffered during transmission. Hence, another stage at output is used to extract the clock that produces a retimed baseband waveform. This clock signal at the receiver will sample the binary pulses at a specific time (see Figure 2.9). If at the sampling time the amplitude of the arriving signal is lower than the amplitude threshold, a 0 bit will be detected. On the other hand, if at the sampling time, the amplitude of the arriving signal is higher than the amplitude threshold, a 1 bit will be identified.

The quality of the received signal will depend on the accuracy of the detection process, which will require the signal to be distinguished from the distortion and noise suffered during transmission. In general terms, channels with wide bandwidths will relax the timing and threshold requirements. However, a wider bandwidth allows more noise into the receiver; therefore, a compromise has to be considered.

Modulation

The processing of a signal to make it suitable for sending over a transmission medium is called *modulation*. The following are the reasons for using modulation:

1. Frequency translating (e.g., when an audio frequency baseband signal modulates a radio frequency carrier)

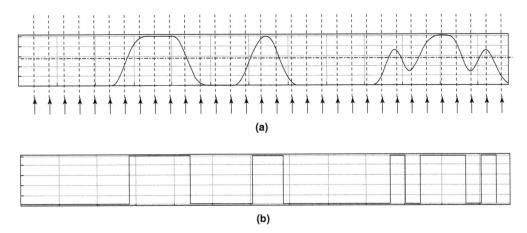

Fig. 2.9 (a) Binary signal after the channel (filter); (b) reconstructed signal after the receiver threshold.

2. Improving SNR by increasing the bandwidth (e.g., using frequency modulation)
3. Multiplexing (i.e., enabling many baseband channels to share the same wideband transmission path)

Modulation schemes can be classified into two major categories, analogue and digital, based on the nature of the data.

Analogue Modulation

Modulation is performed by causing the baseband modulating signal to vary a parameter of a carrier wave. An analogue carrier is defined by three parameters: amplitude, A_c, frequency, f_c, and phase, θ.

$$s(t) = A_c \sin(2\pi f_c t + \theta) \tag{2.5}$$

Thus, there are three basic modulation methods:

1. Amplitude modulation (AM)
2. Frequency modulation (FM)
3. Phase modulation (PM)

When modulation is employed, a modulator is needed at the sending end of a channel, and a demodulator at the receiving end recovers the baseband signal from the modulated carrier. The combination of modulator and demodulator at a terminal is often referred to as a *modem*.

Amplitude Modulation

The simplest form of modulation is amplitude modulation. The modulator causes the envelope of the carrier wave to follow the waveform of the modulating signal, and the demodulator recovers it from this envelope.

If a sine carrier is modulated to a depth, *m*, by a modulating signal, *x(t)*, the AM wave can be mathematically given as

$$s(t) = (1 + mx(t))\sin(2\pi f_c t + \theta) \tag{2.6}$$

In modern communication systems, AM is not a preferred modulation mechanism. It makes inefficient use of the transmitted power because information is transmitted only in the sidebands but the majority of the power is contained in the carrier.

More efficient variations of AM eliminate the carrier and generate only the sidebands. These schemes are known as double sideband suppressed carrier modulation (DSBSC) and single sideband suppressed carrier (SSBSC). The SSBSC signal requires the minimum possible bandwidth for transmission. Consequently, this method is used whenever its complexity is justified by the saving in bandwidth signal.

Frequency and Phase Modulation

Frequency modulation (FM) and phase modulation (PM) are both forms of angle modulation. The angle-modulated signal can be mathematically represented as

$$s(t) = A_c[\sin(2\pi f_c t) + \theta(t)] \tag{2.7}$$

Frequency modulation follows equation 2.7 with the condition that the derivative of the phase $\theta'(t)$ is proportional to the information signal *x(t)*, which can be mathematically represented as

$$\theta'(t) = \beta x(t) \tag{2.8}$$

where β is a constant, known as the *frequency modulation index*.

The instantaneous frequency of the carrier wave is hence changed according to the amplitude of the information signal. FM is more resistant to noise because random interference is more likely to affect amplitude than frequency of the signal.

In phase modulation (PM), the phase of the modulated signal, $\theta(t)$, is directly proportional to the information signal, *x(t)*. Mathematically, the relationship can be represented as

$$\theta(t) = \beta_p x(t) \tag{2.9}$$

where β_p is a constant, known as the *phase modulation index*.

In angle modulation, the information is conveyed by the instantaneous phase of the signal. Consequently, phase distortion in the transmission path causes attenuation distortion of the received signal. The differential delay of the transmission path must therefore be closely controlled over the bandwidth required to transmit the signal. However, because the information is not conveyed by the amplitude of the signal, the receiver can contain a limiter to maintain constant signal amplitude. Consequently, nonlinear distortion in the transmission path does not cause distortion of the demodulated output signal; nor does attenuation/frequency distortion.

In PM, the frequency deviation is proportional to the frequency of the modulating signal as well as to its amplitude. Consequently, for signals such as speech that have the major proportion of their energy at the lower end of the baseband, PM makes inefficient use of the transmission path bandwidth compared with FM. Moreover, to demodulate a PM signal, the receiver must compare the phase of the incoming carrier with that of a locally generated carrier, which must be very stable. FM is therefore preferred to PM for the transmission of analogue signals. It is used whenever sufficient bandwidth can be provided. However, PM is widely used for the transmission of digital signals, as described in the next section.

Digital Modulation

A digital signal can modulate an analogue carrier by using amplitude modulation, frequency modulation, or phase modulation. Some of the basic techniques are described below.

Amplitude Shift Keying

In Amplitude Shift Keying (ASK) or On-Off Keying (OOK), a binary 1 is represented by the presence of a carrier and a binary 0 is represented by the absence of it. Mathematically,

$$s(t) = \begin{cases} A\sin(2\pi f_c t), & \text{for binary 1} \\ 0, & \text{for binary 0} \end{cases}$$

(2.10)

Frequency Shift Keying

If a digital signal is used to modulate a carrier in frequency, this is known as Frequency Shift Keying (FSK). Binary FSK is equivalent to applying ASK to two carriers of frequencies, f_1 and f_2, one being switched on when the baseband signal is 0 and the other when it is 1. Mathematically, binary FSK can be given as

$$s(t) = \begin{cases} A\sin(2\pi(f_c + k)t), & \text{for binary 1} \\ A\sin(2\pi(f_c - k)t), & \text{for binary 0} \end{cases}$$

(2.11)

The spacing between the frequencies used for representing bits 0 and 1 (also known as tone distance) is an important parameter in the design of FSK systems.

Phase Shift Keying

If phase modulation is used to transmit a digital signal, this is known as Phase Shift Keying (PSK). If the signal is binary, the phase of the carrier is shifted between two positions that are 180° apart. Mathematically, binary PSK is represented as

$$s(t) = \begin{cases} A\sin(2\pi f_c t + \pi), & \text{for binary 1} \\ A\sin(2\pi f_c t), & \text{for binary 0} \end{cases}$$

(2.12)

PSK signals can have more than two levels. For example, a 4-level PSK signal uses four carrier phases, separated by 90°. PSK is therefore sometimes known as Quadrature Phase Shift Keying (QPSK). To transmit a binary baseband signal, a pair of binary digits is combined to form a 4-level signal (00, 01, 10, 11) corresponding to the four phases transmitted. At the receiver, each of the four output values is used to generate two consecutive binary output digits. By this means, the digit rate on the transmission path and the required bandwidth are halved.

PSK requires a synchronous demodulator, and synchronism is often difficult to maintain, particularly at high carrier frequencies. The need for accurate carrier recovery can be avoided by using Differential Phase Shift Keying (DPSK). This does not require the generation of a local carrier at the demodulator. In DPSK, the information is conveyed by changes in phase between digits instead of by the phase deviation of each digit from a reference carrier. If a greater number of phase differences, say 4, 8, or 16, are used, then the corresponding systems are called DQPSK, 8-DPSK, and 16-DPSK, respectively. $\pi/4$-DQPSK, another modulation technique, is a variant of DQPSK. In $\pi/4$-DQPSK an additional phase shift of $\pi/4$ radians is inserted in each symbol.

Another hybrid modulation scheme is Quadrature Amplitude Modulation (QAM), where both amplitude and phase are varied to represent bits. The spectrum utilization is greatly improved in QAM. Several higher-level QAM schemes (16-QAM and 64-QAM) are more bandwidth efficient.

Attenuation

Attenuation is an inherent property of any channel that makes the signal power decrease with increasing distance. For a general transmission medium (guided or unguided), attenuation can be defined as

$$A = \frac{P_T}{P_R} \qquad (2.13)$$

where A is the attenuation, P_T is the transmitted power, and P_R is the received power. Equation 2.13 can be expressed in dB:

$$A_{dB} = 10\log_{10}\frac{P_T}{P_R} \qquad (2.14)$$

For transmission lines, coaxial cables, and waveguides, received power is generally represented by the following equation:

$$P_R = e^{-2\alpha d}P_T \qquad (2.15)$$

where α is the attenuation coefficient (dependent on the medium) and d is the distance between the transmitter and receiver (channel length).

It is important to mention that attenuation varies with frequency depending on channel characteristics. For example, if an optical fiber is used as the transmission medium, an optical signal frequency of 193 THz will be attenuated less than 1 at 225 THz.

For wireless (free-space) transmission, attenuation is dependent on the nature of the terrain as well as on the radio frequency used and the distance between the transmitter and receiver. The received power based on a free-space propagation model can be given by the following equation:

$$P_R = P_T G_T G_R\left(\frac{\lambda}{4\pi d}\right)^2 \qquad (2.16)$$

where G_T and G_R are the transmitter and receiver antenna gains, respectively, in the direction from the transmitter to the receiver, and λ is the wavelength of the signal.

For long-distance optical fiber links, the received signal can become very small because of the attenuation (see Chapter 6). Intermediate repeaters or amplifiers can be used to overcome the attenuation losses in the channel. Boosters can also be used to preamplify the signal before the receiver. However, special care must be taken at the receiver not to overload the decision circuitry.

Delay Distortion

All signals suffer from delay after being transmitted through a channel. A perfect channel will provide at the output an identical delayed version of the input signal. However, distortion occurs because of the different speeds of the signal frequencies, causing signal spreading. Specifically for digital data, this spreading will cause adjacent bits to overlap in time. To overcome this distortion effect, an equalization stage can be implemented just after the channel output. Figure 2.10 illustrates these stages in the receiver block

Delay distortion can be considered a type of noise because of the deformation effect it produces on the signal transmitted. Different kinds of noise and their effect on the decision process will be explained in the next subsection.

Noise and Interference

Data transmission from transmitter to receiver is subject to much impairment that causes errors in the received data. These undesirable impairments can occur at the transmitter, in the channel, or at the receiver itself. They can be broadly classified into interference and noise.

Two main forms of interference are adjacent-channel interference and cochannel interference. In adjacent-channel interference, signals in adjacent frequency bands have components outside their allocated ranges, and these components can interfere with a wanted signal. This can be avoided by introducing guard bands between the allocated frequency ranges. A guard band is a small frequency band used to separate two adjacent frequency bands. Cochannel interference, sometimes also referred to as narrowband interference, is due to other nearby systems (say, AM/FM broadcast) using the same transmission frequency. This can be minimized by dynamic channel allocation techniques.

The most common noise impairment in data transmission is the linear addition of wideband noise with a constant spectral density (expressed as watts per hertz of bandwidth)

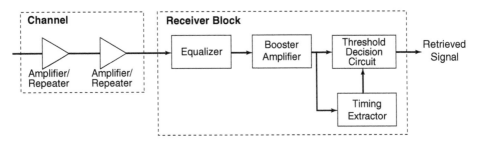

Fig. 2.10 Regenerator at the receiver block, including the equalizer.

and a Gaussian distribution of amplitude. Wideband Gaussian noise (also called white noise) comes from many natural sources, such as the thermal noise or shot noise.

Thermal noise arises from the random agitation of electrons because of temperature variations in the electronic device.

Shot noise is generated at the receiver and is caused by the random varying velocity of electron movement under the influence of externally applied potentials or voltages at the terminals or electrodes. It is similar to thermal noise in that it has a Gaussian distribution and a flat power spectrum. It differs, however, in that it is not directly affected by temperature. Its magnitude is proportional to the square root of the direct current through the device and thus can be a function of signal amplitude.

Other common types of noise include impulse noise and intermodulation noise.

Impulse noise is the effect of an external electromagnetic disturbance, faults, or flaws in the communication system. This type of noise will happen suddenly for a short period of time and will be represented as a spike in the transmitted signal. Impulse noise consists of short spikes of power having an approximately flat frequency response over the spectrum range of interest. It is considered to be a voltage increase of 12 dB or more above the r.m.s. noise for a period of 12 ms or less. This type of noise can be from a number of sources, such as the switching of relays in electromechanical telephone exchanges. Although these "clicks and pops" are annoying to the human ear, data is reasonably tolerant to it. However, impulse noise can cause many serious error rate problems on data or digital circuits.

The effects of impulse noise can be alleviated by the use of a wideband clipping circuit followed by a band-limiting filter. This procedure first reduces the amplitudes of the spectral components and then reduces the number of components.

Intermodulation noise is due to the presence of the products of intermodulation. If a number of signals of different frequencies are passed through a nonlinear device, a series of second-, third-, and higher-order sum and difference frequencies are produced, which are known as intermodulation products. These components can be inside or outside the frequency band of interest for the device. For most analogue systems with many multiplexed channels, the addition of the very large combination of signals results in an output noise spectrum, which is approximately flat with a frequency across a narrowband of about 4 kHz. Intermodulation noise differs from thermal noise because it is a function of the signal power at the point of nonlinearity.

For a given bitstream, the bit error rate (BER) at the receiver strongly depends on the SNR of the input signal, which is generally expressed in dB. For zero-mean Gaussian noise, the probability of making an erroneous decision is given as

$$P_e = \frac{1}{\sigma\sqrt{2\pi}} \int_a^\infty e^{\frac{-x^2}{2\sigma^2}} dx \qquad (2.17)$$

where α is the voltage difference between the received signal and its associated decision threshold at the decision time and σ is the r.m.s. noise voltage. This is the well-known area under the Gaussian tail. The analytical solution to this formula has to be approximated. Figure 2.11 shows an example of a BER graph for a two-level signal under Gaussian noise.

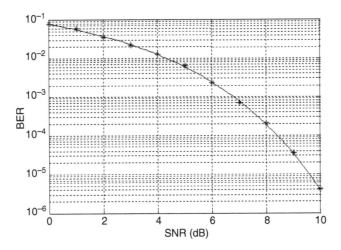

Fig. 2.11 Bit error rate performance in the presence of Gaussian noise.

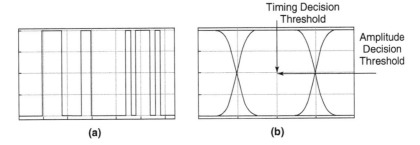

Fig. 2.12 (a) Random binary pattern; (b) eye diagram.

Transmission Performance

In communication systems, performance measurements are required to evaluate the efficiency of the system. The main reasons for transmission errors are channel noise, echoes, signal fading, and errors due to loss of synchronization at the receiver. The following subsections introduce different performance measures commonly used in communication systems.

Eye Diagram

The eye diagram, which is also known as the eye pattern, is a time domain representation of a digital data signal from a receiver that is repetitively sampled and applied to the vertical input, while the datarate is used to trigger the horizontal sweep. Figure 2.12 shows a random binary sequence of data and the corresponding eye diagram.

The binary pattern displayed in Figure 2.12(a) is noise free and without any jitter distortion. From Figure 2.12(b) it can be observed that the eye is completely "open" and correct

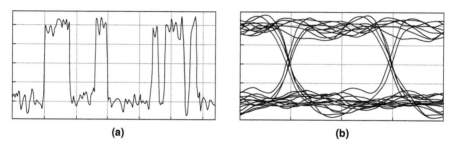

(a)   **(b)**

Fig. 2.13   (a) Random binary pattern with noise; (b) eye diagram with noise.

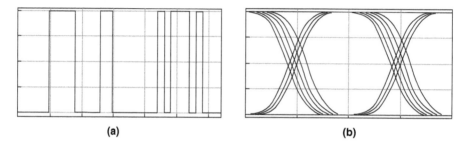

(a)   **(b)**

Fig. 2.14   (a) Random binary pattern with jitter; (b) eye diagram with jitter.

detection of binary sequences can be achieved if the amplitude decision threshold and the timing are located just in the middle of the opened eye. When noise is filtered in the transmission system, the eye becomes "closed." Figure 2.13 displays the binary pattern and its eye diagram with noise.

From the eye diagram of Figure 2.13(b) it is clear that the noise effects are mainly in the superior and inferior part of the eye, closing the diagram and complicating the decision at the receiver. ISI can also be detected on these areas when the pulses are shaped after transmission over a lowpass channel. Another source of distortion that affects system transmission is jitter, which can be observed in the eye diagram as well. Figure 2.14 shows the effect of jittering in a binary pattern and its eye diagram.

From Figure 2.14(b) it can be observed that the eye diagram becomes "closed" because of jitter or variations in time of the binary pattern. This "closure" is observed in the lateral sides of the eye diagram and results in nonoptimum sampling times. Furthermore, the slower the rising and falling edges of the pulses, the more closed the eye will be, which can result in more errors at the receiver.

Signal-to-Noise Ratio

The energy per bit-to-noise power spectral density ratio is a normalized measure of the SNR. The energy per bit of a binary digital data transmitted at a bitrate, R, is equal to the signal power, S_p, times the bit transmission, T_b. N_0 refers to the power spectral density of the noise measured at the detector input. Then,

$$\frac{E_b}{N_0} = \frac{S_p T_b}{N_0} \tag{2.18}$$

The noise can be calculated as $N_0 = kT$, where k is the Boltzmann's constant ($1.380650 \times 10^{-23} J/K$) and T is the receiving-system noise temperature. Equation 2.18 can be related to SNR by considering the noise, N, in a signal with bandwidth, B, being equal to $N = N_0 B$. Thus,

$$\frac{S}{N} = \frac{E_b}{N_0 B T_b} \tag{2.19}$$

E_b/N_0 is used as a measurement to compare the bit error rate performance of different digital modulation schemes without taking bandwidth into account. E_b/N_0 is useful in the design of a transmission link because, knowing this parameter, the corresponding BER can be approximated and the parameters in the formula can be obtained. It is important to mention that the higher the bitrate, the higher the signal power must be to obtain the same E_b/N_0.

Error Ratio

In general terms, error ratio is the number of symbols, codes, or characters incorrectly received in a transmission system compared to the total number of symbols, codes, or characters received. If a binary system is used, the *bit error ratio* denotes the number of bits received in error divided by the total number of bits sent in a specified time interval. The BER is affected by the rate of data transmission and the sensitivity of the system to the signal (by "sensitivity" is understood the minimum power accepted by the receiver to generate an appropriate decision).

The BER is usually measured over long periods of time. The measurement can vary per unit time and, therefore, the mean BER is the quantity required for some types of system. However, for other types the burstiness of the errors is more important—that is, the distribution of errors in time. This term is used because of the possibility of many errors at a specific interval of time and the chance of no or small amounts of errors at any other interval.

Different distributions can be used to estimate the amount of errors in a system. Furthermore, if a system is simulated, specific distributions can be used to calculate the number of errors in a transmission configuration to minimize the amount of computation required. The most common distribution for digital systems is the Gaussian distribution. In some specific optical systems where spontaneous noise is considered, Chi square distributions produce results closer to the error estimation.

Some typical objectives for BER are presented in Table 2.1. Databit error rates of the order of 10^{-9} mean one error in 10^9 bits sent. This measurement will depend on the requirement of the link of error-free transmission. In some cases, such as voice and ADPCM, the restriction is less due to the methods used for error prediction.

Latency

Latency is a measurement of the time it takes a signal to completely arrive at the destination from the moment the first symbol (bit) was delivered by the source. This parameter is very important if a transmission network is considered for the propagation of a signal. In a

Table 2.1 Typical BER Objectives

Digital Transmission Service	Transmission Rate	BER Objective (approx.)
Data	16 kbit/s–2500 Mbit/s	10^{-9}
Voice: ADPCM	32 kbit/s	10^{-4}
Voice: Log-law PCM	64 kbit/s	2×10^{-5}
Video: Linear PCM	60 Mbit/s	2×10^{-7}

network the transmission is usually executed by sending noncontinuous messages, such as packets of information, which can be a set of bits. In this way, the transmission is more bandwidth efficient. To achieve a proper latency calculation, parameters like propagation time, transmission time, and queuing time have to be considered.

Propagation time refers to the time it takes a bit of information to travel from the source to the target receiver. This time will correspond to the ratio of the distance to the speed of propagation in the transmission network.

Transmission time relates to the time to transfer a message. The messages (packets) consist of a set of bits that establish a message size. The transmission time, therefore, is the ratio of the packet size to the bandwidth of the channel.

Queuing time is a parameter in the transmission network that is required because of several factors. One is the need to keep the packet stored while its destination prepares to receive it. Another is the need to avoid internal collisions of packets that are sent to the same destination. Queuing in a transmission system is a topic by itself and is treated elsewhere.

Bandwidth Delay Product

The bandwidth delay product is a measurement of the transmission link, which represents the maximum amount of information that can be stored in a point-to-point connection. This parameter is the multiplication of the total bandwidth of the link multiplied by the time it takes the transmitted data to reach the destination. When a binary system is used, the bandwidth delay product is expressed in bits per second.

Timing and Synchronization

System timing and synchronization make up one of the essential foundations of all transmission equipment. A synchronized clock signal with minimal jitter is a fundamental parameter for proper data transmission.

Clock Requirements

Transmission systems need an efficient decision process at the receiver. Regenerators use timing recovery to extract a synchronization signal from the transmitted data. This is known also as clock extraction, and the most advanced receivers must include this function to decrease the number of possible errors. If the extraction is successful, the system becomes self-timed and it avoids the need to send timing information via a separate channel.

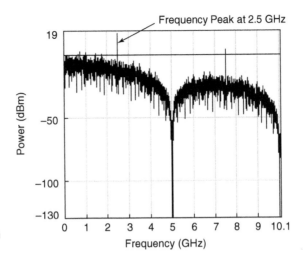

Fig. 2.15 Spectrum of a 2.5-GHz signal.

For efficient clock extraction, coding is needed in the data signal to guarantee a clock component in the signal to be transmitted. The clock component must present frequency, phase, and amplitudes sufficiently consistent for recovery at the receiver. Therefore, this signal must be free of distortions and interference. In some cases, when a long transmission link must be designed, intermediate regenerators are used to recover the attenuated and out-of-phase signal. Clock extraction is executed at these stages to regenerate the signal and reach the destination in the transmission link.

Clock extraction is also useful to decrease the amount of jitter in the signal, which is caused by the time fluctuations in the data patterns being transmitted. Jitter accumulation can be present when several repeaters are used. Therefore, correct clock extraction for data recovery will provide an improvement of the data received.

All sequences of digital pulses present symmetry in their amplitude and phase when they are observed in the frequency domain. This symmetry is independent of the data sequence received. Figure 2.15 shows the spectrum of a 2.5-Gbit/s transmission link at the receiver point. Return-to-zero coding is used (different methods of codification have been discussed elsewhere). As observed from the spectrum, a clear peak at 2.5 GHz is produced from the binary data signal. This peak can be chosen by a clock-selecting filter, which will pass the clock at the data frequency and will reject as much as possible of the pattern-related spectrum and noise. Methods for extraction of this frequency component include *Phase Lock Loop* (PLL) circuits or a *bandpass filter*.

When using a bandpass filter, the selection of the filter must be in agreement with the wanted quality, Q, of the circuit element. Q has two definitions. One definition is concerned with the ratio of the total energy stored in a circuit to the energy dissipated per cycle. The second definition relates to the frequency selectivity of a circuit, which is the ratio of the resonant frequency, f_0 to the 3-dB bandwidth, B. With the first definition it can be said that the larger the value of Q, the more efficient the energy storage element. With the second definition, the larger the value of Q, the better the frequency selectivity of the element due to the reduction of the bandwidth range.

In burst-mode transmission, the clock requirements become stricter because the timing recovery must be carried out as soon as the burst signal reaches the receiver. Some

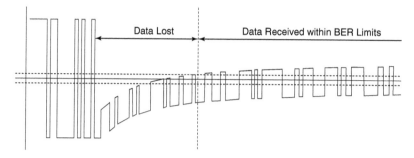

Fig. 2.16   Baseline drift effect in consecutive bursts due to AC coupling.

techniques send a preamble before the data signal, which contains clock information. Edge detection techniques (detection of only the rise and fall time of the data pulses) can also be used to almost instantly lock to incoming bursts. Special attention must be considered if the system is capacitively coupled (AC coupled) because of the adaptation of the bursts to 0-DC level, which can result in data lost at the beginning of the packet or burst. The amplitude and phase levels of the consecutive bursts are also affected by this baseline drift originated by the AC coupling. Figure 2.16 describes this effect in continuous bursts with different amplitude levels. The extraction of the clock will also be affected by the burst adaptation to the zero-DC level.

Synchronization

At the receiver, the transmitted digital signal must be interpreted correctly (i.e., the bits received must correspond exactly to the bits sent). If the clock is faster or slower, a mismatch can occur and the receiver will misunderstand the signals. *Synchronization* in digital signals refers to the process where the receiver will sample the transmitted bits at the correct moment and the correct time. Synchronization can be subdivided into *bit synchronization* and *frame synchronization*. With bit synchronization, the timing sampling is generated on each bit at the receiver. Therefore, an agreement on the bit duration in between the transmitter and receiver must exist to avoid errors. These errors can originate more bits than the transmitted ones if the bit-sampling duration is shorter at the receiver (see Figure 2.17(a)). On the other hand, if the sampling duration of the bit is longer at the receiver compared with the bit-transmitting duration, the receiver will miss some bits and the data transmitted will be incomplete (see Figure 2.17(b)). Other problems arise when a long trail of 1s or 0s is transmitted. This difficulty can impair the receiver's synchronization. To avoid this last issue, *scramblers* can be used, which add extra information to the bits in a random manner. DC components are also eliminated by the use of scramblers.

Bit synchronization distortion can arise whenever the sampling drifts in time with respect to the correct sample period. The synchronization will present timing jitter, and errors will occur at the receiver. Error propagation can appear whenever scrambling is being used and one bit in error is present. This error will derive more bits in error, and the only solution is to fill the scrambler with full correct bits.

Frame synchronization refers to the timing per frame in the transmission. Whenever a transmission starts, the receiver must know at which point a signal is present (or has

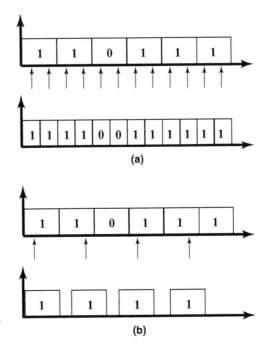

Fig. 2.17 (a) Oversampling and (b) undersampling at the receiver of a digital signal.

arrived). No identification of the signal frame will occur in the errors that are because of the interpretation of noise as random bits. Therefore, the frame synchronization starts with identifying the beginning of the frame. Another important aspect is to recognize subdivisions or frames within the transmitted signal. To do that, specific bit patterns are used as *prefixes*. The repetition of this prefix will allow the receiver to synchronize to the adequate frame by acquiring its bit synchronization signal.

Applications of synchronization are observed in video, digital audio, and digital telephony, for example. In the Synchronous Digital Hierarchy (SDH) standards, synchronization is very important because of the need to synchronize different multiplexed data frames, which usually hold different clock patterns. For this reason, several clock sources are used for synchronization. As priority one, an atomic cesium clock or a clock derived from a satellite can be used to synchronize the data frames. If the first priority is lost, the frames lock to a priority-two clock, which can be derived from another element within the network. If no other clock is available, the SDH network element clocks are locked to themselves, falling in a stage of "holdover"; meanwhile, a higher-priority clock is found in the network.

Jitter

The decision-making process at the receiver requires optimum timing of the digital signal (i.e., at the maximum eye opening of the demodulated bitstream). Variations from the optimum timing might exist, which are known as jitter. The effect of jitter on the decision-making process is illustrated in Figure 2.18. From this figure it can be seen that clock edge jitter can cause decisions to be made at less favorable times. Therefore, the probability of error can increase. Incoming noise and data-dependent timing effects

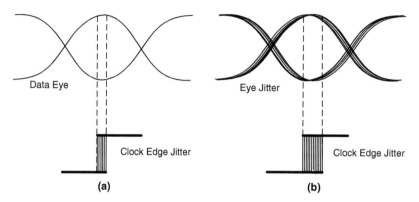

Fig. 2.18 Timing jitter and data eye: (a) clock edge jitter; (b) effect of jitter accumulation.

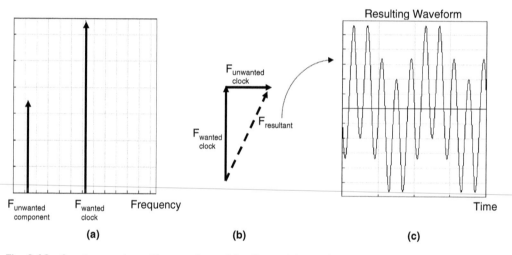

Fig. 2.19 Spectrum and resulting waveform of the jitter originated from an interfering spectral component: (a) spectrum of the desired clock frequency; (b) phasor representation of (a); (c) amplitude and phase modulation of resulting waveform.

that arise from imperfections and limitations in the clock extraction process are the main causes of jitter.

Timing jitter can be caused by spectral components close to the desired clock frequency. The combination of wanted and unwanted spectral components produces a resultant output clock that exhibits both amplitude and phase modulation. From Figure 2.19(a) it can be seen that the spectrum of the wanted clock and the unwanted component are at different frequencies. Figure 2.19(b) is the phasor representation of these two components and their resultant, where the resultant is amplitude and phase modulated. The resulting waveform in the time domain is represented in Figure 2.19(c).

Amplitude variations can be removed by a zero-crossing detector provided that the clock component is large enough. In general, an acceptable low-jitter performance can be achieved if a proper combination of the following elements exists: careful transmission coding (to guarantee a tolerable clock component relative to the continuous spectrum) and received noise limiting and an appropriate choice of Q for the clock selecting filter. When timing variations are present at frequencies lower than 10 Hz, the jitter effect is known as *wander*, and it is caused by propagation delay and phase changes. According to the ANSI T1.403 standard, wander can be classified as long term and short term. Long-term wander includes variations within 24 hours due to temperature changes. Short-term wander refers to the variations produced within 15 minutes.

Jitter Accumulation

Overall growth of jitter will appear because of the jitter arising from the clock extraction and retiming process in a regenerator, which will be transmitted to subsequent regenerators where it might merge with any locally generated timing variations. Two different tools are used to assess the effect of this jitter accumulation in a self-time system: eye jitter and clock edge jitter (see Figure 2.18(b)). Eye jitter is a straight result of any timing variations already present on the received data signal. Clock edge jitter occurs as illustrated in the preceding section (see Figure 2.18(a)). However, a contribution will also be made to the clock edge jitter by the incoming data signal.

The decision-making process at the regenerator requires that the extracted clock follow any phase variations in the incoming data. The most important is the relative timing of a particular decision instant, as far as the instantaneous SNR and the decision error probability are concerned—that is, the difference in the time position at the clock edge with respect to the most favorable decision time for the particular incoming data symbol. However, jitter will still accumulate from one regenerator to another even when jitter tracking is supporting the decision-making process at a particular regenerator. At the end of a link it might be necessary to use a stricter network clock to alleviate the jitter situation. Large data buffers might be also required to decrease data slips (loss or gain of bits). Nevertheless, the most important cause of trouble is the data pattern–dependent effect due to the reinforcement of each at subsequent repeaters. PLLs carefully designed are preferred in long chains of regenerators with extreme accumulation of jitter, rather than bandpass filters in the clock selection circuits. Data scramblers can also be used at intermediate regenerators to break up the data patterns.

Jitter Specifications

Standardization is needed to specify the maximum jitter allowed in a system. These specifications can be in terms of Unit Intervals (UI) versus frequency for different bitrates. UI is defined as one cycle of the clock frequency. The jitter measurement will then be the fraction of the variation with respect to the ideal period of the clock. This kind of metric allows comparison of different bitrates because it scales with the clock frequency. Other metrics include absolute units such as picoseconds, degrees, or radians. For example, if a clock rate of 622.08 MHz is considered, 1 UI will be equal to the period of the signal, which will be $1/622.08 \times 10^6 = 1.60$ ns = 1 UI. These 1.6 ns correspond to a phase of 360°. If 100 ps of jitter are present in the signal, the equivalent in UI will be 0.062208 UI, and the jitter phase will be 22.394°.

A typical specification for jitter limits for a digital transmission system comprising many regenerators is given in Figure 2.20. In this figure, peak-to-peak jitter is expressed in clock periods and is plotted against jitter frequency. This representative example is based on figures given in ITU-T Recommendations G.823 and G.824. The figure serves to

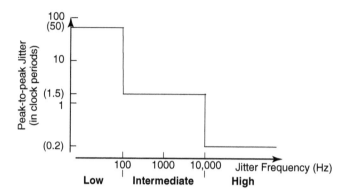

Fig. 2.20   Typical jitter limits for a digital transmission system.

highlight the low-frequency timing jitter (or wander), which can amount to many clock intervals.

Trends in Data Transmission

Data transmission communications are evolving according to demand and advances in technology. With the aim of interconnecting everybody to everywhere, complex networks at different levels have appeared. Some examples of these types of data networks are Ethernet and 10G Ethernet, IP over DWDM, and video codecs.

10G Ethernet

Ethernet is clearly the most widely understood LAN technology. The speed of Ethernet data transmission has grown enormously during the past years. Gigabit Ethernet delivers the scalable performance of Ethernet technology. Currently 10-Gbit/s Ethernet (10G Ethernet) is the highest standardized transmission speed in current use, although there is significant effort to raise this to 100 Gbit/s. 10G Ethernet links adopt full-duplex operation and use a fiber optic medium. SONET/SDH equipment can be used to provide long-distance data communications (see Chapter 10). Potentially carrier-class Ethernet (see Chapter 14) can simplify the architecture of communications networks.

Applications like multimedia (videoconferencing and real-time backup) require a high amount of bandwidth from the LAN, which will demand far higher speeds. Ethernet local-area networks (LANs) usually operate from 1 Mbit/s to 10 Gbit/s, and the trend is to continue increasing the speed due to its cost benefit. With such evidence, the only questions are how quickly 10G Ethernet will be adopted and to what extent it will displace alternative technologies. Early adoption of 10Gb Ethernet is happening where there is a critical need for additional bandwidth.

IP over DWDM

Data packet transmissions using the Internet Protocol (IP) are becoming the most practical transmission method used in data networks. The reason is the ability to transmit information through a channel, optimizing the channel's resources to the maximum. In other words, because of the frame format of IP, a channel can be used by many users simultaneously and without discontinuities in the transmission. On the other hand, advances in lasers and fiber optic devices permit very high speeds (on the order of THz) in the network. Dense Wavelength Division Multiplexing (DWDM) is the process of multiplexing signals of different wavelength onto a single fiber. IP over DWDM is the concept of sending data packets over an optical layer using DWDM for its capacity and other operations. This creates a vision of an all-optical network where all management is carried out in the photonic layer. The optical network is proposed to provide end-to-end services completely in the optical domain, without having to convert the signal to the electrical domain during transit. In addition, employing IP over DWDM saves the use of SDH equipment and multiplexing. In this case, Generalized Multiprotocol Label Switching (GMPLS), which is a derivation of Multiprotocol Label Switching (MPLS), forms the basis for a control plane for all-optical networks (for details of MPLS, see Chapter 13).

Multimedia Applications

A *video codec* is a type of codec that compresses or decompresses a video signal. This compression can be carried out via a device or software. The type of compression will depend on the application. For example, if the application is audio, video, or images, the compression can be lossy. On the other hand, if texts or data files are required to be compressed, a lossless compression must be used (see Chapter 3).

Compression/decompression of a video signal became possible thanks to the introduction of digital television, which acted as an improvement over analogue television. Digital television now offers High-Definition Television (HDTV) as well as standard-definition TV, an increased number of programs, reception in moving means of transportation, and distribution over the Internet and telecommunications networks.

New video codec applications have evolved rapidly. For example, MPEG-1 was developed for video storage on CD-ROM, and MPEG-2 was developed for video broadcast. H.263 was intended for videoconferencing. The trend in video codecs is expected to lead to a reduction in data speeds due to compression. For example, a reduction of approximately 3 Mb/s is expected for good-quality standard-definition video under MPEG-2. The transmission of video through Internet Protocol data is going to be an important visual service in the future, demanding more bandwidth because of the increased number of users. In this case, it is expected that several Mbit/s of IPTV data traffic will be on demand in the future as well. Finally, mobile video will be a service with increasing demand in the future. Compressions between 128 kbit/s and 64 kbit/s are expected.

Acknowledgment   The authors would like to acknowledge the contribution of Professor J. E. Flood in describing modulation.

Resources

Carlson, B., Crilly, P. B., and Rutledge, J. (2002) *Communication Systems,* Fourth Edition, McGraw-Hill, 850.
Couch, L. (2001) *Digital and Analogue Communications Systems,* Seventh Edition, Prentice Hall, 756.
Dunlop, J., and Smith, D. G. (1994) *Telecommunication Engineering,* Third Edition, Chapman and Hall, 593.

Forouzan, B. (2006) *Data Communications and Networking,* Fourth Edition, McGraw-Hill, 1134.

Glover, I., and Grant, P. M. (2004) *Digital Communications,* Second Edition, Pearson/Prentice Hall.

Kennedy, G. (1985) *Electronic Communication Systems,* Third Edition, McGraw-Hill, 741.

Mazda, F. (2001) *Telecommunications Engineer's Reference Book,* Second Edition, Focal Press.

Murthy, C., Ram, S., and Manoj, B. S. (2004) *Ad Hoc Wireless Networks: Architectures and Protocols,* Prentice Hall Communications Engineering.

Nyquist, A. (1928) Certain Topics in Telegraph Transmission Theory, *Trans. AIEE,* 47:617–644.

Stallings, W. (2007) *Data and Computer Communications,* Eighth Edition, Pearson/Prentice Hall, 878.

Stremler, F. (1990) *Introduction to Communication Systems,* Third Edition, Addison Wesley, 757.

3 Digital Video Transmission

John Charles Woods
University of Essex

The transmission and reception of digital video is a complex problem compounded by the idiosyncrasies of the transmission media and the increased value of bits within the stream as a result of compression. To fully understand the implications of transporting digital images and video, the coding and compression processes underlining them must first be understood.

Digital Representation of Video

The use of moving images in a computer requires the use of a discrete, digital representation. Subsequent processing inside the digital domain allows a variety of processing techniques to be applied.

Monochrome Images

Monochrome digital video resides in memory as an array of picture elements (pixels) of size $A \times B \times C$. Each pixel is quantized to D bits, or 2^D brightness levels, so progressive monochrome (black and white) video has a bitrate of $ABCD$ bits/s. The brightness level associated with each pixel is the value that is stored in computer memory and used in subsequent video analysis. The stored data is scanned out to create a moving sequence. Still frames are stored as two-dimensional arrays, ABD; this can be seen in Figure 3.1.

Color Images

When images are represented in color, two additional arrays are required. For high-quality images such as for studio applications, the three arrays can be represented by the primary colors gathered by the camera: red, green, and blue (RGB), respectively. However, more

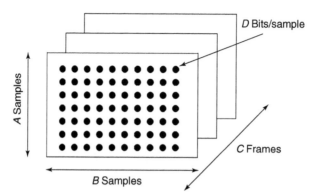

Fig. 3.1   Digital video representation.

commonly a new color space is created consisting of the luminance (black and white) component and two other chrominance (color) components as seen in the NTSC, PAL, and SECAM standards. The amount of color component compared to the luminance categorizes the image format. Image formats have a variety of options—for example: (1) CCIR-601 or 4:2:2 format where the number of horizontal chrominance pixels is half the luminance, and (2) SIF or 4:2:0 format where the horizontal resolution of the chrominance is halved and the vertical is divided by 4 (Ghanbari, 1999). Reducing the amount of color component clearly reduces the quality of the image, but such is the nature of the human eye (Pearson, 1975) that the renditions are generally indistinguishable from the full-resolution RGB.

Image Size

Images come in a variety of sizes depending on whether they use the NTSC or the PAL standard (ISO/IEC, 1995). They range from small thumbnails up to super-high definition. For example, sub-QCIF, QCIF, and CIF images have sizes 128 × 96, 176 × 144, and 352 × 288, and they find widespread use in multimedia applications. When it comes to entertainment systems, there is a variety of images sizes. With standard broadcast size, the PAL standard uses 720 × 576 and NTSC employs 720 × 480 (Ghanbari, 2003). With higher-definition entertainment systems, picture formats are often proprietary and await convergence of the standards.

So-called *HD-ready* pictures are 1024 × 768, *720p* is 1280 × 720, *HD-ready 1080p* is 1280 × 1080, *HDV 1080i* is 1440 × 1080, *FULL-HD* is 1920 × 1080, and *2160p* or *quad HDTV* (super-high definition) is 3840 × 2160. This list is not exhaustive. Other innovations shortly to be offered include 3 DTV where the bandwidth required is up to twice that for standard monoscopic displays.

Raw Digital Video: Motivation for Compression

Table 3.1 illustrates the datarates required to represent monochrome images for various different imaging devices. The addition of color information will increase these bitrate values between 1.5 and 3 times.

Consider a digitized broadcast-quality DVD video used in the UK: 720 pixels/line with 576 lines/frame, 25 pictures/sec, 8 bits/sample, and 4:2:2 color format. This would

Table 3.1 Raw Monochrome Bitrates for Imaging Devices

System	A	B	C	D	ABCD (Mbits/s)
Low-quality videophone	64	64	8	6	0.2
CIF videoconferencing	352	288	30	8	24
Digital broadcast TV	720	576	25	8	83
HDTV	1920	1150	50	8	883
SHDTV	2048	2048	60	10	2500

require 166 million bits per second or 2600 telephone lines and fill a standard 4.7-GB DVD in less than 4 minutes. If this were HDTV this would be 664 million bits per second and would fill the DVD in less than 1 minute. Even with the smallest pictures, SQSIF (Ghanbari, 1999), which is 120 pixels per line and 96 lines per frame at 10 frames per second using 4:2:0 color format, the required channel capacity is still three times that of normal telephony. Clearly, digital video on a DVD lasts longer than this, and so the volume of data has in some way been reduced or compressed. The same is true when the information is transmitted using digital terrestrial broadcast, satellite, or networked-over-IP networks, and so on.

Compression Overview

Video and still images consume large amounts of memory when stored and require high bandwidth when transmitted over networks. Therefore, methods are required that reduce the bitrate requirement of the raw (PCM) data.

Pulse Code Modulation

Pulse code modulation (PCM) is a digital representation of an analogue signal. The analogue amplitude is sampled and then quantized into a number of discrete bands; for example, 8-bit resolution would constrain the signal level between 0 and 255. In video the PCM values form the starting point for subsequent analysis and compression.

Lossless and Lossy Compression

Two types of digital video compression schemes exist: lossless and lossy. A lossless codec has the ability to perfectly reconstruct the original image without degradation (Sayood, 2002). The compression efficiency or gain is very limited, and the high bitrates make lossless codecs unsuitable for the majority of purposes except high-resolution images. A lossy codec acknowledges that higher compression gains are required and achieves this at the expense of image quality (Ghanbari, 1999). A lossy codec discards information, resulting in an imperfect reconstruction or facsimile of the original image; however, it is generally suitable for entertainment purposes.

Many years of research have culminated in the modern generation of hybrid codecs that compromise bitrate and fidelity (Ghanbari, 1999). These codecs include: MPEG-1, MPEG-2, H.261, H.263, H.26L, and MPEG-4.

These codecs exploit the following:

- *Spatial redundancy*, recognizing that pixels inside a picture are similar
- *Temporal redundancy*, recognizing that successive frames are similar
- *Statistical redundancy*, recognizing that more frequently occurring events should be assigned shorter codewords

It should be noted that image coding exploits spatial and statistical redundancy, whereas video coding uses exploit spatial, statistical, and temporal redundancy.

Compression Ratios

A compression ratio is the ratio of the raw bitrate compared to the bitrate after compression, as seen in Figure 3.2. For any given technique, the compression ratio, K, is given by

$$K = B_1/B_2 \text{ for still images} = R_1/R_2 \text{ for video}$$

The use of compression can have profound effects for the application in question. Figure 3.3 shows the possible ranges for typical video applications both with and without the use of compression. The upper limit of each block corresponds to the raw or uncompressed bitrate, R_1, and the lower limit to the compressed rate, R_2. Because the scale is logarithmic, the height of the block, $\log K = \log R_1 - \log R_2$. Compression of two orders of magnitude is achievable for some applications.

Good overall performance is obtained by a combination of efficient source coding and efficient channel coding (modulation), as seen in Figure 3.4. $K = R_1/R_2$ is the data compression ratio. $M = R_2/B$ in bits/Hz is a measure of the channel coder efficiency.

We are interested in the cascaded effect of analogue-to-digital conversion, compression, and modulation. We measure overall performance by the ratio of output to input bandwidth, which we term the bandwidth expansion ratio, E. Therefore:

$$E = \frac{B}{W} = \frac{R_2}{MW} = \frac{R_1}{KMW} = \frac{2N}{KM} \qquad (3.1)$$

Video Compression Standards

Considerable research in still-picture and video coding has resulted in a set of standards that can reduce the bitrate requirements by orders of magnitude while still maintaining subjective quality.

MPEG

Commonly used video compression standards are MPEG-1, MPEG-2, MPEG-4, H.261, H.263/263+, H.264, and H.26l.

Fig. 3.2 Finding the compression ratio.

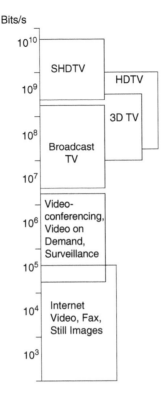

Bits/s

Fig. 3.3 Bitrates for various applications with and without compression.

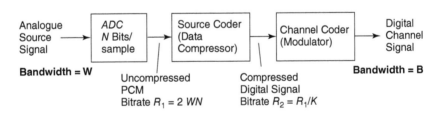

Fig. 3.4 Sampling, compression, and modulation.

MPEG, the Moving Picture Experts Group, standardized MPEG-1 in 1992. The system was designed for VHS-quality video and at 1.2 Mbit/s could hold 75 minutes on a 640-Mbyte compact disk. It had SIF-size images (352 × 288) at 25 Hz and offered editing facilities along with the first example of software coding. It was also suitable for streaming over IP. It used I, B, and P pictures, half-pixel motion estimation, and a slice structure to limit the effects of error propagation. It did not support interlaced video and was effectively super-seded by MPEG-2.

The MPEG-2 standard was published by the ISO/IEC in 1994 and was designed to provide high-quality video encoding suitable for transmission over computer networks (ISO/IEC, 1995). MPEG-2 serves a wide range of applications, such as HDTV, digital storage media, and television broadcasting. It is based on motion-compensated block-based transform-coding techniques, with target bitrates between 4 and 9 Mbit/s. It uses intraframe compression to remove redundancy within frames and interframe compression to reduce redundancy contained over sequences of many pictures.

An MPEG-2 sequence contains three types of frames, I, P, and B, and supports variable picture sizes; in most cases, the pictures are interlaced. MPEG-2 supports scalability (Ghanbari, 2003) by permitting the division of a continuous video signal into two or more coded bitstreams representing the video at different resolutions, picture qualities, or picture rates. Additional frame buffering is required at the encoder for prediction of B frames.

MPEG-3 was originally intended for HDTV, but it was realized that MPEG-2 could be modified to accommodate this and so it was abandoned.

MPEG-4 was standardized in 1998 and represented a departure from standard thinking about video in that it attached importance to objects within the image. This allows objects to be mixed together—for example, cartoons and real objects—and support for the Virtual Reality Modeling Language (VRML) was also offered. MPEG-4 added to MPEG-1 and -2 integration of natural and synthetic support for 2D and 3D content and interactivity. Coding is at very low rates and supports management and protection of intellectual property.

MPEG-4 is also compatible with major existing standards: MPEG-1, MPEG-2, ITU-T H.263, and VRML. The MPEG-4 standard is very extensive, and some regard the specification as a "wish list" because it is so large. Profiles are declared, which define subsets of the standard for individual implementations. The chief difficulty associated with MPEG-4 is the requirement to segment objects from a scene, and consequently MPEG-4 only finds widespread use in the "block-based" profile.

MPEG-7 was formalized in 2000 and produced a standardized description of the multimedia. An ability to describe the content facilitates efficient searching and browsing. The standard does not code video but rather describes it. MPEG-7 is formally called the *Multimedia Content Description Interface*.

MPEG-21 defines a standard for permissions and digital rights for audio and video content. It contains machine-readable license information. The numbering of the MPEG standards remains an enigma, with various stories associated with the missing numbers. However, no clear pattern emerges.

H.26x

The H.261 standard was published by the CCITT (now ITU-T) and was designed for videoconferencing and videophone applications over ISDN telephone lines. It has traditionally utilized 64-kbit/s baseline ISDN and uses multiples of 64 kbit/s (ITU-T, 1993). Buffering is used to smooth out variations in bitrate generated by the video encoder; to achieve a constant bitrate, a simple feedback loop is employed to report on the state of the buffer to the encoder. If the buffer becomes too full, the quantization scale factor is increased to reduce the data at the expense of video quality.

The coding technique in H.261 removes temporal redundancy by interframe prediction, and spatial redundancy by transform coding. There are no B frames, and motion compensation is optional; the codec makes a decision whether to inter- or intracode macroblocks. H.261 defines two resolutions, 352×288 and 176×144, commonly known as CIF and QCIF formats, respectively.

H.263 was published by ITU-T and was particularly aimed at coding video for low bitrates (typically 20–30 kbit/s) (ITU-T, 1998). The H.263 standard is based on techniques common to many current video-coding standards: interpicture prediction to utilize temporal redundancy and transform coding to reduce spatial redundancy. Motion compensation is based on one motion vector per macroblock, coded with half-pixel precision, and the decoder has motion compensation capability. The H.263 standard supports five picture formats: sub-QCIF, QCIF, CIF, 4CIF, and 16CIF. The luminance component of the picture is sampled at these resolutions, while the chrominance components are subsampled by a factor of two in both horizontal and vertical directions.

H.263+ is an extension of H.263 which provides new negotiable modes and additional features (ITU-T, 1998; Côté et al., 1999). These features improve compression performance over packet-switched networks and support custom picture size. Recently, a new video codec, H.264/AVC (Advanced Video Coding), has been standardized. The codec has been in use since May 2003 and is intended for coding at very low bitrates. It has been recommended by the ITU-T Video Coding Experts Group (VCEG) and the ISO/IEC Moving Picture Experts Group (MPEG) (ISO-TEC, 2003). The codec is reported to have better performance than existing codecs, with enhanced compression performance and provision of network-friendly video representation. More literature about H.264 is available from Ghanbari (2003) and Wiegand and colleagues (2003).

JPEG

The Joint Photographic Experts Group (JPEG) standard was developed by the ISO in collaboration with ITU-T for encoding, transmitting, and decoding still images (Ghanbari, 2003; Wallace, 1991). The JPEG coding scheme compresses still images, including both gray scale and color, using any color space, and can be applied to most image sizes. Compression ratios of up to 15:1 for full-color images are achievable. JPEG specifies lossy and lossless classes of compression and decoding, employing the Discrete Cosine Transform (DCT) and Differential Pulse Code Modulation (DPCM), respectively.

Several encoding modes are available with the JPEG lossy compression method: baseline sequential mode, progressive mode, and hierarchical mode (Wallace, 1991).

In baseline mode, an image is partitioned into 8×8 blocks of pixels. A forward DCT is applied to each 8×8 block to generate DCT coefficients, transforming the image from the spatial to the frequency domain. The resulting DCT coefficients are quantized and are ordered in a zigzag scan and entropy-coded for transmission or storage.

JPEG progressive encoding is used to allow a coarse version of an image to be transmitted at a low rate and then progressively improved by subsequent transmissions. This is a useful characteristic, specifically when browsing, because the user is presented with a representation of the content before the entire page is loaded.

Hierarchical (pyramidal) encoding is used when a high-resolution image is to be displayed on a low-resolution display device. JPEG uses pyramidal techniques where an image is down-sampled by a factor of two in each direction. The smaller image is then encoded using another encoding method (progressive or baseline). The encoded image is decoded and up-sampled; the difference between the up-sampled and the original image is encoded using other encoding methods (progressive or baseline). Other modes of image encoding supported by JPEG are entropy encoding and lossless encoding.

A recently standardized JPEG2000 codec (Skodras, Christopoulos, & Ebrahimi, 2000) has the goal of providing better performance by generating fewer and less visible artifacts. These techniques have also been adopted in the later generations of video codecs. With the emergence of new applications that require high-quality images, it is important that coding systems have the ability to provide lower bitrates and better subjective image qualities than existing standards. There are numerous publications available about JPEG2000, such as Christopoulos, Skodras, and Ebrahimi (2000).

Spatial, Temporal, and Statistical Redundancy

Within a still image it is possible to exploit spatial redundancy or statistical redundancy, whereas in a moving sequence you can use spatial, temporal, and statistical redundancy. This section details these approaches.

Spatial Redundancy Reduction

A technique used for a number of years that still has applications in high-quality images is Differential Pulse Code Modulation, or DPCM (Kondoz, 1994). DPCM codes the difference between the pixel of interest and a prediction. The prediction is usually based on the neighborhood of the pixel of interest, and it is the error in the prediction that is coded rather than the value of the incoming pixel. The operation is straightforward and illustrated in Figure 3.5. The output of the system, which represents the difference between the prediction and the incoming pixel, does not in itself represent a data reduction, but with a good predictor the output DPCM will have a small dynamic range.

Data reduction/compression can be achieved with the addition of an entropy coder that assigns shorter codewords to commonly occurring symbols.

If the prediction strategy is appropriate, then compression is achieved, but otherwise expansion can result. A DPCM encoder is shown in Figure 3.5, and the decoder is shown in Figure 3.6.

A similar process but in reverse is applied at the decoder, allowing the original bitstream to be recovered. If a quantizer is included in the path, then coefficients will be forced to reside in discrete levels; there is the potential for increased compression, but the

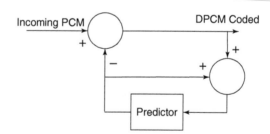

Fig. 3.5 Simple DPCM encoder.

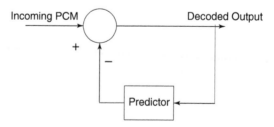

Fig. 3.6 Simple DPCM decoder.

original information can no longer be reconstructed; only a facsimile is possible. The complexity of the prediction and the spatial correlation of the image ultimately dictate compression ratios, which are usually well below 10:1.

It should be noted that even if the predictor is complex and high quantization levels are employed, it is impossible to achieve less than 1 bit per pixel, and therefore the efficiency of this system is bounded. At this level of compression for standard broadcast video, 720×576, 25 frames/sec, and YUV 4:2:2 color format, a DVD would only be capable of holding just over 30 minutes of video. Clearly a different compression method is required that can achieve higher rates than 1 bit per pixel.

Transform Coding

As is often the case in engineering, a problem can seem insoluble until it is subjected to a transformation. The widespread deployment of the two-dimensional DCT in image processing is such a transform. The image is divided into blocks, typically 8×8, on which the DCT is applied. The transformation is similar to Fourier analysis (Körner, 1988) in that it converts from the time domain into the frequency domain and back again.

It should be noted that a block of 8×8 pixels in the time domain results in an 8×8 block of coefficients in the frequency domain, so the transformation itself nets no gain; in fact, it could be regarded as an expansion because the frequency domain coefficients are larger than the time domain coefficients. The compression comes from the way in which the coefficients themselves are quantized and owes a lot to the detailed studies done on the spatial frequency responses of the human eye (Pearson, 1975).

The graph in Figure 3.7 shows the contrast sensitivity of the human eye. The horizontal axis represents the frequency; the vertical axis, the sensitivity. This graph shows us that the eye has a bandpass characteristic with maximum sensitivity around 8 Hz. The relative insensitivity to frequencies above and below this point can be clearly seen. What the graph shows us is that the human eye is relatively insensitive to high and low spatial frequencies. Note particularly that frequencies beyond 50 cycles per degree are difficult to perceive.

With knowledge of human visual response, a codec can be designed to take advantage this phenomenon. Because the DCT is applied to the image and a frequency domain representation is available, emphasis can be applied to the parts of the spectrum that are humanly

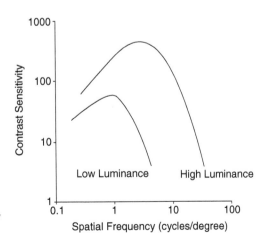

Fig. 3.7 Human spatial frequency response.

important. The application of nonlinear quantization of the coefficients allows the quantity of information conveyed to be reduced while still maintaining subjective quality.

Discrete Cosine Transform

The Discrete Cosine Transform (DCT) separates the image into parts (or spectral subbands) of differing importance with respect to the image's visual quality. The DCT is similar to the discrete Fourier transform: It transforms a signal or image from the spatial domain to the frequency domain. With an input image, **A**, the coefficients for the output image, **B**, are

$$\mathbf{B}(k_1,k_2) = \sum_{i=0}^{N_1-1}\sum_{j=0}^{N_2-1} 4 \cdot \mathbf{A}(i,j) \cdot \cos\left[\frac{\pi \cdot k_1}{2 \cdot N_1} \cdot (2 \cdot i + 1)\right] \cdot \cos\left[\frac{\pi \cdot k_2}{2 \cdot N_2} \cdot (2 \cdot j + 1)\right] \qquad (3.2)$$

The input image is N_2 pixels wide by N_1 pixels high; $\mathbf{A}(i,j)$ is the intensity of the pixel in row i and column j; $\mathbf{B}(k_1,k_2)$ is the DCT coefficient in row k_1 and column k_2 of the DCT matrix. All DCT multiplications are real. This lowers the number of required multiplications as compared to the discrete Fourier transform. The input to the DCT transform is an 8×8 array of integers containing each pixel's gray-scale level and is typically 8 bits or 0 to 255. The output array of DCT coefficients contains integers; these can range from -1024 to 1023. For most images, much of the signal energy lies at low frequencies; these appear in the DCT's upper left corner. The lower right values represent higher frequencies and are often small—small enough to be neglected with little visible distortion. Figure 3.8 shows the basis functions, which are the individual coefficients reconstructed in the pixel domain.

Significance of the Coefficients

For each 8×8 block of pixels contained in the image, the DCT is applied, producing an 8×8 grid of coefficients representing the amount of each individual basis function the block contains. If all of the coefficients are retained, the transformation can be reversed to arrive

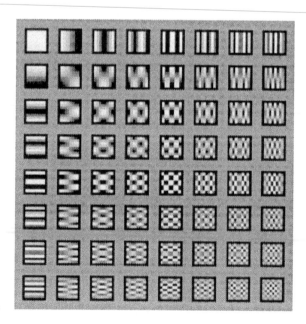

Fig. 3.8   DCT basis functions.

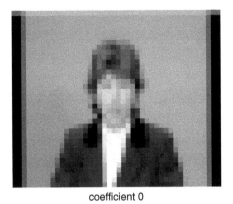

coefficient 0   coefficients 0, 1, and 8

Fig. 3.9   Reconstructions of the Claire image.

at the original pixels; in other words, the image has full fidelity. If, however, some of the coefficients are discarded and some of the frequency information is missing, degradations are introduced into the image reconstruction.

This is shown in Figure 3.9, where the Claire test sequence has been transformed using the DCT and the image has then been reconstructed back into the pixel domain without using all the available coefficients. The figure on the left has been reconstructed using only coefficient 0 (i.e., the dc component), and the figure on the right uses coefficients 0, 1, and 8. Note there is an increase in fidelity when more coefficients are used and that if all 64 were used, the image would have full fidelity. Compression is generally achieved in DCT codecs by quantizing or leaving out some of the DCT coefficients.

Two-Dimensional Variable-Length Coding

Variable-length coding is used to code the DCT coefficients and is discussed later in this chapter. The result of applying the DCT transform to an 8×8 block of pixels is a block of 8×8 numbers or coefficients that represent the amount of each basis function (from Figure 3.8), which go to make up the block of pixels. Subjectively, the important information is generally contained in the top left of the figure, where the low-frequency components reside. Given that we wish to code this information, we can use variable-length coding but have the additional difficulty of needing to code a 2D array of numbers. To solve this, the 8×8 grid needs to be scanned. This is done in a zigzag fashion starting at the top left and descending to the bottom right, as shown in Figure 3.10.

This action produces an array of 64 coefficients. You will recall from Figure 3.7 that the eye is not very responsive to high frequency and therefore the components in the top left corner of Figure 3.10 are of particular importance. When the grid of coefficients has been zigzag-scanned there remains a 1D array of coefficient values. The values at the beginning of the array have the lowest spatial frequency.

Quantization of Coefficients

Transforming a block of pixels using the DCT transform gives rise to a block of coefficients. Performing the inverse DCT on this block of coefficients produces the original block of pixels; however, it should be noted that in performing the DCT transform no compression has taken place. Compression is almost universally achieved in image codecs by the imposi-

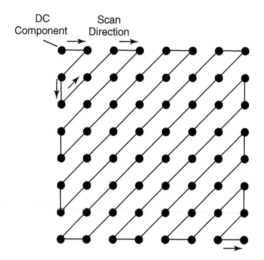

Fig. 3.10 Zigzag scanning of DCT coefficients.

Table 3.2 Quantization Step Size Q = 16

Coefficient	−107	23	12	−15	14	187	−91	3	17
Final value	−6	1	0	0	0	11	−5	0	1

tion of quantization. Figure 3.9(a) and (b) shows the relationship between fidelity and the number of coefficients used.

If the coefficients are subjected to quantization, a reduction in bitrate can be achieved, but this is inevitably at the expense of image quality. A quantization scheme needs to consider a compromise between the bitrate and the resultant distortion. This is known as rate distortion (Ghanbari, 2003) and was studied by researchers prior to implementing the codecs. Therefore, most modern codecs have made this decision for us and have been designed to also take into consideration the behavior of the human visual system.

Typically, bitrate is controlled by adjusting the quantizer step size. The step size is often controlled by the output buffer in an attempt to control the rate of data generation. The quantizer operates on the raw coefficients, reducing both the dynamic range and the number of coefficients to be coded. Table 3.2 shows an example of this where some arbitrary values have been quantized and indexed using a step size of $Q = 16$. Note particularly the presence of 0s produced by the quantizer. For real-world images, typically the final values to be coded consist of long runs of 0s.

Therefore, a technique known as run length coding is used where two values, the number of 0s to the next coefficient followed by the value itself, are coded. This information is fed to a Variable-Length Coder (VLC) (Ghanbari, 2003) to exploit the statistical redundancy still contained within the bitstream.

Temporal Redundancy

Modern codecs exploy the spatial correlation (local similarity) within a single image and use knowledge of human visual response to reduce the overall bitrate. With moving images

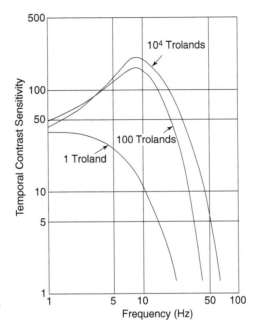

Fig. 3.11 Temporal response of the human visual system.

it is possible to exploit not only the spatial information present but also the temporal (motion), as the codecs have access to images from the past, present, and future.

The temporal response of the human eye, shown in Figure 3.11, also has much to do with the design of modern codecs.

You will note from the figure that the eye is very susceptible to a frequency of around 10 Hz and so frame rates around this frequency should be avoided. Also note that at frequencies around 50 Hz the eye has a very low sensitivity. This is one of the factors governing the frame and field rates in domestic broadcast television.

A pair of consecutive frames taken from a video sequence are generally very similar to one another, allowing codec designers to exploit temporal redundancy. If there were no motion at all, a message could be sent to the decoder to redisplay the previous frame. If a small amount of motion were present, only the changes would need be sent. If content could be found in both frames that had simply moved, then the decoder could be signaled to identify the area and how much it had moved.

Motion Estimation
With these basic premises, a method can be defined that uses the redundancy in coding two consecutive frames of video (ITU-T, 1998). The image is divided into a series of small blocks or macroblocks (e.g., 8×8) and then a search is conducted for content in the current frame using a shifted block from the previous frame. When the point of lowest error has been found, the x and y offsets represent the motion vectors for the particular block.

With reference to Figure 3.12 for the Mean Squared Error (MSE), this can be formalized as

$$M(i, j) = \frac{1}{N^2} \sum_{m=1}^{N} \sum_{n=1}^{N} (f(m, n) - g(m + i, n + j))^2, \, -w \le i, j \le w \qquad (3.3)$$

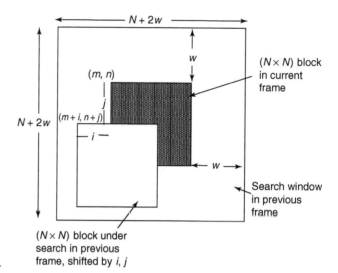

Fig. 3.12   Block search estimation.

In other words, search up, down, left, and right by the width of the search block. The point with the lowest error is the best match.

Motion Compensation

When the motion has been determined for each individual block, it can be used to make a prediction of the next picture. Not all macroblocks within a picture will use motion compensation. A decision is made as to whether the prediction significantly lowers the error between a pair of frames. If it does, then motion compensation will be used; otherwise, DCT coding will be used. Note that this can be coded with respect to the current frame (intra) and with respect to the previous frame (inter). The motion vectors are coded and sent as side information (ITU-T, 1998).

Statistical Redundancy

Having exploited the spatial and temporal redundancy contained in the image, there are still further economies to be had in the redundancy contained inside the stream of data generated from the coding process. The transform coefficients and the motion vectors are subjected to a type of coding that assigns short codewords to frequently occurring values and long codewords to infrequent one. The earliest example of this was Morse code. Therefore, for a data source the minimum average number of bits required per codeword or the entropy $H(x)$ can be calculated as

$$H(x) = -\sum_{i=1}^{n} p_i \log_2 p_i \qquad (3.4)$$

Assigning different-length codewords to data is known as Variable-Length Coding (VLC). There are two types of VLC used in video: Huffman coding and arithmetic coding.

Huffman Coding

The simplest and most commonly seen form of variable-length coding is Huffman; its use is not restricted by patents as arithmetic coding is. To generate a Huffman code requires adhesion to three simple rules:

1. Rank-order the symbols according to their probability of occurrence.
2. Merge the two symbols with the least probability to form a composite symbol and recompute the rank ordering.
3. Trace a path from the root to each leaf node, noting the direction taken at each node.

Table 3.3 shows the symbol set EASYZ in the first column, and the probability of occurrence in the second. The table is used to create the tree shown in Figure 3.13. When the tree has been created, move from right to left and write down a 1 or a 0 depending on which path is taken at each fork.

As can be seen in this simple example, symbols that have low probability are often assigned long codewords, leading to inefficiencies. Techniques to overcome this include a modified Huffman code (Ghanbari, 1999) or a 2D/3D version of Huffman coding.

Arithmetic Coding

The principal difficulty of Huffman coding is that it becomes very inefficient if the probability of the symbol becomes high in part because of its ability to assign only an integer number of bits. Arithmetic coding creates a code string that represents a fractional value, meaning it is not bound by the integer restrictions of Huffman and can achieve theoretical entropy for a given data source. Arithmetic coding is heavily bound by patents, which often restrict its use. Using the same data contained in Table 3.3, the probabilities are arranged into an

Table 3.3 Characters for Coding and Their Probabilities

Symbol	Probability
E	0.4
A	0.32
S	0.2
Y	0.06
Z	0.02

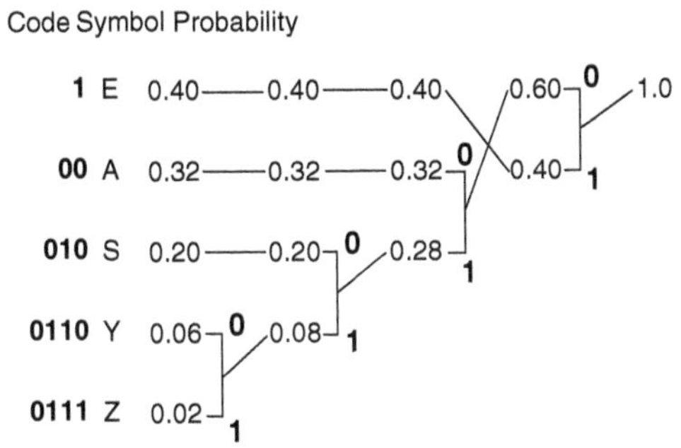

Fig. 3.13 Huffman tree created from Table 3.3.

Fig. 3.14 Cumulative density function for the symbol set in Table 3.3.

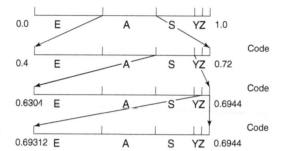

Fig. 3.15 Coding of the codeword sequence ASZ.

interval between 0 and 1; in other words, a cumulative density function of the probabilities is constructed, as seen in Figure 3.14.

Imagine now that we wish to send the sequence of codewords ASZ. Arithmetic coding achieves this by scaling each interval after the generation of each individual codeword. This is shown in Figure 3.15.

The rules are straightforward: When you code a word, the new interval is the interval occupied in the previous iteration according to the symbol probability. For example, at the first step when coding the word A, you will notice from Figure 3.14 that it occupies the interval 0.4 to 0.72; therefore, the interval in the next iteration of coding is 0.4 to 0.72. To subsequently code the word S, given that the interval is now only 0.4 to 0.72, the next interval must lie somewhere in between. Therefore, the lower value is 0.4 + 0.72 × (0.72 − 0.42) = 0.6304, and the upper is 0.4 + 0.92 × (0.72 − 0.42) = 0.6944.

The same rule continues: Calculate the width of the interval and work out the relative positions of the symbols within it at each stage. Finally, conveying a number between 0.69312 and 0.6944 uniquely identifies the code sequence ASZ. The sequence can be calculated by the decoder if it has a copy of the original cumulative density function. In practice, binary rather than floating-point arithmetic is used (Ghanbari, 2003).

Hybrid Codecs: Exploiting Spatial, Temporal, and Statistical Redundancy

Hybrid codecs employ spatial temporal, and statistical redundancy. Spatial and statistical redundancy is employed in image codecs, such as JPEG. Spatial temporal, and statistical redundancy is employed in standard video codecs, such as MPEG-1, MPEG-2, MPEG-4, H.261, H.263, H.264, and H.26L. The block diagram of a hybrid spatial temporal and statistical DCT codec is shown in Figure 3.16.

Bitrate and Quality Variations

Great progress has been made in the last few decades in techniques for the data reduction or compression of digital video signals. The challenge has been to find techniques that provide high compression ratios without introducing significant distortion. Figure 3.16 shows a block diagram of a hybrid video coder. It relies on interframe prediction, followed

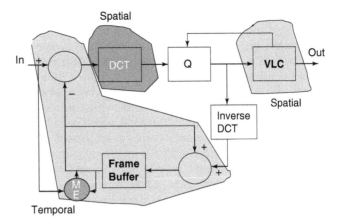

Fig. 3.16 Generic DCT based hybrid codec.

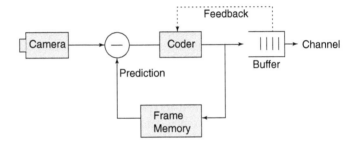

Fig. 3.17 Simplified codec with buffer feedback to control quantization settings.

by coding of the prediction error, typically using a DCT. Therefore, when there is a lot of unpredictable movement in the scene the coder generates a high bitrate, but when there is little or no movement it generates a low bitrate. The output of the coder is thus a variable bitrate (VBR) and is suitable for transmission over most packet-based networks. If the codec is required to be used over a constant bitrate (CBR) channel, then the addition of a smoothing buffer on the output is required; this is connected to the quantizer as shown in Figure 3.17.

As the buffer overflows, the quantizer step size is increased so the coding process produces fewer bits. This feedback introduces a time-varying impairment into the video. Thus, these types of compression schemes, when operated at substantial compression ratios, introduce either bitrate or quality variations.

Structure of the Coded Information

The previous sections have discussed the use of compression techniques to reduce the overall bitrate required to code video. Having undergone heavy compression by exploiting spatial, temporal, and statistical redundancy, each remaining bit is worth N bits from the original uncompressed bitstream, where N is the compression ratio. The result is a stream of bits with an increased value. If this stream were transmitted in this basic form, a single error within the stream would make all subsequent information difficult or impossible to decode. This can be likened to reading a book where all of the letters have shifted by an unknown amount (e.g., a *c* becomes an *e* and a *d* becomes an *f*); even though they have only shifted by two places, clearly the content is meaningless to the decoder (the reader).

As a consequence, video codecs tend to arrange the bitstream so that if an error does occur its effects will be short-lived and can be recovered from at a later date. This is the structure of the coded bitstream, and this structure exists at the micro as well as the macro level. By providing structure of this kind, errors are prevented from propagating. Using the book example again, if the error occurs anywhere on a single page but is reset at the beginning of the next, the reader will have lost some understanding but will be able to regain the context when the alignment is restored. This restoration could occur in a piece of writing at the level of the chapter, page, paragraph, sentence, and finally word. In video coding it occurs at the group of pictures, at the individual picture, and finally at the slice level. In adopting this strategy when packet or bit loss occurs the degradation only continues until the next "alignment," where the spatial and temporal information is resynchronized.

Types of Pictures

Video compression can be categorized as spatial, temporal, and statistical. Statistical redundancy is used for both still- and moving-picture data reduction and is implied in most codecs. Coding relative to spatial information has another name: intraframe coding. Coding relative to temporal information has another name: interframe coding. As a consequence, a minimum of two types of frame can be identified in video codecs: intracoded frames or I frames, which are coded relative to themselves, and intercoded frames, or P frames (predicted), which are coded relative to other pictures. In the MPEG standard a further B-type picture is found, which is a predicted picture based on future as well as past information.

Intracoded or I Pictures

An intrapicture is independent from other pictures. In a hybrid DCT codec, no motion compensation is employed, and each macroblock is DCT-coded, quantized, and run-length–coded in the usual way. The I picture forms a marker within the bitstream, a point at which no assumptions have been made about the past or future. Therefore, if other types of predictive frames have been used within the video stream and errors have occurred in those, the arrival of an I-frame has the effect of resetting the coder back to normal. However, it should be noted that other frames use the I-frame as the basis of their predictions, and therefore errors in or loss of the I-frame can have cumulative effects for a number of frames to follow. An intracoded frame also has a high bitrate associated with it, making it costly in network terms.

Intercoded or P Pictures

An interframe-coded picture is a predicted, or P, picture that codes the difference between itself and other neighboring pictures within a sequence. In being dependent on information from other pictures, it will be affected by not only loss in its own information but also loss that has occurred in associated I frames; the P frame assumes the existence at the decoder of the information the prediction is based on. The predicted picture may or may not use motion compensation and will have selected the option that yields the lowest coding expense. The difference between the actual and the predicted picture is DCT-coded in the usual way.

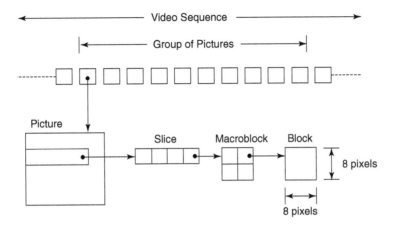

Fig. 3.18 Picture structure.

MPEG-1 took this idea one step further by introducing a bidirectional picture, or a B picture. A B picture can use past, present, and future information as the basis of its predictions. The extreme research suggests that there is merit in buffering hundreds or even thousands of frames for use in motion compensation. It should be noted that while not resilient to error, such predictive frames are cheap to encode.

Video Structure

A video sequence consists of a sequence of pictures, which are presented at a rate of N frames per second. As seen earlier, predictive pictures are based on previously coded frames, and therefore within a sequence some frames are related to one another. The sequence of frames is first divided into a group of pictures that describes how the sequence is made up of I, P, and B pictures and how they are related to each other. Below this is the individual picture, which can be either I, P, or B. The picture is subdivided into slices to reduce the propagation of errors. The slices themselves are made up of macroblocks. A macroblock is usually made up of four 8×8 blocks of pixels, with the finest granularity being the pixel itself. This is clarified in Figure 3.18.

Video Transport over the Internet

Video is usually accompanied by audio, and when transporting multimedia over the Internet a number of operations are required, including deriving the source media, compressing the content, controlling the rate of generation, packetizing the data, smoothing bursts, network interfacing, multiplexing with other traffic, scheduling, and protocol. These operations are usually transparent to the user and integral to the codec, the operating system, and the network.

Video Traffic Types

Real-time video traffic is where the video is delivered as it is generated and displayed at the receiver as quickly as possible. The delay between generation and final display is of the order of tens of milliseconds, and two-way communication is also possible. Little or no packet buffering is used, and there is generally no mechanism to recover lost packets; the application has to deal with late or lost packets and the associated degradation caused. A typical example is the videoconferencing software Skype.

Streamed video is the simultaneous delivery and playout of multimedia; a fundamental example is television. Archived or near real-time video is held by the sender (server) and delivered to the receiver (client) by default or by request. Often when using digital media over the Internet, the receiver buffers the data and only plays the video when the buffer is sufficiently full. This allows time so that lost packets can be retransmitted and late packets can be used. The media player examines the status of the buffer and can even adjust the playout speed to avoid an empty buffer and the associated break in provision. Video can be played at a frame rate and quality that might not otherwise be achieved given current network conditions. A typical example is the video-share site YouTube.

Video file transfer is where a server has data that is requested by the client, and that data is delivered in packets to the receiver. The transfer has to complete before the file is of use to the receiver. This is the type of thing that happens when you request an MPEG file or an AVI file from a Web site.

Data Transfer

Data transferred across any medium is subject to the possibility of loss. In modern networks the primary cause of data loss is the finite resource, which has to be shared between multiple users. There is considerable variation in the types of traffic on the Internet, and the interactions between them can often be complex and difficult to predict. Nevertheless, a communication can be simply defined by its probability of loss, its delay, and its jitter.

The main task of a communication network is to enable end-to-end information transfer. The Internet is a ubiquitous communication medium, and a simple diagram of communication is shown in Figure 3.19.

The Internet uses the TCP/IP reference model, which has four readily identifiable layers—namely, application, transport, Internet, and network interface (Pearson, 1975). The application layer consists of protocols such as HTTP, FTP, and Telnet, while the transport layer provides two services, Transmission Control Protocol (TCP) and User Datagram Protocol (UDP). TCP offers a reliable, connection-oriented transfer for exchanging data and uses retransmissions to handle packet loss in the network. UDP, on the other hand, provides best-effort connectionless transfer of individual messages and offers no error recovery or control capabilities.

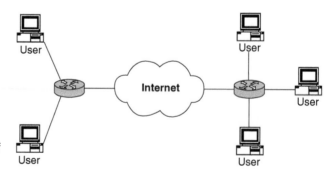

Fig. 3.19 Simplistic view of end-to-end communication.

While TCP offers a reliable delivery service, UDP is more suitable and mostly used for transmission of real-time media such as video and audio because there are no retransmissions to increase the round-trip delay in the packets. The Real-Time Transport Protocol (RTP) (Schulzrinne et al., 1996) can be used on top of UDP to provide a more robust end-to-end delivery service. RTP services include payload identification, sequence numbering, and timestamping. The payload identifier specifies payload type as well as the encoding/compression scheme used, enabling the receiving application to interpret the payload. The sequence number allows the ordering and detection of lost packets. The timestamp helps the receiver to play the video at the correct rate.

The Internet layer manages transfer across multiple networks. Information moves from one network to another through routers and gateways, where networks are identified by their unique IP (Internet Protocol) addresses (Schulzrinne et al., 1996; Peterson & Davie, 2000).

The behavior of packets on the Internet is widely documented (Baldi & Ofek, 2000), and the general conclusion is that the Internet is not an ideal medium for the transport of real-time traffic. The current Internet offers a single-best-effort level of service, where all packets are independently routed to the next node without guarantee of delivery and might be subject to delay or jitter.

Packet Loss

Packet loss is when packets go missing in a network. The principal cause of packet loss is when sources overload network elements. As a network becomes congested, forwarding nodes cannot process the data at the rate at which it arrives, and it is therefore discarded by the overflowing buffers. Other loss due to equipment failure or bit error caused by interference can occur but is in general rare. Bursty traffic is particularly adept at causing buffer overflow, and the convergence of a number of VBR traffic sources at a network element (such as a router) can soon overload the buffer capabilities. Packets can also be discarded if they have been delayed. An IP packet has a Time-to-Live (TTL) stamp, which when decremented to 0 results in the packet being discarded (Peterson & Davie, 2000; Tanenbaum, 2003). This can actually be desirable because for real-time traffic the playout time at the receiving codec is likely to have passed.

Packet loss can result from the protocols employed by the transmitting sources themselves (Baldi & Ofek, 2000). For example, in the TCP protocol, if packets are lost because of congestion, the TCP control mechanism will supply a negative acknowledgment and a retransmission will follow. Acknowledgment and retransmission packets interact with current data packets, causing delay, loss, and increased network congestion. If the video is sent using UDP, it still has to interact with other traffic, which is likely to be TCP, and therefore the same argument applies (Baldi & Ofek, 2000).

The end-to-end behavior of the Internet has been extensively studied, and losses can be summarized as bursty and typically between 2 and 10 percent (Sanghi et al., 1993). Bolot, Crepin, and Vega-Garcia (1995) showed that the number of consecutively lost packets is low provided the network load is not high.

Packet loss causes severe effects on video traffic, affecting overall Quality of Experience (QoE) (Woods & Siller, 2006). The type of distortion is dependent on the type of information that has been lost (i.e., does it belong to an inter- or intracoded frame?). Degradations include parts of the image missing, the image becoming blocky, the reconstruction not resembling the actual, and abrupt/jerky motion. Some works (e.g., Woods & Siller, 2006) have suggested the selective discarding of packets during times of congestion: B frames first followed by P and then finally I frames.

Delay and Jitter

Packet delay is the time required to transmit the unit of data from source to destination. The delay includes packetization, propagation, queuing, and synchronization (Tanenbaum, 2003). Packetization itself does not create significant delay, but waiting for the packet to fill with the coded information does (Peterson & Davie, 2000). Propagation delays are generally predictable for a given path and depend only on distance. Queuing delays at the buffers of network elements are difficult to quantify because there are variations caused by the differing link capacities, scheduling mechanisms, and current levels of network congestion (Schulzrinne et al., 1996). At the receiver, packets might have become reordered, and it might be necessary to wait for late arrivals; these have to be synchronized before playout can occur.

Jitter is the variation in delay between consecutive packets. It is a result of packets taking different paths across the network and variable queues at network elements, which combine to cause the interarrival times to vary. This means that packets might arrive out of order, and the decoder will need to buffer a number of frames to overcome these difficulties. Real-time multimedia applications have strict requirements in terms of maximum delay and delay variation (jitter). The delay and jitter must be kept within the tolerances of human perception.

The end-to-end delay requirement varies from one application to another; for example, real-time interactive applications involving video and audio have a one-way delay limit of 100 to 150 ms (Baldi & Ofek, 2000). Synchronization between audio and video to maintain lip sync dictates variations of below 80 ms for one-way communication (Baldi & Ofek, 2000).

One of the most common effects of delay in real-time traffic is when a packet arrives too late to be of use to the application. The application has the constraint that it must play the video in a timely fashion, so even though a packet does eventually arrive, it is ignored/discarded by the decoder; it might as well not have been sent at all.

Some applications, particularly VBR (variable-bitrate) video, are very bursty in nature and can temporarily consume network bandwidth and penalize other users of the resource. It seems to be a general observation of the Internet that those causing congestion are not necessarily the ones who suffer from packet loss.

Video Transmission

A generic list of tasks to transmit multimedia over a network includes compression, traffic control, packetization, multiplexing, and QoS (Quality of Service) control.

Compression mechanisms, as discussed previously, reduce the quantity of information transmitted, albeit increasing the value of each bit.

Traffic control is where the source controls the bitrate generated by the codec, adjusting the rates according to available channel capacity (Bolla et al., 2000). The coding process adjusts the quantizer step size, frame rates, picture size, and so forth, according to the prescribed budget.

The resulting bitstreams from the codec are placed into packets, which are then fed via a socket to the network. The policy adopted in the *packetization* will affect the performance of the media. Placing a large amount of data in a single packet makes for good utilization of resources (with the addressing overheads being proportionately small), but a single packet loss results in a considerable loss of data. Conversely, constructing small packets reduces the impact of loss but at the expense of poor network utilization; the addressing mechanism is larger now as a proportion of the packet.

Multiplexing refers to time sharing within the Internet. This can occur at the source, where, for example, audio, video, and hypertext are transmitted simultaneously using only a single network socket. It also occurs at network forwarding nodes, which have to multiplex multiple sources to a single connection.

QoS is a general term referring to the way in which some datastreams can be provided with higher priorities than others. This is almost universally employed at the input buffers of forwarding nodes where packets can be labeled for preferential treatment using techniques such as DifServ and IntServ, which give priority to packets and flows, respectively.

Network Queuing

When a stream reaches a network element such as a router, it must be multiplexed with other traffic taking the same path. The network element only has a finite resource, and packets are queued prior to forwarding to their destination according to the address contained in the packet header. If the incoming traffic has a CBR, then, as shown in the top half of Figure 3.20 the output is predictable as NR bits/s. For VBR, the data is statistically multiplexed, and therefore a burst in one stream can result in a loss of resources for another; packet loss occurs as a result of bursts in VBR traffic.

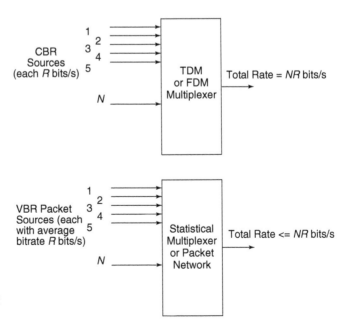

Fig. 3.20 CBR and VBR multiplexing.

Different packets can have different levels of importance, and if a network permits, they can be classified and placed in different queues. These can then be scheduled to the output of the network element according to a number of queue-scheduling techniques. The technique of increasing the probability of delivery of a particular stream is described generically as QoS. The techniques that are described in Woods and Siller (2006) include First-in-First-out (FIFO), Priority Queuing (PQ), Fair Queuing (FQ), Weighted Fair Queuing (WFQ), and Weighted Round-Robin (WRR), also known as Class-Based Queuing (CBQ).

FIFO is the most common queue-scheduling scheme, where all packets are treated equally by placing them in a single queue; they are serviced in the order in which they arrive at the queue. PQ provides a simple way of supporting differentiated services, where packets are classified and placed into different priority queues. Packets are serviced in order of decreasing priority, provided that the higher-priority queue is empty. FQ was designed to ensure that each flow has fair access to network resources and prevents bursty flows from overconsuming network bandwidth; flows are serviced in a round-robin fashion. WFQ allows a fair distribution of bandwidth by supporting flows with different bandwidth requirements. In WRR, packets are classified according to service class: real-time, interactive, file transfer, and so on, and then serviced using round-robin.

To provide service differentiation to enhance QoS on the Internet, the Internet Engineering Task Force (IETF) defined two models: Integrated Services (IntServ) and Differentiated Services (DiffServ) (Blake et al., 1998). IntServ bases its functionality on resource allocation for individual flows by marking the IP Type of Service (ToS) field to provide QoS guarantees. IntServ uses the RSVP signaling protocol (Woods & Siller, 2006), which sets up and reserves network resources.

Despite considerable effort having been put into differentiated services, IPv4, the current Internet protocol, ignores the ToS field and RSVP is challenging to implement in large networks. Packets incident at network elements are forced to compete with other traffic, with the stark reality that those who make a small contribution to the congestion often have their packets discarded as a result of large bursts caused by larger congestors. The Internet is inherently unfair and awaits a policy that does not penalize responsible users. Recent work has suggested the use of congestion-based pricing to solve these problems (Woods & Siller, 2006).

Packet Loss Statistics

Internet Protocol (IP) allows three basic communication models: unicast, broadcast, and multicast (broadcast is not used on the Internet). With no ubiquitous form of prioritization, all packets are equiprobably lost. The current Internet Protocol (IPv4) ignores the ToS field, and so increased delivery probability can only be achieved using overhead information. ATM networks have an inherent method of prioritization, but they are not widely utilized, and IPv6 purports prioritization but has yet to find widespread deployment.

It should be noted that in any network there is no such thing as truly guaranteed delivery. The traditional "best-effort" service of the Internet offers no guarantee of delivery or delivery in the order the packets were sent. Expected rates over the Internet can range from zero in times of congestion to tens of kilobits, although the rates are strictly governed by the network topology. High bandwidth might be possible across a few hops or within a single institution, but as the topological distance increases the available bandwidth generally decreases. Real-time traffic needs to be able to cope with potentially serious loss.

There are various higher-level protocols for dealing with real-time media and for reserving resources to guarantee certain QoS. The most popular of these are the Real-Time Transport Protocol (RTP), currently implemented at the application layer for IPv4 and the Reservation Protocol (RSVP), which are expected to be provided at the transport layer within IPv6. The RTP protocol provides facilities such as sequence numbering and timestamps, which are required by real-time media for synchronization, loss detection, and so forth. However, RTP itself does not provide any mechanisms for QoS or real-time delivery. The design and implementation of robust media coders warrants a detailed understanding of the characteristics of the network used.

If an Internet-type network becomes congested, packets are queued at the routers. IP packets contain a TTL stamp that is effectively a count of the number of routers through which the packet has passed. If the TTL figure is exceeded, the packet will be discarded; this mechanism prevents packets from circulating the network indefinitely. Another implication of queueing is the interarrival time; that is, the source–destination journey time is variable. If a packet is delayed, the client will need to buffer all other data that has arrived as the delayed packets are anticipated.

This has implications for the hardware required, interactivity requirements, and audio–video lip synchronization. When the maximum acceptable delay has been exceeded, all late packets are then regarded as lost. Packets damaged during transit will be discarded by a node if they do not pass parity checks. Hence, packets received will either be correct, erroneous but having passed parity checks, or simply absent.

Internet-type networks are very prone to packet loss and more particularly burst loss. Under the current version of IPv4, no facility exists for prioritization, and all packets at a router are queued and treated equiprobably. Under the UDP protocol:

1. Most packet losses are isolated.
2. Packet loss periods are generally short in duration, assuming the network has not failed.
3. The probability of a loss period duration N decreases geometrically with N.
4. Packet arrival and delays follow a Poisson process.
5. Out-of-order delivery is common, although MBONE routers are more consistent.
6. Geographic location is very important.

The quality of the media delivered to the destination relies essentially on the number of lost packets and the delay variation (i.e., jitter between successive packets). From experiment, packet loss and delay characteristics vary with the time of day and geographical location of the site. Many different measures can be used to characterize the loss process; the most obvious measure is mean packet loss. Experimental results show mean packet loss rates are typically between 1 and 30 percent for the majority of locations. An alternative way to categorize loss is the amount of burstiness—in other words, the number of consecutively lost packets. Figure 3.21 shows a histogram for consecutively lost packets.

Most loss periods are less than two packets in duration. This suggests that such loss might be accommodated using Forward Error Correction (FEC) techniques; however, FEC requires the use of overhead information, ultimately increasing the bitrate and further loading the network elements. Error concealment by the application is the preferred method for video because it results in no overhead.

Packet loss is often modeled using simple models, and while incapable of modeling the self-similarity of the traffic in the Internet, these models are often fit for purpose. Multicast UDP packet losses have been successfully modeled (Clark & Fang, 1998; Bolot et al., 1995)

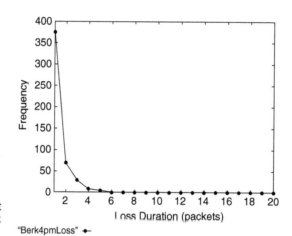

Fig. 3.21   Consecutive packet loss between eagle.essex.ac.uk and bmrc.berkeley.edu.

"Berk4pmLoss"

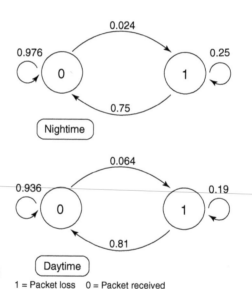

Fig. 3.22   Packet loss modeling using a two-stage Gilbert model.

1 = Packet loss   0 = Packet received

using a simple two-stage Gilbert loss model, shown in Figure 3.22. The statistics for the model were obtained using 384,000 packets transmitted over a 24-hour period. You will note considerable variations in delivery probability according to the time of day.

On the Internet, delays are highly variable. Figure 3.23 shows a typical round-trip delay between eagle.essex.ac.uk and bmrc.berkeley.edu at 4 p.m. GMT on a weekday. The playout instant for media must strike a compromise between delay and interactivity when dealing with round-trip variations. Shorter delays introduce higher losses, as delayed packets are assumed to have been lost; longer delays reduce the interactivity and increase the buffering requirements.

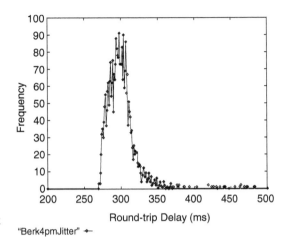

Fig. 3.23 Interarrival times or jitter between eagle.essex_ac_uk and bmrc.berkeley.edu.

"Berk4pmJitter" ✦

Self-Similarity of Traffic

Published work (SIGMETRICS, 1996) has shown that Internet loss statistics are fractal in nature and that packet loss profiles possess self-similarity when examined across different lengths of time. Knowledge of such behavior is useful in predicting loss statistics and has been highlighted here to encourage the reader to investigate further, but it remains outside the scope of this chapter.

Packet Loss Effects

The impact of packet loss on a piece of video is largely dependent on the sequence of events leading to it and the types of packet that have sustained loss. Figure 3.24(a) shows a situation where predictive information, or P packets, have been lost in an H.263 codec; the effect is to propagate errors because coding is relative. Figure 3.24(b) shows the catastrophic effect of the loss to I packets; the image fails to render subsequent predicted packets, which use these as their basis, so the compound errors.

Countering Loss Effects

Because of complexity and the time required to repair losses in the video stream, codecs generally have to accept that loss of data is inevitable and recover from it as best they can.

Forward Error Correction

The Forward Error Correction technique can be used to help codecs recover from bit errors. To provide a mechanism that allows recovery from error requires that additional information be inserted into the bitstream. This is an overhead or additional expense that will require extra data. In its simplest form, this could be sending multiple

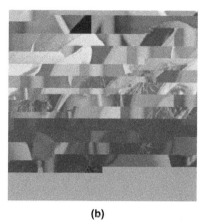

(a)   (b)

Fig. 3.24 Off-air pictures with packet loss.

copies of the same packet; more usually, block or convolutional coding is employed (Lee, 1997). While the techniques are often applied in wireless applications and used for video storage in DVD, FEC is usually avoided in real-time Internet-type traffic because of the overhead.

TCP Retransmission Request

IP networks permit the inclusion of Transmission Control Protocol (TCP) (Tanenbaum, 2003) packets. TCP provides a method of a connection orientation where packets are assured delivery provided the network remains in an operational state. This is achieved by keeping a record of the packet numbers at the sending and receiving sides. If a packet delivery fails, a request is made for retransmission. The difficulty with this approach is the round-trip time—the time taken for the retransmission request (negative acknowledgement) and the retransmissions. This delay means the video display must also be delayed, and this will impact the feeling of interaction in a real-time two-way communication.

Congestion-Driven Rate Control

If the receiver sends side information back to the transmitter to explain the level of loss and hence the congestion it is experiencing, then the rate of generation can be increased or decreased according to network conditions. This not only improves the display of the user, but, if adopted by all, it should achieve better utilization of the network.

Error Concealment

The key to making video transport over the Internet possible is the ability of the codec to recover from packet loss. Error cancealment can be applied to any of the conventional hybrid DCT codecs and awaits standardization (Ghanbari, 2003). In its simplest form when a codec recognizes that a macroblock has been lost, content from the previous frame is rendered in its place; motion compensation can also be employed if available. This remarkably simple solution dramatically improves the subjective quality of the video in the event of loss, although it can produce blocking artifacts against neighboring macroblocks that have been updated. Subsequent filtering of the image can alleviate these effects.

Conclusions

This chapter provided a brief overview of digital video, its transport over packet-based networks, and the difficulties encountered. The high-bitrate nature of digital video dictates the use of high compression; otherwise, networks would have insufficient bandwidth. Compression uses a variety of techniques that have resulted from decades of active research. The act of compression has left the bitstream vulnerable to loss because the absence of a single packet can have catastrophic effects on the reconstruction.

The Internet itself offers no guarantee of delivery, and the interactive nature of two-way video dictates that applications must tolerate packet loss and delay. For noninteractive applications, overheads in the form of TCP and FEC can be applied but increase the bandwidth and reduce the available resource for others. Video transport over the Internet is a complex problem fraught with difficulty; that it works at all is testament to the efforts of universities and standardizing bodies.

References

Blake, S., Black, D., Carlson, M., Davies, E., and Wang, Z. (1998) Architecture for Differentiated Services, IETF Technical Report, RFC 2475.

Bolla, R., Iscra, A., Marchese, M., and Zappatore, S. (2000) A Flow Control Algorithm for Multimedia Network Applications, *IEEE Packet Video Workshop*, 289–298.

Bolot, J., Crêpin, H., and Vega-Garcia, A. (1995) Analysis of Audio Packet Loss in the Internet, *Proceedings NOSSDAV*, Springer, 163–174.

Charles Lee, L. H. (1997) *Convolutional Coding: Fundamentals and Application*, Artech House.

Christopoulos, C., Skodras, A., and Ebrahimi, T. (2000) The JPEG2000 Still Coding System: An Overview, *IEEE Trans. Consum. Electron.* 46(4):1103–1127.

Clark, D., and Fang, W. (1998) Explicit Allocation of Best Effort Packet Delivery Service, *IEEE/ACM Trans. Netw.* 6(4):362–373.

Côté, G., et al. (1999) H.263+: Video Coding at Low Bit Rates, *IEEE Trans. Circuits Syst. Video Technol.* 8(7):849–866.

Ghanbari, M. (2003) *Standard Codecs: Image Compression to Advanced Video Coding*, IEE.

Ghanbari, M. (1999) *Video Coding: An Introduction to Standard Codecs*, IEE.

ISO/IEC and ITU-T (2005) *ITU-T Recommendation H.264, Advanced Video Coding for Generic Audiovisual Services.* Joint Video Team (JVT) of ISO/IEC and ITU-T VCEG.

ISO/IEC (1995) *Generic Coding of Moving Pictures and the Associated Audio Information.* ISO/IEC 13818-2, MPEG-2.

ITU-T (1998) *Recommendation H.263+, Video Coding for Very Low Bit-rate Communication.*

ITU-T (1993) *Recommendation H.261, Video Codec for Audiovisual Services at $p \times 64$ kbit/s.*

Kondoz, A. M. (1994) *Digital Speech*, John Wiley & Sons.

Körner, T. W. (1988) *Fourier Analysis*, Cambridge University Press.

Pearson, D. E. (1975) *Transmission and Display of Pictorial Information*, Wiley.

Peterson, L., and Davie, B. (2000) *Computer Networks: A Systems Approach*, Morgan Kaufmann.

Sanghi, D., Agrawala, A., Gudmundsson, O., and Jain, B. (1993) Experimental Assessment of End-to-End Behaviour on the Internet, *Proceedings IEEE INFOCOM*, 867–874.

Sayood, K. (2002) *Lossless Compression Handbook*, Academic Press.

Schulzrinne, H., et al. (1996) *RTP: A Transport Protocol for Real-Time Applications*, IETF Technical Report, RFC 1889.

SIGMETRICS (1996) Self-Similarity in World Wide Web Traffic: Evidence and Possible Causes, *Proceedings SIGMETRICS '96*.

Skodras, A., Christopoulos, C., and Ebrahimi, T. (2000). JPEG2000: The Upcoming Still Image Compression Standard, *Proceedings 11th Portuguese Conference on Pattern Recognition*, 359–366.

Tanenbaum, A. (2003) *Computer Networks*, Pearson Education International.

Wallace, G. (1991) The JPEG Still Picture Compression Standard, *IEEE Trans. Consum. Electron.* 38(1).

Wiegand, T., et al. (2003) Overview of the H.264/AVC Video Coding Standard, *IEEE Trans. Circuits Syst. Video Technol.* 13:560.

Woods, J. C., and Siller, M. (2006) Using an Agent Based Platform to Map Quality of Service to Experience in Conventional and Active Networks, *IEE Commun.* 153(6):828–840.

4 Coding

M. D. Macleod
QinetiQ Ltd and University of Strathclyde

The Need for Error Control Coding

In a digital communications system, information is represented as a sequence of digits (numbers), also known as symbols. To transmit them, the digits are first converted to an analogue (continuous) form, such as voltage on a wire or light intensity in a fiber; this is called *modulation*. The received signal is then converted back into digits by *demodulation*. The communication channel, which is analogue, introduces noise and distortion, however, and these cause the corruption or loss of some digits at the receiver. The system designer reduces the probability of these errors by appropriate design of the analogue parts of the communication system, but it is usually either impossible or too costly to achieve a sufficiently small probability of error in this way. A better solution is usually to use Error Control Coding (ECC).

Principles of ECC

Error Control Coding is the controlled addition of redundancy to the transmitted digit stream in such a way that errors in the received digits can be detected, and in certain circumstances corrected, in the receiver. It is therefore one aspect of channel coding, so called because it compensates for imperfections in the channel. The other form of channel coding is transmission (or line) coding, which has different objectives, such as spectrum shaping of the transmitted signal. The benefit of ECC is that it reduces the probability of error in the digits output by the receiver. The added redundancy means, however, that extra digits have to be transmitted over the channel, so either the channel transmission rate must be increased or the rate of transmission of digits from input to output must be reduced.

The controlled addition of redundancy in ECC contrasts with source coding (data compression), in which redundancy is removed from the source signal. Examples of the latter include MP3 and other algorithms for data compression of speech and music.

The functional blocks used for ECC are a coder that precedes the modulator in the transmitter and a decoder that follows the demodulator in the receiver. The decoder can be designed to detect digit errors or to correct them. These functions are known as error detection and error correction, respectively.

Exactly the same principles apply to the encoding of information for storage (e.g., on CD, DVD, or magnetic disk) and the subsequent correction of errors on readout.

Shannon, in 1948, derived a theoretical "channel capacity" (in bits/sec) based on the bandwidth of a channel and the Signal-to-Noise Ratio (SNR) at the receiver. Shannon showed that data could be transmitted at almost this limiting rate with arbitrarily low probability of error using ECC, provided sufficiently long codewords were used. The challenge is how to design powerful codes that can be decoded with feasible complexity and quickly enough. Two revolutionary developments in the 1990s (Turbo and LDPC codes) now make it possible to achieve very nearly the theoretical channel capacity.

Types of ECC

There are two main types of ECC: block coding and convolutional coding. In block coding, the input is divided into blocks of k digits. The coder then produces a block of n digits for transmission, and the code is described as an *(n,k) code*. Each block is coded and decoded entirely separately from all other blocks. In convolutional coding, the coder input and output are continuous streams of digits. The coder outputs n output digits for every k digits input, and the code is described as a *rate-k/n code*.

If the input digits are included unmodified in the coder output, the code is described as *systematic*. The additional digits introduced by the coder are then known as *parity* or *check* digits. As well as the conceptual attractiveness of systematic codes, they have the advantage that a range of decoder complexities is made possible. The simplest decoder can simply extract the unmodified input digits from the coded digit stream, ignoring the parity digits. A more sophisticated decoder can use the parity digits for error detection and a full decoder for error correction. Unsystematic codes also exist, but they are less commonly used.

Forward and Feedback Error Correction

In Forward Error Correction (FEC) the decoder applies error correction to the received codeword, and it can also detect some errors that it cannot correct. However, no return path from the receiver to the sender is assumed. Either block or convolutional codes can be used for FEC.

In feedback error correction, for which only block codes can be used, the receiver only attempts to detect errors and sends return messages to the sender, which cause repeat transmission if any errors are detected in a received block. In the OSI model for packet data networks, this function is carried out within the data link layer by the return to the sender of a positive or negative acknowledgement (ACK or NAK) on receipt of a data block; this is known as *stop-and-wait ARQ* (Automatic Repeat Request). In *go-back-N ARQ*, receipt of a NAK by the transmitter makes it retransmit the erroneous codeword and the $N - 1$ that follow, where N is chosen so that the time taken to send N codewords is less than the round-trip delay from transmitter to receiver and back again. This obviates the need for a buffer at the receiver. In *selective-repeat ARQ*, only the codewords for which NAKs have been returned are retransmitted. Performance analysis (Lin & Costello, 1983) shows that this is the most efficient system, although it requires an adequately large buffer in the receiver.

Arithmetic for ECC

In a digital communication system, each transmitted digit is selected from a finite set of M values and is described as an M-ary digit. For example, binary digits (bits) have one of two values, which can be represented as 0 and 1. (The actual values of the physical signal used to transmit the digits—for example, voltages—are irrelevant here.) If the message digits in the transmitter do not use the same value of M as the transmitted digits, they are first converted to M-ary digits before coding.

The analytical design of coders and decoders for M-ary digits requires the use of Galois field arithmetic, denoted GF(M). If M is a prime number (including binary digits, where $M = 2$), Galois field arithmetic is equivalent to arithmetic modulo-M. GF(M) arithmetic also exists when M is equal to a power of a prime—for example, for quaternary (4-valued) and octal (8-valued) digit systems—but then GF(M) arithmetic is not equivalent to modulo-M arithmetic.

Modulo-2 addition is equivalent to the logical exclusive OR (XOR) function, and multiplication is equivalent to the logical AND function. In modulo-2 arithmetic, addition and subtraction are equivalent. Some textbooks only discuss binary coding and therefore treat all subtractions as additions. However, for values of M other than 2 addition and subtraction are not equivalent, so the distinction matters. For nonbinary systems, multiplication and addition operations can be implemented using simple logic.

Types of Error

If the physical cause of digit errors is such that any digit is as likely to be affected as any other, the errors are described as random. A typical cause of random errors is thermal noise in the received signal. Other types of interference, however, might make it likely that when an error occurs, several symbols in succession will be corrupted; this is known as a *burst error*. A typical cause of burst errors is interference. Although the true behavior of the channel can be more complex than either of these simple models, the random error and burst error models are simple, effective, and universally used for describing channel characteristics and ECC performance.

A channel used for transmitting binary digits is known as a *binary channel*, and if the probability of error is the same for 0s and 1s, the channel is called a *binary symmetric channel*. The probability of error in binary digits is known as the *Bit Error Rate* (BER).

Coding Gain

Coding gain is a parameter commonly used for evaluating the effectiveness of an error correcting code and hence for comparing codes. It is defined as the saving in energy per source bit of information for the coded system, relative to an uncoded system delivering the same BER, and is expressed as a ratio in dB. The effects of both error correction and the increase in transmission rate by a factor of n/k must be included when calculating the coding gain.

Criteria for Choosing a Code

The primary objective of ECC is to achieve a desired end-to-end probability of either uncorrected or undetected digit errors. The choice of code will depend on the error characteristics of the channel (particularly the random and burst error probabilities). The other important factors are likely to be the value of n/k (the increase in transmission rate over the channel) and the implementation complexity and cost of the coder and decoder.

Block Coding

In an (n,k) block code, blocks of k input digits are mapped to blocks of n digits for transmission.

Single-Parity Checks

The simplest block coder appends a single-parity digit to each block of k message digits. This produces a systematic $(k+1,k)$ code known as a *single-parity code*. For binary digits, the parity can be the modulo-2 sum of the message bits; this is known as *even parity* because the sum of the $k+1$ bits of the codeword (including the parity bit) is 0. If the polarity of the parity bit is inverted, that is known as *odd parity*.

The decoder forms the sum of the bits of the received codeword. A correct codeword produces a sum of 0, so a sum of 1 means that there has been an error, although there is no way to deduce which bit(s) are in error. Thus the error cannot be corrected. Also, if more than one error occurs, and the number of errors is even, the sum will be 0 and so the code will not detect them. This is therefore a single-error–detecting (SED) code.

Linear Block Codes

The even-parity code is the simplest example of a powerful class of codes called linear block codes. For such codes, the block of k message digits is represented as the k-element row vector, **d**, and the n digit codeword produced by the coder is represented by the n-element vector, **c**. The function of the linear block coder is described by equation 4.1, where **G** is the $(k \times n)$ generator matrix.

$$\mathbf{c} = \mathbf{dG} \tag{4.1}$$

The multiplications and additions in this equation are carried out in GF(M) arithmetic (i.e., modulo-2 for binary digits). The codewords generated by the equation are called *valid* codewords. Because there are 2^k possible datawords, only 2^k of the 2^n possible n-digit words are valid codewords. Systematic linear block codes are produced by a generator matrix of the form shown in equation 4.2, where $\mathbf{I}_k$ is the $(k \times k)$ unit matrix.

$$\mathbf{G} = [\mathbf{I}_k | \mathbf{P}] \tag{4.2}$$

When **G** has this form, the codeword **c** has the form of equation 4.3. In other words, the first k digits of the codeword equal the dataword and the last $n - k$ digits are parity digits.

$$c = [d|d\ P] \tag{4.3}$$

Let $\mathbf{r}$ be the received codeword in the absence of errors $\mathbf{r} = \mathbf{c}$. The decoder performs the operation of equation 4.4, where $\mathbf{H}$ is the $((n - k) \times n)$ parity check matrix, to produce the $(n - k)$ element syndrome, $\mathbf{s}$.

$$s = rH^T \tag{4.4}$$

$\mathbf{H}$ is chosen so that all valid codewords produce a 0 syndrome; the syndrome then plays a crucial role in error correction. For the systematic code given before, the optimum form of parity check matrix is as in equation 4.5, so $\mathbf{H}$ and $\mathbf{G}$ are very simply related to each other.

$$H = [P^T|I_{n-k}] \tag{4.5}$$

Construction of the parity check matrix is less straightforward for unsystematic codes.

Distance and Code Performance

The Hamming distance between two codewords is merely the number of bit (or digit) positions by which they differ. Assume the receiver demodulates each digit in turn to give a received codeword, $\mathbf{r}$. This is called hard-decision decoding.

If the distance between two codewords, c_1 and c_2, is d, and c_1 is transmitted, then at least d errors would have to occur for $\mathbf{r}$ to equal c_2. More generally, if the minimum Hamming distance between codeword c_1 and any other valid codeword is d_{MIN}, and c_1 is transmitted, then, provided no more than $d_{MIN} - 1$ errors occur, $\mathbf{r}$ will not be a valid codeword. The decoder could therefore detect up to $d_{MIN} - 1$ errors.

Alternatively, when codeword $\mathbf{r}$ is received, the distance, d_i, between it and each valid codeword, c_i can be calculated. If one such distance, d_j, is less than all the others, then (if the errors are random) it is more likely that the transmitted codeword was c_j and no other. (This is because the probability that the number of errors will be between 1 and d falls as d increases). The decoder could therefore output the "most probably correct" codeword, c_j; this is known as maximum-likelihood or minimum-distance error correction.

Such error correction is only possible if the number of errors is less than $d_{MIN}/2$ because otherwise the distance to a incorrect codeword could be less than or equal to the distance to the correct one.

In general it is possible to trade off error detection and correction ability. A code that is required to be able to correct n_C errors and detect a further n_D errors must have a minimum distance given by equation 4.6.

$$d_{MIN} = 2n_C + n_D + 1 \tag{4.6}$$

Determining the minimum distance of a code by comparing every pair of codewords would be time consuming for large codeword lengths. However, for a linear block code only the 2^k valid codewords themselves need to be checked because the minimum Hamming distance between codewords is equal to the minimum Hamming weight among its nonzero codewords. (The Hamming weight of a codeword is the number of 1s in it.)

Hard- and Soft-Decision Decoding

In hard-decision decoding, as noted earlier, the receiver demodulates each digit in turn before decoding. For binary signals, demodulation (slicing) is a simple thresholding operation. The decoder then performs error detection and correction using this (possibly corrupted) received digit stream.

In soft-decision decoding, the input to the decoder is the unsliced (analogue) sample stream. Because the decoder implementation is usually digital, the sample stream has to be digitized before input to the decoder, but it has been found that in practice very low-resolution digitization (for example, to only 8 or 16 levels) is often adequate. Soft decision decoding is computationally more demanding than hard-decision decoding. It has long been used with convolutional codes (see the Convolutional Codes section) to give extra coding gain, typically of about 2 dB. Its use for block codes was less common until recently, but the most recent block codes, that is, Turbo and LDPC (see the Turbo and LDPC Codes section) usually use soft-decision decoding.

Some codes, such as Reed–Solomon codes (see the Reed–Solomon Codes section), can handle erasures (i.e., digits that the receiver can detect and flag as erroneous) as well as unknown errors; this is an additional form of soft-decision decoding.

Hard-Decision Decoding of Linear Block Codes

For hard-decision decoding, the detection and correction of errors is based on the syndrome. For (n, k) linear block codes. The $(n - k)$-bit syndrome of valid codewords is 0 as in equation 4.7.

$$\mathbf{s} = \mathbf{cH}^T = \mathbf{0} \tag{4.7}$$

If the syndrome of the received codeword, $\mathbf{r}$, is 0, the decoder assumes that it is correct and the output dataword is extracted directly from it (very easily, for a systematic code).

A nonzero syndrome is a certain indication of errors. If the transmitted codeword, $\mathbf{c}$, is corrupted by the (modulo-2) addition of the n-bit error pattern, $\mathbf{e}$, so $\mathbf{r} = \mathbf{c} + \mathbf{e}$. The syndrome is given by equation 4.8

$$\mathbf{s} = (\mathbf{c} + \mathbf{e})\mathbf{H}^T = \mathbf{eH}^T \tag{4.8}$$

which is a function only of the error pattern, $\mathbf{e}$. There are only $2^{(n-k)} - 1$ different nonzero syndromes, but there are $2^n - 1$ different error patterns, so every syndrome (including 0) can be generated by many different error patterns.

However, as mentioned before, error patterns with small numbers of errors are more likely than those with larger numbers of errors, so the most likely cause of a given nonzero syndrome is the corresponding error pattern with the fewest 1s. This is why the decoder assumes no errors if the syndrome is 0.

One way to implement the decoder is to store in a lookup table (called the standard array) the chosen error pattern for each syndrome. The decoder then calculates the syndrome, reads the corresponding error pattern from the table, and subtracts it (using bit-by-bit XOR) from the received codeword. Finally, the output dataword is extracted from the corrected codeword.

Types of Block Code

The most important linear block codes now in use are cyclic codes, including BCH and Reed-Solomon codes (see the Cyclic Codes section) and Turbo and LDPC codes (see the Turbo and LDPC Codes section). Many other linear and nonlinear block codes, some now of mainly historical interest, exist. They include the following:

2-D parity codes. These are created by arranging the message bits on a rectangular array and calculating parity bits for each row and column. Changing any one message bit changes one row parity and one column parity as well (a total of three changes), so this is a distance-3 code, which can be used for single-error correction or dual-error detection. The decoder forms the row and column parity sums. If all are 0, the codeword is assumed to be correct. If only a row sum or only a column sum is 1, the error is assumed to be in the corresponding parity bit. If one row sum and one column sum are 1, the error is assumed to be in the databit at the intersection of that row and column, which is therefore corrected. If two or more row sums or two or more column sums are 1, a larger number of errors have occurred, which cannot be corrected.

Hamming codes. These are also distance-3 linear block codes, but more efficient than 2D parity codes. For binary Hamming codes, the codeword length is given by $n = 2^r - 1$, where r is the number of parity bits; the number of message bits is therefore given by $k = n - r$. The first four Hamming codes, for example, are (3,1), (7,4), (15,11), and (31,26). Error correction can be achieved by syndrome decoding, as described in the Hard-Decision Decoding of Linear Block Codes section. There are cyclic forms of Hamming code, described in the Cyclic Codes section.

Hadamard codes. The codewords are the rows of a Hadamard matrix, which is a binary $n \times n$ matrix (n even) in which each row differs from any other row in exactly $n/2$ positions; if the matrix elements are denoted +1 and −1, the rows are also orthogonal. Using this matrix and its complement (i.e., the matrix formed by exchanging the +1 and −1 elements), a blocklength-n, distance-$n/2$ code, with $2n$ codewords, can be formed.

Golay code. The Golay code is a (23,12) triple-error–correcting (TEC) code.

Constant-ratio codes. Also known as m-out-of-n codes, these have blocklength n, and each codeword has m bits set. These are nonlinear codes of primarily historical interest.

Shortened Codes

If no convenient code exists with exactly the required value of the dataword length, k, a code having a larger value of k, say k', can be shortened. The number of check (parity) bits is unaltered by shortening, but the dataword and codeword lengths are both reduced by the same amount, so the (n',k') code becomes $(n' + k - k',k)$. An example would be the use of a (15,11) Hamming code, shortened to (12,8), for the byte-by-byte transmission of data. Shortening may or may not increase the minimum distance, and hence the error control ability, of the code.

At its simplest, shortening the code consists simply of encoding k' databits, of which k are the required data and $k' - k$ are 0. The main advantage arises when a systematic code is used, because then the first $k' - k$ bits in the codeword can be set to 0 and need not be transmitted. In principle, the received codeword can then be lengthened to n' bits again by the reinsertion of $k' - k$ 0s prior to decoding with the (n',k') decoder.

In practice, however, the decoder should be modified because errors cannot occur in untransmitted bits. Consider a shortened Hamming code. The Hamming code can normally correct any single error. However, if on receiving a shortened codeword the decoder calculates a syndrome that would normally imply an error in one of the untransmitted bits, that cannot be the true cause. The next most likely cause is a double-error pattern in the transmitted bits. If there were only one double-bit error pattern that could produce that syndrome, the decoder could apply the corresponding double-bit correction. If, however, more than one double-bit error pattern could have produced that syndrome, the decoder cannot guarantee to correct the error.

In decoders for shortened cyclic codes, further modifications are required (see Shortening Cyclic Codes section).

Extended Codes

An odd-distance code can be extended by adding an overall (even) parity bit; for binary codes this is just the modulo-2 sum of the codebits. This increases the code distance by 1. If the original distance is $2t + 1$, the code is t-error correcting; the extended code can correct up to t errors and detect the case in which there are $t + 1$ errors, which is often useful. An example of an extended code is the (24,12) extension of the Golay (23,12) code, which has a convenient rate, k/n, of exactly 0.5.

Interleaved and Concatenated Codes

Interleaving (also known as interlacing) is one technique used to combat burst errors, such as those that occur in fading radio channels. As an illustration, consider arranging 110 databits as ten 11-bit rows and encoding each row using a Hamming (15,11) coder.

Assume that the resulting codewords are stored as the ten rows of a 10×15 bit array, but that the array is then transmitted column by column. This means that the first bit of each codeword will be transmitted, then the second of each, and so on. The receiver regroups the received bits into the original pattern of ten 15-bit codewords before decoding.

In this example, if there is a burst error of up to ten bits duration, no more than one bit in each codeword will be affected. Because the code is single-error correcting, the receiver can correct these errors, so the use of 10-way interleaving allows correction of 10-bit burst errors.

The extension of this technique to more powerful codes is straightforward and permits correction of combinations of random and burst errors.

Concatenated coding means the application of one error-correcting code to the input data, followed by the application of another code to codewords output by the first coder, and so on. Concatenated coding is valuable when the different codes can combat different types of error (e.g., burst versus random). It is used either for convenience of implementation or when the overall efficiency (k/n) of the concatenated code is higher than that of any single code with the same error-correcting ability.

Cyclic Codes

Cyclic codes are a very important class of linear block codes for two reasons: first, cyclic codes are available for a wide range of error-detecting and error-correcting requirements, including burst error correction; second, the encoders and decoders can be implemented efficiently using feedback shift registers.

The mathematical theory of cyclic codes is based on the "algebra of polynomials over GF(M)." In this framework the digits of a dataword or codeword are treated as the coefficients of a polynomial. For example, the 4-bit word (1011) is represented by the polynomial of equation 4.9.

$$d(x) = 1 \cdot x^3 + 0 \cdot x^2 + 1 \cdot x^1 + 1 \cdot x^0 \tag{4.9}$$

When adding, subtracting, multiplying, or dividing these polynomials, the arithmetic that is carried out on the coefficients is done in GF(M) (i.e., modulo-2 for binary cyclic codes). The variable x never needs to be evaluated; it serves only a place-keeping role.

A cyclic code is completely defined by its generator polynomial, which for an (n,k) code is a factor of $x^n - 1$ and is of degree $r = (n - k)$. Hence it is of the form given by equation 4.10 and has $r + 1$ coefficients.

$$1 \cdot x^r + g_{r-1}x^{r-1} + \ldots + g_0x^0 \tag{4.10}$$

Published tables are available that give, for each code, the values of n and k, the coefficients of $g(x)$, and the distance of the code and hence its error detection and correction ability. In such tables it is common practice to represent the binary coefficients of the generator polynomial in octal notation. For example, if the generator of a (7,3) cyclic code (which therefore has $7 - 3 + 1 = 5$ coefficients) is written as 35 in octal, the binary coefficients are (11101) and the generator polynomial is given by equation 4.11.

$$g(x) = 1 \cdot x^4 + 1 \cdot x^3 + 1 \cdot x^2 + 0 \cdot x^1 + 1 \cdot x^0 \tag{4.11}$$

In systematic cyclic codes, in which the first k bits of the codeword are equal to the data word, equation 4.12 defines the parity check polynomial, $p(x)$, which defines the r parity bits. $p(x)$ is computed as the remainder polynomial after $x^r d(x)$ is divided by $g(x)$.

$$c(x) = x^r d(x) + p(x) \tag{4.12}$$

Given the received codeword, represented as a polynomial $r(x)$, the decoder calculates a syndrome polynomial, $s(x)$, which is used for error detection and correction. $s(x)$ is also calculated as the remainder after a division. These polynomial division operations are efficiently implemented in the way described in the next section.

Implementation of Cyclic Coders and Decoders

Cyclic coders and decoders are implemented using Linear Feedback Shift Registers (LFSRs). The general form of a systematic cyclic coder for the code is shown in Figure 4.1, with generator $g(x)$ given by equation 4.13.

$$g(x) = 1.x^r + g_{r-1}x^{r-1} + \ldots + g_0x^0 \tag{4.13}$$

All multiplications and additions are in GF(M) (i.e., modulo-2 for binary digits). The rectangular boxes are latches that store digits; there are r multipliers and latches. The latches

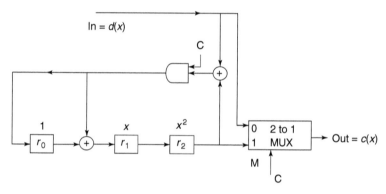

94   Chapter 4   Coding

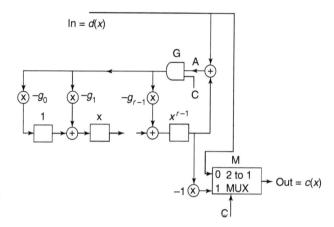

Fig. 4.1 general form of a systematic cyclic coder.

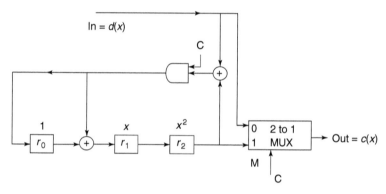

Fig. 4.2 Systematic binary cyclic coder.

are initially cleared. During the first k cycles of operation, control line C is ON and so AND gate G is open and the k digits of the dataword are fed through multiplexer M to the output. In the last r cycles of operation, C is OFF and so gate G is closed and the r values stored in the shift register are sequentially output through M.

The function of this circuit is equivalent to the polynomial division $x^r d(x)/g(x)$; the contents of the shift register after the kth cycle are the remainder.

For binary coders, multiplication by 0 is equivalent to removal of the corresponding path and adder, and multiplication by −1 or +1 is equivalent to a direct connection to the adder input. Each modulo-2 adder is an XOR gate. Hence, the systematic binary coder for the (7,4) code with the generator polynomial given by equation 4.14 (which would be represented in tables as 1011 in binary or 13 in octal) is as shown in Figure 4.2.

$$g(x) = x^3 + x + 1 \tag{4.14}$$

The decoder calculates the syndrome of the received (possibly corrupt) codeword, using an LFSR circuit exactly as in the coder. The contents of the shift register after the nth cycle are the remainder (i.e., the required syndrome). Other forms of syndrome are sometimes used, but the decoder operation is very similar in either case.

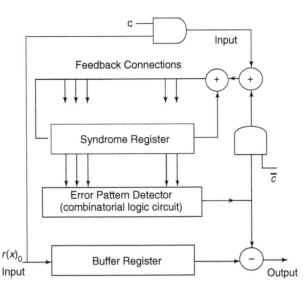

Fig. 4.3 General form of a Meggitt, or error-trapping cyclic error-correcting decoder.

A nonzero syndrome means there are one or more errors. In an error-correcting decoder, further processing is required to correct the error. The general form of a "Meggitt," or "error-trapping," decoder is shown in Figure 4.3.

During the first n cycles, control line C is ON and the syndrome is calculated as already described; while this is happening the received codeword is stored in a buffer register. During the second n cycles, C is OFF and the received codeword is output from the buffer register. The error pattern detector circuit outputs a 1 (in the binary case) whenever the corresponding bit of the received codeword has been deduced to be in error. This inverts, and hence corrects, the output codeword. Because the code is systematic, its first k digits are the required dataword.

It can be seen from Figure 4.3 that the signal from the error pattern detector circuit is also fed back to the LFSR input. This produces an "updated" syndrome; after the whole codeword has been corrected, the remaining syndrome in the shift register should be 0. If it is not, it indicates that more errors have occurred than can be corrected by the code. Depending on the system requirements, this condition can be detected and flagged.

The internal details of the error pattern detector are simple for single ECCs. In this case, the syndrome corresponding to an error in the first (most significant) bit of the codeword is calculated. The error pattern detector is then simply a circuit that outputs a 1 when it detects this pattern in the shift register. For multiple error-correcting decoders, as discussed in the following section, the design of the error pattern detector is more complex.

BCH Codes

BCH codes are the most extensive and powerful family of error-correcting cyclic codes. Because of their mathematical structure, they allow decoders to be implemented reasonably easily, even for multiple-error correction. Both binary and nonbinary BCH codes exist; there are two classes of binary BCH code: primitive, which have a block length of $n = 2^m - 1$, and nonprimitive, where n is a factor of $2^m - 1$.

For any positive integers, m and $t(t < 2^{m-1})$, there exists a primitive BCH code with the parameters n and r given by equations 4.15 and 4.16 and the minimum distance given by equation 4.17.

$$n = 2^m - 1 \qquad (4.15)$$

$$r = n - k \le mt \qquad (4.16)$$

$$d_M \ge 2t + 1 \qquad (4.17)$$

Such a code can therefore be used as a t-error–correcting code.

Both the algebraic basis of BCH codes and the decoder algorithms are described in many textbooks. Primitive BCH codes include cyclic forms of the Hamming SEC codes, and nonprimitive BCH codes include a cyclic form of the (23,12) Golay TEC code.

The two main procedures used in decoding are

1. Peterson's direct solution, suitable for up to about 6-error correction.
2. The Berlekamp–Massey algorithm, an iterative algorithm applicable to any BCH code.

Some decoder algorithms have also been developed for specific BCH codes—for example, the Kasami algorithm for the (23,12) TEC code.

Reed–Solomon Codes

Reed–Solomon (RS) codes are an important subclass of nonbinary BCH codes. RS codes have a true minimum distance between their codewords, which is the maximum possible for a linear (n,k) code, as in equation 4.18. They are therefore examples of maximum-distance-separable codes.

$$d = n - k + 1 \qquad (4.18)$$

For decoding RS codes, both Peterson's method and the Berlekamp–Massey algorithm can be used. The latter is also known as the FSR synthesis algorithm because it is equivalent to the generation of the coefficients of a certain LFSR. A further technique for decoding RS codes is called transform decoding, which uses a finite field analogue of the Fourier transform.

A particularly important ability of RS codes (and nonbinary BCH codes in general) is their ability to perform error and erasure decoding. This is of value if the receiver can under some circumstances signal the loss of a received digit. For example, in a CD player the demodulator can often detect invalid (corrupt) digits. If such a code has distance d, it can correct combinations of 1 errors and s erasures provided that $2l + s < d$.

While RS codes are naturally suited to nonbinary digits, they can sometimes be used very effectively for binary channels by treating groups of m bits as 2^m-ary digits. It has been shown that such codes outperform binary codes with the same rate and block length at low output error rates. Also, when used in this way, RS codes have a natural burst error–correcting ability because, for example, a burst of up to m bit errors will affect at most 2 "digits."

Majority Logic Decodable Codes

These cyclic codes are slightly inferior to BCH codes in terms of error correction, but they have simple decoder implementations in which the error pattern detector circuit is a combination of XOR and majority logic gates.

Burst Error–Correcting Codes

Fire codes are a widely used class of algebraically constructed burst error–correcting cyclic codes, which require a minimum of $3b - 1$ parity bits to correct bursts of up to b bits in length. Other, more efficient burst error–correcting cyclic codes, found by computer search, are also available.

Shortening Cyclic Codes

Cyclic codes can be shortened; the design of a systematic cyclic coder is unaffected by this, except that only k bits are clocked in rather than k' before the control line is switched to output the parity check bits. In the decoder, the same approach of simply reducing the number of clocks can be used for syndrome polynomial calculation; however, if error correction is to be carried out, the error pattern detector must be modified to deal with the missing leading bits and the fact that there cannot be errors in those missing bits.

Convolutional Codes

A convolutional coder converts a continuous stream of source data symbols into a continuous stream of encoded symbols. Like block codes, convolutional codes can be systematic or unsystematic, and can be used for error detection only or correction as well. However, where only error detection is required, block codes are simpler and are almost always used.

By contrast with the analysis and design of block codes, the behavior of convolutional codes is harder to analyze, and some of the best codes are found by computer search. A rigorous mathematical treatment is, however, available in Forney (1970). Tables of good convolutional codes for various requirements are available in reference books, together with measures of their performance, such as their coding gain at different output BERs.

Convolutional Coding

A convolutional coder takes M-ary input digits in groups of k at each timestep and produces groups of n output digits. Because the input and output datarates are k and n digits per timestep, the code is known as a *rate-k/n code*. The coder contains k parallel shift registers of maximum length $L - 1$, where L is known as the *constraint length* of the code. Two common pictorial representations of such a coder (with $k = 1$, $n = 2$, and $L = 4$) are shown in Figure 4.4.

The output digits are formed as weighted sums of the input digit(s) and the digits in the shift register(s), with arithmetic in GF(M) as for block codes. In binary, the only possible weights are 0 and 1, so the output bits are formed by selected modulo-2 sums of the input bit(s) and register contents. After the output digits have been formed, the input digit(s) are shifted into the shift register(s).

An alternative representation of a convolutional coder is as a single shift register of length $\leq(L - 1)k$ digits. In each timestep the n output digits are calculated and then the k

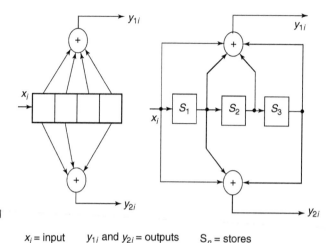

Fig. 4.4 Two commonly used representations for convolutional coders, for the example of a binary rate ½ constraint.

x_i = input y_{1i} and y_{2i} = outputs S_n = stores

input digits are loaded sequentially into this register. Sometimes Lk is called the *constraint length* (a conflicting definition with the previous one).

The digits in the shift register(s) (s_1 to s_3 in Figure 4.4) define the state of the coder at each timestep; there are therefore $M^{(L-1)k}$ possible coder states. The possible transitions from a given coder state at one timestep to another state at the next timestep are determined by the shift register arrangement. The actual output code digit stream is determined by both the shift register arrangement and the particular summations chosen to form the output digits.

There are M^k patterns of k input digits, so from each state there are M^k possible next states. The behavior of the coder can therefore be described using a trellis diagram, as shown in Figure 4.5. In this, the possible states of the coder are represented as $M^{(L-1)k}$ nodes in a vertical column, and the state-to-state transitions at each timestep are represented by a rightward step of one column. From each node at timestep i there are M^k branches to successor states at timestep $i+1$. In a worthwhile code no two different patterns of input digits give the same transitions.

Viterbi Decoding

The error-correcting power of a convolutional code arises from the fact that only some of the possible sequences of digits are valid outputs from the coder; these sequences correspond to possible paths through the trellis. The job of the decoder is to find which valid digit sequence is closest to the received digit sequence. This is analogous to the job of a block decoder, but because the input digit sequence is continuous the coder must operate continuously and must have an acceptably small delay between the arrival of particular input digits and the output of the corresponding decoded digits.

The optimum (in the sense of maximum-likelihood) decoding of convolutionally coded sequences can be carried out using the Viterbi algorithm, which can be applied to both hard-decision and soft-decision decoding. Calculating the most likely digit sequence output by the coder is equivalent to calculating the most likely path followed by the coder through

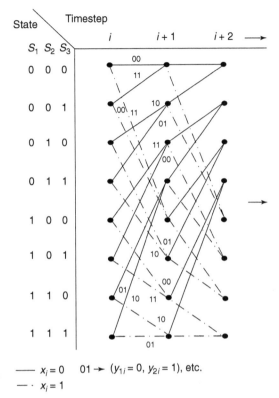

Fig. 4.5 Trellis diagram for the coder in Figure 4.4.

$x_i = 0$ $01 \rightarrow (y_{1i} = 0, y_{2i} = 1)$, etc.

$x_i = 1$

the trellis. In a hard-decision decoder (and assuming a binary symmetric channel for simplicity) the most likely coder path is the one that has the smallest number of disagreements with the received bitstream. In a soft-decision decoder the most likely path is the one with the smallest squared difference from the received signal. These measures of discrepancy between the received signal and possible transmitted signals are referred to as path metrics.

It appears at first sight that the number of candidate paths grows exponentially with each timestep. However, at timestep i the decoder only needs to keep track of the single best path so far to each node; any worse path to that node cannot be part of the overall best path through the node. The number of paths to be remembered therefore remains constant and equal to the number of nodes. For each node the decoder must store the best path to that node and the total metric corresponding to that path. By comparing the actual n received digits at timestep $i + 1$ with those corresponding to each possible path on the trellis from step i to step $i + 1$, the decoder calculates the additional metric for each path. From these, it selects the best path to each node at timestep $i + 1$ and updates the stored records of the paths and metrics.

In theory the decoder has to wait until the hypothetical end of transmission before it can decide the overall maximum-likelihood path from start to finish. Fortunately it can be shown that a fixed finite delay between input and output is all that is necessary to give nearly optimum performance. In particular, it can be shown that if all the reasonably likely paths at timestep *i* are traced backward in time, they will converge to the same path within about 5*L* timesteps. The decoder therefore operates with a fixed delay of δ steps, where $\delta \geq$ 5*L*; at time *i* it selects one of the reasonably likely paths, and the bit from that path at time $i - \delta$ is output as the decoded output. The path stores within the decoder can therefore be implemented as δ-bit shift registers.

Code Performance

A decoding error occurs when the path through the trellis selected by the decoder is not the correct path. The result will almost always be a burst error; it is even possible, for some codes, for the worst error to be infinite in extent. Such codes are clearly unusable. Analysis of the exact statistics of the error bursts (their probability as a function of their duration) is very complex; approximations can, however, be derived using the minimum free distance of the code. This is obtained by considering all possible error paths—that is, incorrect paths that depart from the correct path at one symbol time and rejoin it later. The minimum free distance is the minimum number of symbol differences between any of these error paths and the correct path. Note, however, that calculation of the true error rates involves analysis of many other interacting effects.

In practice, the user of a convolutional code will normally use the published graphs of output BER as a function of SNR (E_b/n_0). The coding gain is also often quoted, but it is important to remember that this varies with SNR.

Turbo and LDPC Codes

Until the 1990s, the most powerful codes for telecommunications, known as RSV codes, were created by concatenating Reed-Solomon and Viterbi-decoded convolutional codes.

Two major advances came with Turbo codes (1993) and LDPC codes (1995). The huge performance benefit of these codes arises because they generate very long codewords but in a way that allows the decoder to converge rapidly to the optimum estimate of the data digits.

Turbo Codes

In 1993 Berrou, Glavieux, and Thitimajshima announced the invention of "Turbo codes." These achieve 3 dB or more of coding gain compared to RSV codes, and as a result they closely approach the Shannon limit. In the original Turbo code, a data block (**d**) is fed into a convolutional encoder, which computes a set of parity bits ($\mathbf{p}_1$). The data is in parallel scrambled by an interleaver, giving a scrambled data block (**d′**) which is fed into a second convolutional encoder, which computes a further set of parity bits ($\mathbf{p}_2$). Finally, the transmitter sends **d**, $\mathbf{p}_1$, and $\mathbf{p}_2$. These arrive, corrupted, at the receiver as data blocks $\mathbf{d}_R$, $\mathbf{p}_{1R}$, and $\mathbf{p}_{2R}$.

At the receiver there are two decoders; the first processes $\mathbf{d}_R$ and $\mathbf{p}_{1R}$, while the second processes $\mathbf{d}_R′$ and $\mathbf{p}_{2R}$, where $\mathbf{d}_R′$ is generated from $\mathbf{d}_R$ by a de-interleaver matching the one in the transmitter. A crucial aspect is that the decoding is iterative, with each decoder taking in not only the analogue signal samples but also information from the other decoder. The first decoder passes estimated values of the data, **d**, with "confidence levels" (actually a quantity called a *log-likelihood ratio* for each estimated digit), to the second decoder, which uses them, together with $\mathbf{d}_R′$ and $\mathbf{p}_{2R}$, to compute revised decisions, each with an associated confidence level. These are passed back to the first decoder, which computes revised decisions, and so on. This iterative process converges to a solution (when little further change occurs typically) after 4 to 10 iterations.

This original type of Turbo code is called a Turbo Convolutional Code (TCC); it is quite complex to implement and is heavily protected by patents on both the encoding and decoding sides. It is nevertheless included as an option in the IEEE 802.16a standard and in 3G mobile telephony standards such as 3GPP (UMTS/W-CDMA) and 3GPP2 (CDMA2000).

Another type of code, known as a Turbo Product Code (TPC), has been shown to provide performance similar to that of the TCC but with lower complexity and smaller encoding delay than either the TCC or RSV decoders. Product codes (first proposed by Elias in 1954) are constructed from 2D (row and column) arrangements of simple linear block codes (typically simple parity or Hamming codes). In TPCs, iterative decoding is again used, with information passed to and fro between the row and column decoders. The principle can be extended to 3D arrangements of codes.

LDPC Codes

Low-Density Parity Check (LDPC) codes were first proposed by Robert Gallager in 1960. Implementing them was not feasible then, and they were largely forgotten until Tanner, in 1981, generalized them and developed a graphical method of representing them (called Tanner or bipartite graphs). They were reinvented in 1995 (MacKay and Neal, 1995), and since then there has been a great deal of work on them. LDPC codes are now in many respects the best codes known.

An LDPC code is based on the use of a parity check matrix, **H** (see the Linear Block Codes section), which contains only a few 1s (hence the term "low-density" or "sparse"). It is "random" (generated using random numbers) but follows specific construction rules. The resulting code is unsystematic. Encoding is performed straightforwardly as described in the Linear Block Codes section, using a generator matrix, **G**, derived from **H**. The decoder is designed by first deriving from equation 4.2 multiple parity check equations, all of which are satisfied by any generated codeword. A network of processing nodes is created, each one handling one of those equations. Decoding proceeds iteratively; information (comprising estimates of some digits, together with confidence levels) is passed between the nodes, which also take in a "soft" (i.e., analogue) input of the received signal. This iterative probabilistic equation solver is of a type known as a *message-passing algorithm, belief propagation algorithm,* or *sum-product algorithm.*

With the flexibility of LDPCs, codes can be constructed to match exactly a particular block size or code rate, though practical implementations can have constraints on block sizes and/or obtainable code rates.

Concatenation to Overcome Error Floors

What is meant by an "error floor" is that the error probability of a given code does not approach 0 as quickly for medium to high SNR as it does at low SNR. Such error floors primarily affect codes with low- (Hamming-) weight codewords, such as LDPC and Turbo codes.

To lower the error floor, the LDPC (or Turbo) code can be concatenated with an outer code, such as a BCH code. Digital Video Broadcast (DVB) has chosen this method of FEC, using LDPC and BCH coding, for its new DVB-S2 standard.

Lattice and Trellis Codes

Although the Term *trellis code* is sometimes used to refer to convolutional codes in general, Trellis-Coded Modulation (TCM) is a combined modulation and coding scheme (Burr, 1993) invented by Ungerboeck (1982). In TCM each multibit data symbol is split into two groups of bits; one group from each symbol (group 1) is then fed into a convolutional encoder, whose output bits determine which of a number of signal constellations (sets of possible signal values) is used for that symbol. The other group of bits (group 2) selects which value from the constellation is transmitted.

A very simple example would be that at each symbol time the choice of symbol is either between 1 and 3 or between 2 and 4. Based on the received values, a convolutional code decoder in the receiver estimates the sequence of constellations and, from that, the values of the group-1 bits. It then determines which of the constellation points was most probably transmitted and, hence, the values of the group-2 bits. If, for example, the decoder knows that the received value is either 2 or 4 at one particular symbol time, then it does not choose 3. The separation between the allowed possible symbol values at each symbol time is thereby increased, which gives rise to the performance gain of TCM. TCM is incorporated in the CCITT V.32 and V.32 modem standards. Lattice-coded modulation is the block code–based equivalent of TCM but is little used.

Space–Time Codes

Multiple Input Multiple Output (MIMO) techniques, invented by Foschini et al. at Bell Labs in 1996, have become an important means of improving the capacity and robustness of wireless systems. They are now being incorporated in several standards (for example, IEEE 802.11n and 802.16e-2005). MIMO systems use multiple antennas at the transmitter or receiver or both, and exploit the fact that the transmitted signals traverse a "multipath" environment (with scattering, reflection, refraction, and so on). Before 1996 it was known that using multiple receive antennas and either selecting the best received signal or optimally combining received signals could improve transmission reliability. This is known as *diversity reception*. Where multiple antennas are used instead at the transmitter, Space–Time Block Codes (STBCs) can be used to encode the datastream across the multiple antennas. At the receiver, which can use one or more antennas, some of the received copies of the data will be "better" than others, and this redundancy can be exploited to estimate the transmitted digits with minimum error. STBCs offer only diversity gain (compared to single-antenna schemes) and not coding gain. One particular STBC is the code invented by Alamouti in 1998 for a two-transmit-antenna system. It is the simplest example of a class called *orthogonal STBCs*, and these are particularly attractive because optimal (maximum-likelihood) decoding can be achieved at the receiver with only linear processing.

MIMO techniques can achieve not only diversity gain but also an increase in capacity. Space–time trellis codes (STTCs) provide both diversity and coding gain by spreading a conventional trellis code over space (multiple antennas) and time (Tarokh et al., 1998).

Encryption

It is sometimes the case that a sender wishes to send a recipient a message that he wants to keep secret from an eavesdropper. This message, which in its usable form is called *plaintext*, must be transmitted over a channel to which the eavesdropper is presumed to have access. The process by which the sender and recipient can achieve secrecy is called *encryption* or *encipherment*.

In encryption, the plaintext is transformed into a message, called the *ciphertext*, in such a way that the recipient can recover the plaintext from the ciphertext while the eavesdropper cannot. The transformation of plaintext to ciphertext and back is controlled by one or more strings of symbols or digits called *keys* (Boyd, 1993).

A further use of encryption is for authentication, when it is necessary to check that you are communicating with the correct person and not an impostor (message authentication codes or digital signatures).

Principles of Encryption

Encryption algorithms have two possible components: substitution, in which plaintext symbols are mapped to different symbols, and transposition, in which the locations of symbols in the ciphertext are altered from the locations of the corresponding symbols in the plaintext. The need for substitution is obvious; many symbols have sufficient significance that an eavesdropper could deduce information from them wherever they are and perhaps alter them. Transposition is desirable to prevent the eavesdropper from deducing information by comparing messages, or corrupting known parts of a message, even if he does not know what they have been altered to.

The most completely secure system is known as the *one-time pad*. This is a symbol-by-symbol substitution system in which the key is as long as the message; hence no deductions about one part of the ciphertext help the eavesdropper to decipher the rest. The problem is in distribution of the key itself (key management), which is now as big a problem as the original problem of sending the message securely.

Practical encryption systems must have manageable keys and are based on algorithms for which it is too difficult, rather than theoretically impossible, to decrypt the ciphertext. Various (worst-case) assumptions are made in designing and analyzing the security of the system; it is normally assumed that the eavesdropper knows the encryption and decryption algorithms but not the key. When a very high degree of security is required, it is also assumed that the eavesdropper has obtained (by other means) the plaintext corresponding to some of the intercepted ciphertext.

In conventional or symmetric cryptosystems, it is easy to deduce the deciphering key from the enciphering key (they might even be the same), so both must be kept secret. In public-key cryptosystems, however, the enciphering key can be made public by the recipient. Despite this, it is believed to be computationally infeasible for an eavesdropper to work out the decipherment key. A further technique, *public-key distribution* (Diffie & Hellman, 1976) applies the same principles to the secure distribution of keys for conventional cryptosystems.

In stream ciphers, a stream of plaintext symbols is enciphered symbol by symbol. Instead of a true one-time pad, the stream of enciphering symbols is generated by a symbol generator under the control of the enciphering key. A simple example for binary data is the use of a PseudoRandom Binary Sequence (PRBS) generator, initialized by the chosen key. This produces a stream of 0s and 1s, which are XORed with the databit stream. Clearly, recovery of the original bitstream requires exactly the same operation, so the decipherment key is the same as the encipherment key. Such systems are widely used in communications; their advantages include ease of encryption and decryption and the fact that single-bit errors in the channel cause only single-bit errors in the decrypted plaintext. Disadvantages include the effort of initial synchronization and resynchronization if synchronization is lost, and the fact that there is no transposition element in this system.

In block ciphers, the message is divided into n-bit blocks, and the blocks are input sequentially to the algorithm. At each stage the key (which is usually constant for the whole message) is also entered, and the encryptor typically uses both substitution and transposition to produce consecutive blocks of ciphertext. In such systems an error in transmission usually results in corruption of the whole received block.

In all these systems, it is essential that the key is changed frequently; the security of an otherwise satisfactory cryptosystem can be compromised if the key is reused.

Specific Cryptosystems

Many cryptosystems have been developed, both by government agencies and commercial organizations. The first public standard system was the National Bureau of Standards Data Encryption Standard (DES), which is a block cipher with a 64-bit block length involving both substitution and transposition under the control of a 56-bit key (NBS, 1977). The original proposal was for a 64-bit key, and there are claims that DES keys have been broken in less than 24 hours and that DES is therefore too insecure for many applications. Triple DES, which consists of applying DES three times in succession, is believed to be secure in practice. DES has now been superseded by the Advanced Encryption Standard (AES), also known as Rijndael (NIST, 2001). AES has a fixed block size of 128 bits and a key size of 128, 192, or 256 bits.

The RSA algorithm (Rivest, 1978) is a public-key encryption algorithm in which the recipient publishes an encipherment key, N, consisting of the product of two primes, each of order 10^{100}, together with another number, E. The sender breaks the plaintext into blocks that can be represented by numbers less than N, then encrypts the blocks using a simple modulo-N arithmetic operation involving E. The recipient recovers the plaintext using further simple modulo-N operations that rely on knowledge of the factors of N. The difficulty of factoring numbers of order 10^{200} means that the eavesdropper cannot work out the two factors of N and cannot decrypt the message.

Spread-Spectrum Systems

Spread spectrum systems (Proakis, 1989; Tsui and Clarkson, 1994) are used in digital communications for

- Combating interference arising from jamming by others, or self-interference due to multipath effects
- Making it difficult to detect the signal to achieve covert operation
- Making it difficult to demodulate the signal to achieve privacy
- Allowing multiple users to share one frequency band ("multiple access") in a way that provides flexibility in the number of users and the allocation of transmission capacity to different users. This is also known as Code Division Multiple Access (CDMA).

Applications

Until 1995, the main application of spread-spectrum techniques was in military systems to achieve the preceding first to third objectives. However, with the introduction of the IS-95 CDMA-based digital cellular system and the subsequent incorporation of spread-spectrum techniques in many standards, their dominant application is now in commercial digital cellular phone and wireless networking systems. The advantage of using spread-spectrum techniques is that they overcome problems due to multipath and that multiple users can be provided with a uniformly acceptable service rather than good service for some and bad for others, as can happen with older Frequency Division Multiple Access (FDMA) systems.

Direct Sequence Spread Spectrum

In Direct Sequence Spread Spectrum (DSSS), the transmitter and receiver contain identical pseudorandom sequence generators producing a pseudonoise (PN) stream of symbols at a rate (known as the chip rate) that is a multiple of the data symbol rate. The ratio of chip rate to data symbol rate is the spreading factor (L). In the transmitter, the input datastream is XORed with the PN signal before transmission, so each data symbol is replaced by L chips. The bandwidth of the transmitted signal is therefore L times greater than that of the datastream. In the receiver the received signal is XORed with the PN stream to recover the original datastream; this is equivalent to correlation with the known PN sequence.

The use of DSSS gives LPI (Low Probability of Intercept) because the total signal power is spread over a wide bandwidth and the signal is noiselike, making it hard to detect. In antijamming (AJ) applications, the transmitter introduces an unpredictable element into the modulation of the signal, known to the receiver but kept secret from opponents, as in stream ciphering. This, together with the wide bandwidth of the transmitted signal, makes jamming more difficult than for conventional signals.

DSSS is used in a large number of applications, including

- CDMA cellular (mobile) phones (IS-95, CDMA2000, W-CDMA, UMTS, FOMA)
- Some cordless phones
- The 802.11b/g Wi-Fi standards
- Satellite navigation systems (GPS, Glonass, and Galileo)
- ZigBee/802.15.4 (for low-power wireless networked controls)

In CDMA applications of DSSS (DS-CDMA), the receiver synchronizes to the received signal and correlates it with its particular PN pattern; this is called *despreading*. Despreading increases the power of the wanted signal by a factor of L relative to the signals from other users (and noise). The transmitters that are sharing the channel use different PN sequences, chosen so that their cross-correlation is low, to reduce the interference between the received signals. The PN sequences must also have sharply peaked autocorrelation functions to help the receiver synchronize correctly to the partially unknown timing of the received signal. Some often used sets of sequences with these properties are called Gold and Kasami sequences (Proakis, 1989).

Frequency Hopping Spread Spectrum

In Frequency Hopping Spread Spectrum (FHSS), the available channel bandwidth, LW, is divided into L slots (channels) of bandwidth W. In any signaling interval the signal occupies only one slot or a few ($\ll L$).

The spreading of the signal spectrum arises because the active slot frequency is "hopped" around pseudorandomly. Because of the difficulty of maintaining phase references as the frequency hops, FSK modulation and noncoherent demodulation are normally used in FHSS, rather than PSK and coherent demodulation.

In block-hopping FHSS, the input signal is first modulated using a binary, *M*-ary FSK, or OFDM (Orthogonal Frequency Division Multiplex) modulator, whose output signal occupies a bandwidth, *W*. This signal is then frequency-shifted to somewhere in the full bandwidth, *LW*, by mixing it with a local oscillator signal derived from a frequency synthesizer controlled by a PN generator. The receiver has a matching PN generator, frequency synthesizer, and mixer, which shift the frequency block down to baseband again, where it is demodulated by a conventional demodulator matching the modulator.

An alternative approach, which has better resistance to some kinds of jamming but requires a more complex demodulator, is called *independent tone hopping*. In this, each digit value is combined with the output of the PN generator to control the frequency synthesizer directly. The resulting frequency separation of the tones corresponding to different digit values at any particular time can be up to the full channel bandwidth.

The terms *fast hopping* and *slow hopping* traditionally described the relationship between the frequency hopping rate and the symbol rate. However, these terms are now used with great license.

Fast hopping can be used to combat a follower jammer, which attempts to detect the signal tone(s) and immediately broadcasts other tones with adjacent frequencies. However, when fast hopping is used, the fact that phase coherence cannot be maintained across hops means that the energy from the successive hops in one symbol must be combined incoherently; this causes a performance loss.

FHSS signals are primarily used in AJ and CDMA ("FH-CDMA") systems, including Bluetooth, and in some cordless phones. Advantages of FHSS over DSSS are that it has less severe timing requirements and is less sensitive to channel gain and phase fluctuations. Bluetooth also includes adaptive FHSS, which improves interference resistance by avoiding the use of crowded frequencies in the hopping sequence. In the United States, use of FHSS is permitted in unlicensed bands around 900 MHz and 2.4 GHz, governed by FCC part 15 rules.

References

Alamouti, S. M. (1998) A Simple Transmit Diversity Technique for Wireless Communications, *IEEE J. Sel. Areas Commun.* 16(8):1451–1458.

Berrou, C., Glavieux, A., and Thitimajshima, P. (1993) Near Shannon Limit Error Correcting Coding and Decoding: Turbo-Codes, *Proceedings of ICC '93*, 1064–1070.

Boyd, C. (1993) Modern Data Encryption, *Electron. Commun. Eng. J.* 5(5):271–278.

Burr, A. G. (1993) Block versus Trellis: An Introduction to Coded Modulation, *Electron. Commun. Eng. J.* 5(4):240–284.

Diffie, W., and Hellman, M. E. (1976) New directions in Cryptography, *IEEE Trans. Inf. Theory*, 22(6):644–654.

Forney, G. D. (1970) Convolutional Codes 1: Algebraic Structure, *IEEE Trans. Inf. Theory* (IT-16): 720–738.

Lin, S., and Costello, D. J. (1983) *Error Control Coding: Fundamentals and Applications*, Prentice-Hall.

MacKay, D. J. C., and Neal, R. M. (1995) Good Codes Based on Very Sparse Matrices, *Cryptography and Coding: 5th IAM Conference, Lecture Notes in Computer Science No. 1025*, Springer-Verlag, 100–111.

National Bureau of Standards (1977) *Data Encryption Standard*, Federal Information Processing Standard (FIPS) Publication No. 46.

National Institute of Standards and Technology (2001) *Federal Information Processing Standards*, Publication No. 197.

Proakis, J. G. (1989) *Digital Communications, Second Edition*, McGraw-Hill.

Rivest, R. L., Shamir, A., and Adelman, L. (1978) A Method for Obtaining Digital Signatures and Public-Key Cryptosystems, *Commun. ACM* 21(2):120–126.

Shannon, C. E. (1948) A Mathematical Theory of Communication, *Bell Sys. Tech. J.* 27:379–423, 623–656.

Tarokh, V., Seshadri, N., and Calderbank, A. R. (1998) Space–Time Codes for High Data Rate Wireless Communication: Performance Analysis and Code Construction, *IEEE Trans. Inf. Theory* 44(2):744–765.

Tsui, T. S. D., and Clarkson, T. G. (1994) Spread-Spectrum Communication Techniques, *Electron. Commun. Eng. J.*, 6(1):3–12.

Ungerboeck, G. (1982) Channel Coding with Multilevel/Phase Signals, *IEEE Trans. Inf. Theory*, 28(1):55–67.

Resources

Michelson, A. M. and Levesque, A. H. (1985) *Error-Control Techniques for Digital Communication*, Wiley-Interscience.

Peterson, W. W., and Weldon, E. J. (1972) *Error-Correcting Codes, Second Edition*, MIT Press.

Part 2
Transmission Media

This part discusses the transmission media that are used in core telecommunications networks. Before 1980 the predominant transmission medium was copper cables (either copper pairs or coaxial cables). A smaller proportion of systems were carried over radio channels. However, since the advent of commercial optical fiber transmission systems in 1980, copper cables for transmission have been displaced either to the edges of the network or to connections within or between equipment. They are therefore not covered in this volume.

Microwave systems, on the other hand, are still widely used, although the proportion of traffic they carry in core networks is small compared to fiber systems. The basic principles of radio propagation are therefore covered here. However, radio remains a key technology for mobility and for local-area network applications, such as in hotspots or home and office networks. These applications will be discussed in more depth in Volume 3 of *The Cable and Telecommunications Professionals' Reference*.

Optical fiber systems, however, have become firmly established as the mainstay of telecommunications transmission. From modest beginnings in the first commercial systems in 1980, with a single channel per fiber, bit rates of 2 Mbit/s and system ranges around 30 km, they progressively increased in bit rate to 2.5 Gbit/s per channel by 1990. At this stage wavelength division multiplexing was not considered economic; each time a higher-capacity system was needed, digital electronic technology was able to increase its speed to meet the requirement. (A commonly applied rule of thumb was that each time speed increased by a factor of 4, cost increased by a factor of 2½).

Further increase of speed from 1990 onward proved increasingly difficult to achieve economically. By 1995 speeds had increased to 10 Gbit/s, but even with these bit rates, long-haul carriers in the United States were beginning to run out of fiber capacity: They were faced with the choice of either laying new cables on long-distance routes or adopting WDM technology. The principles of WDM were well known, and many suitable components had already been developed but until then they had not been at a price that could compete with higher-speed electronics. However, competing with the cost of laying new cables was a different matter and, with further speed increases remaining difficult over the next ten years, fiber capacity increased mainly by augmenting the number of wavelengths. In that period, the number of wavelengths that could be launched into a fiber increased from 1 to 160 for commercial systems. In the same period, system speeds increased more modestly from 10 to 40 Gbit/s. However, there is currently a strong drive to increase system speeds to 100 Gbit/s for Ethernet applications.

In recent years there have been substantial developments in radio technologies in the areas of miniature antenna designs, new types of modulation format, and encryption. Many of these are associated with cellular systems, which were introduced in volume 1, and broadband access, which will be covered in volume 3. While radio provides much less bandwidth than that offered by fiber, it provides the means for mobile connectivity and can be used for broadband access without the need to lay a new cable infrastructure. The chapters in Part 3 of this volume of *The Cable and Telecommunications Professionals' Reference* provide an introduction to the principles of radio propagation.

Chapter 5 describes the fundmentals of light propagation in optical fibers and the fiber designs that have been developed for short-, medium-, and long-haul transmission. Chapter 6 provides an introduction to fiber transmission systems and their properties, including wavelength division multiplexing and optical amplifiers. It then goes on to discuss transmission impairments and methods to mitigate them and ways to optimize system performance. Chapters 7, 8, and 9 provide an introduction to radio transmission, covering the basic theory of electromagnetic waves, propagation of radio waves, and antenna design.

5 Optical Fibers

Takis Hadjifotiou
Telecommunications Consultant

Introduction

Optical fiber communications have come a long way since Kao and Hockman (then at the Standard Telecommunications Laboratories in England) published their pioneering paper on communication over a piece of fiber using light (Kao & Hockman, 1966). It took Robert Maurer, Donald Keck, and Peter Schultz (then at Corning Glass Works–USA) four years to reach the loss of 20 dB/km that Kao and Hockman had considered as the target for loss, and the rest, as they say, is history. For those interested in the evolution of optical communications, Hecht (1999) will provide a clear picture of the fiber revolution and its ramifications.

The purpose of this chapter is to provide an outline of the performance of fibers that are fabricated today and that provide the pathway for the transmission of light for communications.

The research literature on optical fibers is vast, and by virtue of necessity the references at the end of this chapter are limited to a few books that cover the topics outlined here.

The Structure and Physics of an Optical Fiber

The optical fibers used in communications have a very simple structure. They consist of two sections: the glass core and the cladding layer (Figure 5.1).

The core is a cylindrical structure, and the cladding is a cylinder without a core. Core and cladding have different refractive indices, with the core having a refractive index, n_1, which is slightly higher than that of the cladding, n_2. It is this difference in refractive indices that enables the fiber to guide the light. Because of this guiding property, the fiber is also referred to as an "optical waveguide." As a minimum there is also a further layer known as the secondary cladding that does not participate in the propagation but gives the fiber a minimum level of protection. This second layer is referred to as a coating.

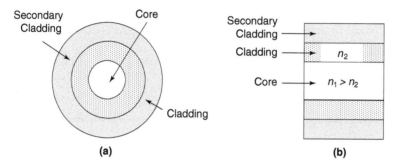

Fig. 5.1   (a) Cross section and (b) longitudinal cross section of a typical optical fiber.

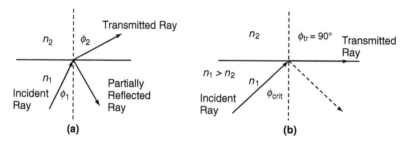

Fig. 5.2   Snell's law.

The basics of light propagation can be discussed with the use of geometric optics. The basic law of light guidance is Snell's law (Figure 5.2a). Consider two dielectric media with different refractive indices and with $n_1 > n_2$ and that are in perfect contact, as shown in Figure 5.1. At the interface between the two dielectrics, the incident and refracted rays satisfy Snell's law of refraction—that is,

$$n_1 \sin\phi_1 = n_2 \sin\phi_2 \tag{5.1}$$

or

$$\frac{\sin\phi_1}{\sin\phi_2} = \frac{n_2}{n_1} \tag{5.2}$$

In addition to the refracted ray there is a small amount of reflected light in the medium with refractive index n_1. Because $n_1 > n_2$ then always $\phi_2 > \phi_1$. As the angle of the incident ray increases there is an angle at which the refracted ray emerges parallel to the interface between the two dielectrics (Figure 5.2(b)).

This angle is referred to as the critical angle, ϕ_{crit}, and from Snell's law is given by

$$\sin\phi_{\text{crit}} = \frac{n_2}{n_1} \tag{5.3}$$

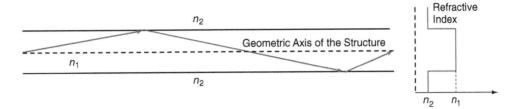

Fig. 5.3 Light guidance using Snell's law.

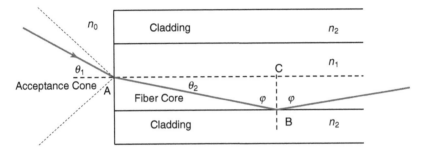

Fig. 5.4 Geometry for the derivation of the acceptance angle.

If the angle of the incident ray is greater than the critical angle, the ray is reflected back into the medium with refractive index n_1. This basic idea can be used to propagate a light ray in a structure with $n_1 > n_2$, and Figure 5.3 illustrates this idea.

The light ray incident at an angle greater than the critical angle can propagate down the waveguide through a series of total reflections at the interface between the two dielectrics. The ray shown in Figure 5.3 is referred to as a "meridional ray" because it passes through the axis of the fiber. For rays not satisfying this condition, see Senior (1992). Clearly, the picture of total internal reflection assumes an ideal situation without imperfections, discontinuities, or losses. In a real-world fiber, imperfections will introduce light that is refracted as well as reflected at the boundary. In arriving at the basic idea of light propagation, it was assumed that somehow the ray has been launched into the fiber.

For a ray to be launched into the fiber and propagated it must arrive at the interface between the two media (with different refractive indices) at an angle that is at minimum equal to ϕ_{crit} and in general less than that. Figure 5.4 illustrates the geometry for the derivation of the acceptance angle. To satisfy the condition for total internal reflection, the ray arriving at the interface, between the fiber and outside medium, say air, must have an angle of incidence less than θ_{acc}, otherwise the internal angle will not satisfy the condition for total reflection, and the energy of the ray will be lost in the cladding.

Consider that a ray with an incident angle less than the θ_{acc}, say θ_1, enters the fiber at the interface of the core (n_1) and the outside medium, say air (n_0), and the ray lies in the meridional plane. From Snell's law at the interface we obtain

$$n_0 \sin\theta_1 = n_1 \sin\theta_2 \tag{5.4}$$

From the right triangle ABC (Figure 5.4), the angle ϕ is given by

$$\phi = \frac{\pi}{2} - \theta_2 \tag{5.5}$$

where the angle ϕ is greater than the critical angle. Substituting equation 5.5 into equation 5.4 we obtain

$$n_0 \sin \theta_1 = n_1 \cos \phi \tag{5.6}$$

In the limit as the incident angle, θ_1, approaches θ_{acc}, the internal angle approaches the critical angle for total reflection, ϕ_{crit}. Then, by introducing the trigonometric relation $\sin^2 \phi + \cos^2 \phi = 1$ into equation 5.4, we obtain

$$n_0 \sin \theta_1 = n_1 \cos \phi = n_1 (1 - \sin^2 \phi)^{1/2} = n_1 \left(1 - \left(\frac{n_2}{n_1} \right)^2 \right)^{1/2} = (n_1^2 - n_2^2)^{1/2} \tag{5.7}$$

This equation defines the angle within which the fiber can accept and propagate light and is referred to as the "Numerical Aperture" (*NA*).

$$NA = n_0 \sin \theta_{acc} = (n_1^2 - n_2^2)^{1/2} \tag{5.8}$$

When the medium with refractive index n_0 is air, the equation for the *NA* of the glass fiber simplifies to

$$NA = \sin \theta_{acc} = (n_1^2 - n_2^2)^{1/2} \tag{5.9}$$

This equation states that for all angles of incident where the inequality $0 \leq \theta_1 \leq \theta_{acc}$ is satisfied the incident ray will propagate within the fiber. The parameter NA expresses the propensity of the fiber to accept and propagate light within the solid cone defined by an angle, $2\theta_{acc}$. The equation for the *NA* can be also expressed in terms of the difference between the refractive indices of core and cladding—that is,

$$\Delta = \frac{n_1^2 - n_2^2}{2n_1^2} \approx \frac{n_1 - n_2}{n_1} \tag{5.10}$$

With these simplifications the NA can now be written as

$$NA = n_1 (2\Delta)^{1/2} \tag{5.11}$$

To appreciate the numbers involved, consider a fiber made of silica glass whose core refractive index is 1.5 and that of the cladding is 1.46. The ϕ_{crit} and the NA of the fiber are calculated to be

$$\phi_{crit} = \sin^{-1} \left(\frac{n_2}{n_1} \right) = \sin^{-1} \left(\frac{1.46}{1.50} \right) = 76.73°$$

$$NA = (n_1^2 - n_2^2)^{1/2} = (1.50^2 - 1.46^2)^{1/2} = (2.25 - 2.13)^{1/2} = 0.346$$

$$\theta_{acc} = \sin^{-1} NA = \sin^{-1} 0.346 = 20.24°$$

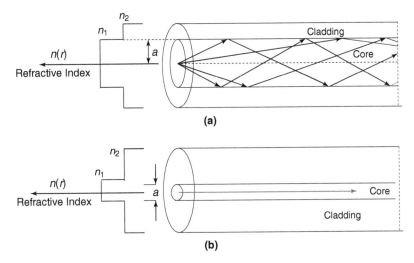

Fig. 5.5 (a) Multimode and (b) single-mode propagation for step index fiber.

If one uses the equation for the approximation of NA, equation 5.11, the result is

$$\mathrm{NA} \approx n_1 (2\Delta)^{1/2} = 1.5 \times \left(2 \times \left(\frac{1.50 - 1.46}{1.5}\right)\right)^{1/2} = 1.5 \times (2 \times 0.0266)^{1/2} = 0.346$$

If the ray does not lie in the meridional plane, the analysis for obtaining the NA is slightly more complex.

The discussion up to now has been based on the launching and propagation of a single ray. Because all the rays with a reflection angle less that the critical angle can propagate, it is expected that a number of rays will propagate provided they can be launched into the fiber. The class of fiber that can support the simultaneous propagation of a number of rays is known as multimode fiber. The term "mode" is used in a more sophisticated analysis of light propagation using Maxwell's equations, and it corresponds to one solution of the equations. Therefore, the concept of a ray and a mode are equivalent, and in the rest of this chapter the term "mode" will be used.

A fiber that allows the propagation of one mode only is called a single-mode fiber. A fiber can be multimode or single mode, and the behavior depends on the relative dimensions of the core and the wavelength of the propagating light. Fibers with a core diameter much larger than the wavelength of the launched light will support many propagating modes, and the propagating conditions can be analyzed with geometric optics. A fiber with a core diameter similar to that of the light wavelength supports only one propagation mode. Figure 5.5 illustrates these concepts.

Material Characteristics of Fibers—Losses

The basic material used in the manufacture of optical fiber for optical transmission is silica glass. A large number of glasses have been developed and studied with the objective of improving fiber transmission properties. There are two parameters of glass that have a substantial impact on its performance: the losses and the changes of refractive index with

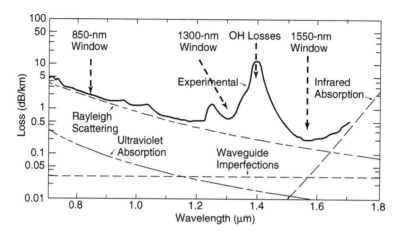

Fig. 5.6 Measured loss versus wavelength for SiO$_2$ fibers.   *Source:* Adapted from Hecht (1999). Used with permission.

wavelength. The basic material used in the manufacture of optical fibers is vitreous silica dioxide (SiO$_2$), but to achieve the properties required from a fiber, various dopants are also used, (Al$_2$O$_3$, B$_2$O$_3$, GeO$_2$, P$_2$O$_5$). Their task is to slightly increase and decrease the refractive index of pure silica (SiO$_2$). Initially the fiber losses were high, but through improvements in the quality of the materials and the actual production process, the losses have been reduced so as to be close to the theoretical expected losses.

In the part of the electromagnetic spectrum where optical fiber transmission takes place, the losses are bracketed between two asymptotes: ultraviolet absorption and infrared absorption. The measured loss–versus–wavelength curve is shown in Figure 5.6.

The actual loss in the window between 0.8 and 1.6 μm is dictated by the Rayleigh scattering losses. The physical origin of the Rayleigh scattering losses is the excitation and reradiation of the incident light by atomic dipoles whose dimensions are much less than the wavelength of the light. The loss can be expressed in terms of decibels per kilometer by the expression

$$\alpha_{\text{scat}} = \frac{A_R}{\lambda^4} \tag{5.12}$$

Where the constant A_R reflects the details of the doping process and the fiber fabrication, and the value of A_R is around 0.95 dB/km $- \mu$m^4. In fiber with a pure silica core, the value of A_R can be as low as 0.75 dB/km $- \mu$m^4. The actual scattering within a fiber is higher than in that of bulk material because of additional scattering from interfaces and in homogeneities in the fiber structure.

The losses discussed up to now are usually referred to as intrinsic because their origin is in the physical properties of the material. The total intrinsic loss can be expressed as the sum of the three—that is,

$$\alpha_{\text{intrinsic}} = \alpha_{\text{ultraviolet}} + \alpha_{\text{infrared}} + \alpha_{\text{scat}} \tag{5.13}$$

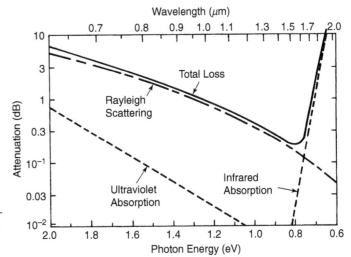

Fig. 5.7 Theoretical intrinsic loss characteristics of germania-doped silica (GeO$_2$–SiO$_2$) glass. *Source:* Adapted from Senior (1992). Used with permission.

The theoretical intrinsic loss has a minimum value of 0.185 dB/km at a wavelength close to 1550 nm. Figure 5.7 summarizes the intrinsic theoretical losses of a silica glass fiber.

In addition to the intrinsic losses, the fiber has additional losses that are referred to as extrinsic. These are associated with the presence of various substances in the glass, the quality of the glass, the processing, and the mechanical imperfections associated with the fiber structure.

These losses can be removed with refinements in the fabrication process and in the quality of glass. The presence of metallic and rare-earth impurities contributes to the extrinsic fiber loss as does the presence of the hydroxyl group OH that enters the glass through water vapors. These contributors to the extrinsic loss are the most difficult to remove (see Figure 5.6).

Another contributor to the extrinsic losses is the micro-bending and macro-bending of the fiber that arises from periodic microbends as a result of the spooling or cabling of the fiber and the bending of the fiber for cabling and deployment (Senior, 1992; Buck, 2004).

When the intrinsic and extrinsic losses are combined, the total fiber loss is as shown in Figure 5.6. The experimental results shown there are from the early 1980s, and the fiber of today has a loss that is very close to the intrinsic loss of the material. This was achieved by removing the OH extrinsic loss and improving the quality of the glass. The total loss of a modern fiber is shown in Figure 5.8.

The evolution of fiber in terms of loss generates three optical windows. The first window was around 850 nm with multimode fiber. The reason for operating at 850 nm was the availability of semiconductor lasers at this wavelength. When the silica zero-dispersion wavelength at 1300 nm was identified, the first single-mode systems operated at 1300 nm, the second window.

When it was possible to reduce the losses at the minimum-loss wavelength, 1550 nm, then single-mode systems started operating at 1550 nm, the third window. Long-haul systems operate in the region of 1550 nm either with a single wavelength or with multiple wavelengths (wavelength division multiplexing—see Chapter 6). The 1300-nm window is still used for short-haul systems.

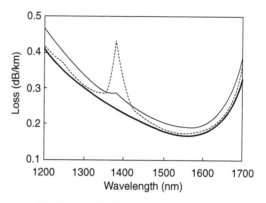

Fig. 5.8 Spectral loss of water loss free PSCF.

Source: Adapted from Chigusa et al. (2005). Used with permission.

—— Water loss free PSCF
------ Conventional PSCF
—— Conventional G.652D fiber
PSCF = Pure Silica Core Fiber

Material Characteristics of Fibers—Dispersion

The variation of refractive index of a material with wavelength is known in optics as dispersion, and it is responsible for resolution of white light into its constituent colors. In the context of fiber light propagation, the dispersion can be divided into two parts. The first part is the dispersion induced on the light by the material used in the waveguide, and this is known as material dispersion. The second part is the impact of the actual waveguide structure, and it is known as waveguide dispersion. This section will address material dispersion only (Senior, 1992; Buck, 2004).

The speed of propagation of monochromatic light in an optical fiber is given by the simple equation

$$u_{\text{phase}} = \frac{c}{n_1(\lambda)} \tag{5.14}$$

This speed is the phase velocity of the light wave, and it is different for each wavelength.

In transmitting a pulse of light through the fiber, the pulse can be expressed as the summation of a number of sine and cosine functions, which is known as the spectrum of the pulse. If the spectrum is centered on a frequency, ω, and has a small spectral width around ω, then a velocity can be associated with this group of frequencies, and this is known as the group velocity.

In conformity with the idea of velocity of propagation for monochromatic radiation, a group index can be defined as corresponding to the group of frequencies around ω. From a detailed analysis of propagation, group velocity is defined by

$$u_{\text{group}} = \frac{c}{n_1 - \lambda(dn_1/d\lambda)} = \frac{c}{N} \tag{5.15}$$

where the group index, N, of the material is defined as

$$N \equiv n_1 - \lambda \frac{dn_1}{d\lambda} \tag{5.16}$$

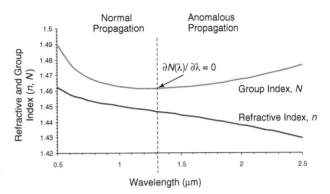

Fig. 5.9 Refractive index and group index for SiO₂ glass.

The packet of frequencies corresponding to the pulse will arrive at the output of the fiber sometime after the pulse is launched. This delay is the group delay, and it is defined as

$$\tau_g = \frac{L}{c}\left(n_1 - \lambda\frac{dn_1}{d\lambda}\right) = \frac{L}{c} \cdot N \tag{5.17}$$

where L is the fiber length. Figure 5.9 shows the refractive index and group index of SiO₂ glass fiber.

For any material, the zero group dispersion is the wavelength at which the curve of the refractive index has an inflection point. For silica this wavelength is at 1300 nm, and it is known as λ_0. From Figure 5.9 one can distinguish two regions associated with the group index curve. The first region is the region with wavelengths less than λ_0. In this region the group index decreases as the wavelength increases. This means that the spectral components of longer wavelength of the pulse travel faster than spectral components of shorter wavelengths. This regime is identified as the normal group dispersion regime and imposes a positive chirp. For wavelengths greater than λ_0 the opposite behavior is observed, and the region is identified as the anomalous group dispersion regime and a negative chirp is imposed on the pulse (Buck, 2004; Agrawal, 2001).

When measurements of these parameters are considered, it is useful to define a new parameter known as the material dispersion parameter, $D_m(\lambda)$. This parameter is defined as

$$D_m(\lambda) \equiv \frac{d(\tau_g/L)}{d\lambda} = \frac{1}{c} \cdot \frac{dN}{d\lambda} \tag{5.18}$$

where $D_m(\lambda)$ is expressed as picoseconds per nanometer of source bandwidth per kilometer of distance (ps/nm-km). Figure 5.10 shows the material dispersion of a pure silica glass fiber.

There are two ways to write in the time domain the spread of a pulse whose spectrum is centered at λ_s. For pulses without initial chirp, the change in pulse width is

$$\Delta\tau_m = -\Delta\lambda_{width}\, D_m(\lambda_s)L \tag{5.19}$$

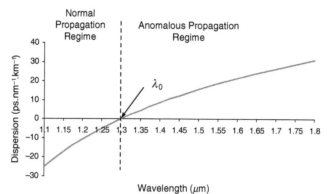

Fig. 5.10 Material (chromatic) dispersion of pure silica.

and in terms of the r.m.s. (Root Mean Square) width

$$\sigma_m = -\sigma_\lambda D_m(\lambda_s)L \tag{5.20}$$

where σ_λ is the r.m.s. width of the optical source. The value of λ_0 depends weakly on the dopants used in the fiber. For example, the use of 13.5 percent of GeO_2 shifts the λ_0 by 0.1 nm with respect to the λ_0 of pure SiO_2. The material dispersion is also referred to as the chromatic dispersion.

Multimode Fibers

As explained in the section titled The Structure and Physics of an Optical Fiber, a multimode fiber can support the propagation of a number of modes (rays). Each of these modes carries the signal imposed on the optical wave. When the modes arrive at the receiver, they create a multi-image of the pulse launched in the waveguide. This multi-image can force the receiver to make a wrong decision regarding the transmitted bit of information.

Before we discuss this feature of multimode transmission, we will first look at the loss of multimode fibers (Senior, 1992).

The loss of multimode fibers should not be different from the spectral loss presented in Figure 5.8 because it is a question of material and processing. However, some manufacturers make available fiber that still has a higher loss, around 1380 nm, but far less loss than that shown in Figure 5.6. The reason for this is the high cost of completely eradicating the residual loss at the OH wavelength. In Figure 5.11 the spectral loss of a multimode fiber is shown; it is important to notice that the loss at the OH wavelength is only around 0.5 dB, compared to 10 dB shown in Figure 5.6.

Light propagating in a multimode fiber is subject to the material dispersion of the fiber. However, there is another form of dispersion that is specific to multimode fibers. Consider a step index multimode fiber with two propagating modes (Figure 5.12).

The first is the axial mode that propagates along the geometric axis of the fiber. The time taken by this mode to reach the end of the fiber is the minimum possible and the output is delayed by that time, which is given by Senior (1992) and Buck (2004):

$$\tau_{\min} = n_1 \frac{L}{c} \tag{5.21}$$

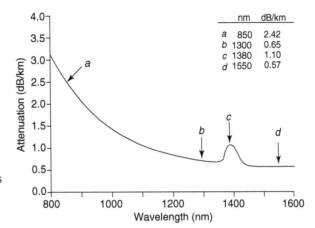

Fig. 5.11 Typical spectral loss for multimode fiber. Notice the low loss at the OH wavelength.
Source: Corning Inc.

nm	dB/km
a 850	2.42
b 1300	0.65
c 1380	1.10
d 1550	0.57

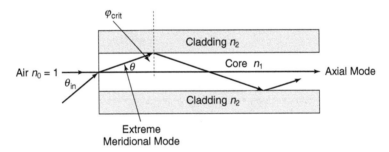

Fig. 5.12 Optical paths of meridional and axial modes.

where the symbols have their usual meaning. Now the extreme meridional mode will reach the output of the fiber after a delay that in this particular geometric arrangement will be the maximum and given by

$$\tau_{max} = \frac{n_1}{c} \cdot \frac{L}{\cos\theta} \tag{5.22}$$

From Snell's law (equations 5.3 and 5.6), we have for the critical angle at the core–cladding interface the following equation:

$$\cos\theta = \sin\phi_{crit} = \frac{n_2}{n_1} \tag{5.23}$$

Then, substituting into equation 5.22 for $\cos\theta$,

$$\tau_{max} = \frac{n_1^2}{n_2} \cdot \frac{L}{c} \tag{5.24}$$

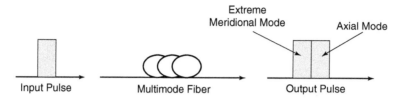

Fig. 5.13 Impact of the multimode fiber delay on the pulse output for a delay equal to the pulse width.

The difference in the delay between the two modes is given by

$$\Delta \tau_s \approx \Delta \frac{Ln_1}{c} \approx (NA)^2 \cdot \frac{L}{2cn_1} \tag{5.25}$$

where NA stands for the numerical aperture of the fiber and Δ is given by equation 5.10. The r.m.s. broadening of the pulse is given by

$$\sigma_s|_{step} \approx \Delta \frac{Ln_1}{2\sqrt{3}c} \approx (NA)^2 \cdot \frac{L}{4\sqrt{3}cn_1} \tag{5.26}$$

To appreciate the impact of this differential delay, let us assume that a pulse of nominal width T is launched into the fiber. If the differential delay is equal to the pulse width, the output consists of two pulses occupying a total width of $2T$. This is illustrated in Figure 5.13.

The receiver will therefore detect two pulses when only one was sent. This effect is called the intermodal dispersion of the fiber, and it is an additional dispersion imparted on the pulse. There are a number of reasons why one has to be cautious in using equation 5.25 for long-haul communications.

The first is that equation 5.25 is a worst-case scenario. In an actual fiber, the power between adjacent modes is coupled back and forth, which leads to a partial mode delay equalization. The second is related to the physics of propagation. Modes that are weakly confined in the core spread out in the cladding, and if they meet higher loss than in the core, then the number of modes propagating is limited to those strongly confined in the core. The additional loss might be present in the cladding because the material used has a higher intrinsic loss or some of the modes might be radiated into the secondary cladding. These two effects limit the spread of the delay per mode, reducing the intermodal dispersion. In addition to pure fiber effects, there is another source of unpredictability in the mode structure of the light source and its statistics, leading to modal noise (Epworth, 1978).

To appreciate the quantities involved, consider a multimode step index fiber of 10 km length with a core refractive index of 1.5 and $\Delta = 2\%$. Then from equation 5.26 the r.m.s. pulse broadening is

$$\sigma_s|_{step} = \Delta \frac{L \times n_1}{2\sqrt{3} \times c} = 0.02 \frac{10 \times 10^3 \times 1.5}{2\sqrt{3} \times 2.998 \times 10^8} = 2.88 \times 10^{-2} \frac{ns}{m} \times 10 \times 10^3 \, m = 288 \, ns$$

The maximum transmission bitrate in terms of the pulse r.m.s. width is given by

$$B_{T\,\text{max}} = \frac{0.25}{\sigma_{\text{pulse}}} \qquad (5.27)$$

Therefore, for $\sigma_s = \sigma = 288$ ns, the maximum bitrate is 868 Kbit/s, which is not a useful value for most modern applications.

The key question now is whether the differential delay for a multimode fiber can be improved. The reason the differential delay between the axial mode and the extreme meridional mode is high is that the meridional mode has to reach the boundary between core and cladding before it is reflected back into the core. If the flight time of a meridional mode is reduced, then the differential delay will also be reduced. This can be achieved with the use of graded index fiber.

The basic concept here is to vary the refractive index from a maximum at the center of the core to a minimum at the core–cladding interface. The general equation for the variation of refractive index with radial distance is

$$n(r) = \begin{cases} n_1(1 - 2\Delta(r/a)^{\alpha})^{1/2} & r < a \quad \text{core} \\ n_1(1 - 2\Delta)^{1/2} = n_2 & r \geq a \quad \text{cladding} \end{cases} \qquad (5.28)$$

where Δ is the relative refractive index difference, r is the axial distance, and a is the profile parameter that gives the refractive index profile. For $a = \infty$ the representation corresponds to the step index profile. Figure 5.14 illustrates the fiber refractive index for various values of the profile parameter a. The step index profile is obtained by setting $a = \infty$.

The improvement in differential delay can be observed by considering the modes of a multimode fiber with the profile parameter a set to 2 (Figure 5.15). Two effects may be observed. First, the axial mode propagates through the section of the fiber core where the refractive index has its maximum value, which implies that the axial mode is slowed down. The meridional modes are bent toward the axis of the fiber, reducing their flight time. Together these two effects reduce the differential delay.

If electromagnetic theory is employed to analyze the differential delay, the value obtained is

$$\Delta\tau_s|_{\text{graded}} = \frac{n_1 L}{c} \times \frac{\Delta^2}{8} = \Delta\tau_s|_{\text{step}} \times \frac{\Delta}{8} \qquad (5.29)$$

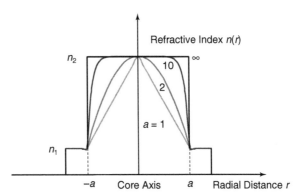

Fig. 5.14 Fiber refractive index for different values of the profile parameter a.
Source: Adapted from Senior (1992). Used with permission.

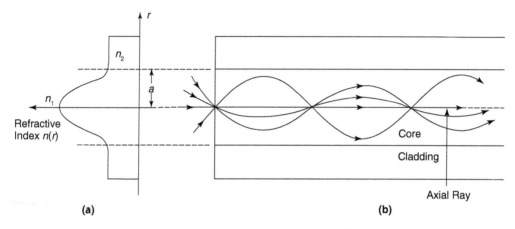

Fig. 5.15   Graded index multimode fiber: (a) refractive index profile; (b) meridional modes within the fiber.   *Source:* Adapted from Senior (1992). Used with permission.

The implication of this equation is to highlight the fact that the reduction of the differential delay for step index fiber is $\Delta/8$. The r.m.s. pulse broadening is now given by

$$\sigma_s|_{index} = \frac{n_1 L \Delta}{2 \times \sqrt{3} \times c} \times \frac{\Delta}{10} = \sigma_s|_{step} \times \frac{\Delta}{10} \qquad (5.30)$$

and there is a $\Delta/10$ reduction in the r.m.s. pulse broadening. It is now straightforward to compare the r.m.s. pulse broadening for step and graded index fibers. From the example studied before, we have a 10-km graded index fiber with a core refractive index of 1.5 and $\Delta = 2\%$. Then, from equation 5.30, the r.m.s. pulse broadening is

$$\sigma_s|_{index} = \frac{n_1 L \Delta}{2 \times \sqrt{3} \times c} \times \frac{\Delta}{10} = \sigma_s|_{step} \times \frac{\Delta}{10} = 288\,\text{ns} \times \frac{0.02}{10} = 0.576\,\text{ns}$$

and the maximum bitrate that can be used with this graded index fiber is now 434 Mbit/s!

Comparing the r.m.s. pulse broadening of the two fibers for a length of 1 km, one obtains

$$\sigma_s|_{step} = 28.8\,\text{ns/km} \quad \text{and} \quad \sigma_s|_{graded} = 0.057\,\text{ns/km}$$

Because of its substantially improved performance, graded index multimode fiber is the clear choice when one wants to exploit the advantages of the multimode fiber with low intermodal dispersion.

Multimode fibers have been fabricated in various core–cladding dimensions addressing specific requirements. For communication purposes two core–cladding sizes have been standardized.

The two fibers differ in the core size, with the first being 50 μm (ITU-I G standard G.651) and the second being 62.5 μm. The cladding for both fibers is 125 μm. At the moment there are no standards for the 62.5-μm fiber. The major advantage of the 50/125 μm fiber is the higher bandwidth it offers compared to that of 62.5/125 μm at 850 nm. An advantage of the 62.5/125 μm is that more power can be coupled into the fiber from an optical device such as a Light Emitting Diode (LED) or laser because of the larger NA (NA = 0.275).

One of the important questions related to the performance of a fiber is that of the bandwidth. In general time and frequency domain techniques can be used to establish the performance, but the industry decided to adopt the bandwidth × length product as the performance index of a multimode-fiber bandwidth.

A simple equation relating the bandwidth, BW, and the length of the fiber is

$$\frac{BW_L}{BW_S} = \left[\frac{L_L}{L_S}\right]^{\gamma} \tag{5.31}$$

where the subscripts L and S correspond to the bandwidth and length for long and short lengths of fiber, and the exponent γ is the length ratio of interest. Usually $\gamma = 1$, but there has long been debate about the value of γ because it reflects the conditions under which the measurements are taken. The value is calculated from the solution of equation 5.31—that is,

$$\gamma = \frac{\log\left[\frac{BW_L}{BW_S}\right]}{\log\left[\frac{L_L}{L_S}\right]} \tag{5.32}$$

In the rest of this section the value of $\gamma = 1$ will be assumed. The first-order estimate of the system length that a fiber can support is calculated from the given bandwidth × length product. For example, for a bandwidth × length product of 1000 MHz × km the system length for a 2.4 Gbit/s system is

$$\text{System length} = \frac{\text{Bandwidth} \times \text{Length}}{\text{Bitrate}} = \frac{1000\,\text{MHz} \times \text{km}}{2.4 \times 10^3\,\text{MHz}} = 401\,\text{m}$$

When this approach is used to estimate the system length, one should understand the technique or techniques used to estimate the bandwidth × length product of the fiber.

Some of the key parameters of the Corning InfiniCor graded index fiber (see Figure 5.11 for losses) are summarized in Tables 5.1 and 5.2. Similar fibers are available from other manufacturers.

The multimode fiber was introduced and discussed up to now with the tacit assumption that there are a large number of modes, but their number was not stated. This question will be addressed now. To derive the number of modes supported by a given multimode fiber,

Table 5.1 Attenuation of Corning InfiniCor Fiber

Wavelength (nm)	Maximum value (dB/km)
850	≤2.3
1300	≤0.6

Source: Corning Inc. Data Sheet, January 2008 Issue.

Table 5.2   Bandwidth × km Performance of Corning InfiniCor Fiber

Corning Optical Fiber	High-performance 850 nm only (MHz × km)	Legacy performance (MHz × km)	
InfiniCor® eSX+	4700	1500	500
InfiniCor® SX+	2000	1550	500
InfiniCor® SXi	850	700	500
InfiniCor® 600	510	500	500

Source: Corning Inc. Data Sheet, January 2008 Issue.

the Maxwell equations must be solved in the context of the fiber structure (cylinder) and the material (dielectric, glass).

From such an analysis a very useful quantity emerges called normalized frequency, V. It combines the wavelength, the physical dimensions of the fiber, and the properties of the dielectric (glass), and is given by

$$V = \frac{2\pi}{\lambda} a \cdot (\text{NA}) = \frac{2\pi}{\lambda} \cdot a \cdot n_1 (2\Delta)^{1/2} \tag{5.33}$$

where a is the core radius, Δ is the relative refractive index difference, and λ is the wavelength.

It can be shown that the number of modes supported by a step index multimode fiber, M_{step}, is

$$M_{\text{step}} \approx \frac{V^2}{2} \tag{5.34}$$

and by a graded index multimode fiber

$$M_{\text{graded}} \approx \left(\frac{\alpha}{\alpha + 2} \right) \cdot \left(\frac{V^2}{2} \right) \tag{5.35}$$

where α is the refractive index profile parameter. For a parabolic profile ($\alpha = 2$) the number of modes is $M_{\text{graded}} \approx V^2/4 = M_{\text{step}}/2$. Consider a multimode fiber operating at 1300 nm with radius 25 μm, $n_1 = 1.5$, and $\Delta = 2\%$. Then the step index fiber supports

$$V \approx \frac{2\pi}{\lambda} \cdot a \cdot n_1 (2\Delta)^{1/2} = \frac{2 \times \pi}{1300 \times 10^{-9}} \times 25 \times 10^{-6} \times (2 \times 0.02)^{1/2} = 24$$

and

$$M_{\text{step}} \approx \frac{V^2}{2} = \frac{(24^2)}{2} = 288 \qquad \text{guided modes}$$

With a parabolic refractive index profile the same fiber will support 156 guided modes. The number of modes and their fluctuations can impair the performance of a multimode fiber system because of modal noise.

After a pulse propagates in a fiber the r.m.s. pulse broadening s given by

$$\sigma_{\text{total}} = (\sigma_{\text{mat}}^2 + \sigma_{\text{w}}^2 + \sigma_{\text{mod}}^2)^{1/2} \tag{5.36}$$

where σ_{mat} is the material (chromatic) dispersion, σ_{w} is the waveguide dispersion (discussed in the next section), and σ_{mod} is the intermodal dispersion (for step index or graded index fiber). For multimode transmission the material and waveguide dispersion terms are negligible compared to the intermodal dispersion, and they are usually ignored. Therefore, the pulse in the output of the fiber has an r.m.s. width given by

$$\sigma_{out} = (\sigma_{inp}^2 + \sigma_{mod}^2)^{1/2} \qquad (5.37)$$

where σ_{inp} is the r.m.s. width of the input pulse.

Single-Mode Fibers

In the section The Structure and Physics of an Optical Fiber, the concept of the single-mode fiber was introduced, and the most important physical feature of a single-mode fiber is its small core. For a useful image of how a single-mode fiber operates, one can use the idea that only the axial mode propagates in the fiber. The bandwidth of single-mode fiber is so large compared to that of the multimode fiber that single-mode fiber is used in all long-haul communications today, terrestrial and submarine. Concepts used to arrive at a reasonable understanding of the propagation features of multimode fiber cannot be used in single-mode fibers, and the solution of Maxwell equations in the context of the geometry and materials is required. Only the results of the mathematical analysis based on Maxwell equations will be presented in this section.

The spectral loss of a single-mode fiber is that shown in Figure 5.8. This is a research result, but we will see that single-mode fibers are available with these loss characteristics as products for terrestrial and submarine applications.

The most striking difference between multimode and single-mode fibers is in the dispersion. In multimode fibers the total dispersion is dominated by the intermodal dispersion. In single-mode fibers only one propagation mode is supported, so the total dispersion consists of three components: material dispersion, waveguide dispersion, and polarization mode dispersion.

The reason waveguide dispersion and polarization mode dispersion became important is the low value of the material dispersion. The normalized frequency parameter, V, plays an important role in single-mode fiber. The parameter V is given by

$$V = \frac{2\pi a}{\lambda} \cdot n_1 (2\Delta)^{1/2} \qquad (5.38)$$

where the symbols have their usual meaning. Single-mode operation takes place above a theoretical cutoff wavelength λ_c given from equation 5.38 as

$$\lambda_c = \frac{2\pi a}{V_c} \cdot n_1 (2\Delta)^{1/2} \qquad (5.39)$$

where V_c is the cutoff normalized frequency. With the help of equation 5.38, equation 5.39 can be written as

$$\frac{\lambda_c}{\lambda} = \frac{V}{V_c} \qquad (5.40)$$

For a step index single-mode fiber—that is,

$$n(r) = \begin{cases} n_1 & r < a \quad \text{core} \\ n_2 & r \geq a \quad \text{cladding} \end{cases} \tag{5.41}$$

$V_c = 2.405$ and the cutoff wavelength is given by

$$\lambda_c = \frac{V\lambda}{2.405} \tag{5.42}$$

For operation at 1300 nm the recommended cutoff wavelength ranges from 1100 to 1280 nm. With this choice for the cutoff wavelength, operation at 1300 nm is free of modal noise and intermodal dispersion. The material dispersion for silica glass with a step index single-mode fiber is given by

$$D_m(\lambda) \approx \frac{\lambda}{c} \cdot \frac{d^2 n_1}{d^2 \lambda} \tag{5.43}$$

where $d^2 n_1 / d^2 \lambda$ is the second derivative of the refractive index with respect to the wavelength.

The value of $D_m(\lambda)$ is usually estimated from fiber measurements, but for theoretical work a simple equation with two parameters offers a reasonable approximation. The equation is

$$D_m(\lambda) = \frac{S_0 \lambda}{4} \cdot \left[1 - \left(\frac{\lambda_0}{\lambda} \right)^4 \right] \tag{5.44}$$

where S_0 and λ_0 are the slope at λ_0 and the zero material dispersion wavelength, respectively. For standard single-mode fiber the slope at λ_0 is in the range 0.085 to 0.095 ps/nm^2 × km.

The Corning single-mode fiber SMF-28+ is a typical example of high-quality modern fiber whose key parameters are summarized next. For this Corning fiber, equation 5.44 is valid for the range $1200 \leq \lambda \leq 1625$ nm. Tables 5.3, 5.4, and 5.5 summarize the performance of the Corning SMF-28+ fiber. Single-mode fibers of similar performance are also available from other fiber manufacturers.

After material dispersion the next important dispersion component is waveguide dispersion. The physical origin of waveguide dispersion is the wave-guiding effects, and even without material dispersion, this term is present. A typical waveguide dispersion curve is shown in Figure 5.16.

The importance of this term is not so much in the value of the dispersion but in the sign. In silica fiber the waveguide dispersion is negative, and because the material disper-

Table 5.3   Attenuation of Corning SMF-28e+™ Single-Mode Fiber

Wavelength (nm)	Maximum values (dB/km)
1310	0.33–0.35
1383	0.31–0.33
1490	0.21–0.24
1550	0.19–0.20
1625	0.20–0.23

Source: Corning Inc. Data Sheet, December 2007 Issue.

Table 5.4 Single-Mode Fiber Dispersion of Corning SMFe+ in the
Third Optical Window

Wavelength (nm)	Dispersion value [ps/(nm × km)]
1550	≤18.0
1625	≤22.0

Source: Corning Inc. Data Sheet, December 2007 Issue.

Table 5.5 Zero-Dispersion Wavelength and Slope
of Corning SMF-28e+ Single-Mode Fiber

Zero-dispersion wavelength (nm)	$1310 \leq \lambda_0 \leq 1324$
Zero-dispersion slope [ps/(nm^2 × km)]	$S_0 \leq 0.092$
λ_0 typical value (nm)	1317
S_0 typical value [ps/(nm^2 × km)]	0.088

Source: Corning Inc. Data Sheet, December 2007 Issue.

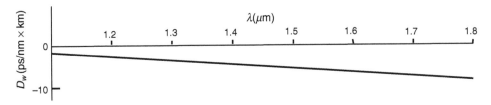

Fig. 5.16 Waveguide dispersion with $\lambda_c = 1.2$ and $\Delta = 0.003$.

sion is positive above λ_0, the value of the total dispersion is reduced. The immediate impact is to shift λ_0 to longer wavelengths. By itself this small shift in λ_0 is of marginal importance, but if the value of the waveguide dispersion is designed to take much higher values, then the shift is substantial and gives rise to two new classes of single-mode fiber: dispersion shifted and dispersion flattened, which will be discussed in the next section.

The third component of dispersion in single-mode fiber is polarization mode dispersion, or PMD. For ease of study it has been assumed tacitly that the fiber is a perfect cylindrical structure. In reality the structure is only nominally cylindrical around the core axis. The core can be slightly elliptical. Also, the cabling process can impose strain on the fiber, or the fiber might be bent as it is installed.

In these situations the fiber supports the propagation of two nearly degenerative modes with orthogonal polarizations. When the optical wave is modulated by the carried information, the two waves corresponding to the two states of polarization are modulated. Therefore, two modes orthogonal to each other propagate in the waveguide. In systems where the receiver can detect both states of polarization, the receiver will function as expected if both modes arrive at the same time; that is, there is no differential delay between the two modes. However, the refractive index of the silica glass can be slightly different along the two orthogonal axes, and consequently there is differential delay. The average Differential Group Delay (DGD) is the PMD coefficient of the fiber.

$$PMD = <DGD> \tag{5.45}$$

To compound the difficulties, the DGD is not the same along the length of the fiber but varies from section to section. Because of this feature, the PMD is not described by a single value but rather by a probability distribution function. The impact of PMD on pulse width is illustrated in Figure 5.17.

The PMD is a complex effect, and its study and mitigation have attracted a very substantial effort. Theoretical studies and experiments have demonstrated that the PMD is proportional to the square root of the fiber length—that is,

$$D_{PMD} = PMD_{coeff}\sqrt{L} \quad ps\sqrt{km} \tag{5.46}$$

Because of its importance for high-speed systems, the PMD fiber coefficient is part of the fiber specifications. For example, the Corning fiber STM-28e+ has a PMD coefficient of ≤0.06 ps/√km .

The calculation of the amount of PMD a system can tolerate is an involved process and is part of the system design. An easily remembered guideline is that the PMD of the link satisfies the inequality

$$D_{PMD} \leq \frac{\text{Pulse width}}{10} \tag{5.47}$$

The impact of PMD on high-speed, systems is quite severe, as can be seen from Table 5.6. From this table it is clear that long-haul high-capacity systems are very vulnerable to PMD, and the selection of the fiber type is a very important decision in system design.

The ITU-T has issued standards for the conventional fiber that are described by the G.652 recommendation. The fiber covered by this recommendation is the most widely installed in the worldwide telecommunications network.

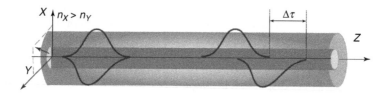

Fig. 5.17   Effect of PMD on a pulse.   *Source:* Corning Inc.

Table 5.6   Impact of PMD on System Length for Typical Bitrates

Bitrate (Gbit/s)	Timeslot (ps)	Link PMD (ps)	System length (km) PMD_{coeff} 0.02 ps/√km	PMD_{coeff} 0.5 ps/√km
2.5	400	40.0	4.0×10^6	6.4×10^3
10.0	100	10.0	2.5×10^5	4.0×10^2
40.0	25	2.5	15.6×10^3	25.0
100.0	10	1.0	2.5×10^3	4.0

Until now the discussion has been limited to what may be called the conventional single-mode fiber. That is a single-mode fiber with $\lambda_0 \approx 1300$ nm, low waveguide dispersion ($D_m \geq D_w$), and a step index profile. Since the introduction of the conventional single-mode fiber, a number of applications have arisen where a fiber optimized for the application can deliver better performance. We are now going to take a very brief look at these special fibers.

Special Single-Mode Fibers

As was mentioned in the Single-Mode Fibers section, waveguide dispersion can be used to shift the λ_0 of a fiber to a direction that the dispersion of the fiber will be substantially altered. Initially, two classes of fiber emerged to exploit this property of waveguide dispersion. The first was the dispersion-shifted fiber in which the value of the waveguide dispersion was large and the λ_0 shifted to 1550 nm. At this wavelength the zero dispersion and the minimum loss value of silica glass coincided, offering the best of both worlds: dispersion and loss. This fiber design favors Electrical Time Division Multiplexed (ETDM) systems, and it was introduced at a time when the only practical way to increase the capacity appeared to be ETDM.

Later, however, a better approach in terms of cost–performance was identified with Wavelength Division Multiplexing (WDM), and another fiber was introduced in which the dispersion presented a plateau enabling the WDM channels to face similar dispersion. Unfortunately, again the fiber could not be employed in large quantities because of the nonlinear behavior of the silica fiber when optical amplifiers are used (see the section on nonlinear fiber effects) demands the presence of substantial dispersion between channels if the nonlinear effects are to be mitigated. Therefore, the two first attempts to break the confines of the conventional single-mode fiber were not very successful. The dispersion characteristics of standard fiber, dispersion-shifted fiber, and dispersion-flattened fiber are shown in Figure 5.18.

The dispersion-shifted fiber standards are addressed in ITU-T recommendation G.653.

The introduction of the erbium-doped silica optical amplifier operating over the band 1530 to 1565 nm and the suppression of the OH loss peak prompted the division of the

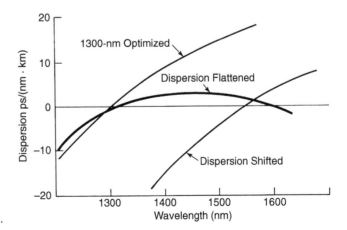

Fig. 5.18 Dispersion characteristics of standard, dispersion-flattened, and dispersion-shifted fibers.
Source: Adapted from Senior (1992). Used with permission.

Table 5.7   Optical Bands Available for Fiber Communications with SiO$_2$ Fibers

Band	Description	Wavelength range
O-band	Original band	1260–1360 nm
E-band	Extended band	1360–1460 nm
S-band	Short-wavelengths band	1460–1530 nm
C-band	Conventional (erbium window)	1530–1565 nm
L-band	Long-wavelengths band	1565–1625 nm
U-band	Ultralong-wavelengths band	1625–1675 nm

Table 5.8   Attenuation of Nonzero Dispersion-Shifted Fiber

Wavelength (nm)	Loss (dB/km) Maximum	Loss (dB/km) Typical	
1310 nm	≤0.40	≤0.35	The maximum attenuation in the range 1525–1625 nm is no more than 0.05 dB/km greater than the attenuation at 1550 nm.
1383 nm	≤0.40	≤0.25	
1450 nm	<0.26	≤0.25	
1550 nm	≤0.22	≤0.20	
1625 nm	≤0.24	≤0.21	

Source: Copyright © 2005 Furukawa Electric North America, Inc.

Table 5.9   Dispersion of Nonzero Dispersion-Shifted Fiber

Optical band	Dispersion ps/(nm × km)	
C-band 1530–1565 nm	5.5–8.9 ps/nm-km	Zero dispersion wavelength: ≤1405 nm
L-band 1565–1625 nm	6.9–11.4 ps/nm-km	Dispersion slope at 1550 nm: ≤0.045 ps/nm^2-km
S- and L-bands 1460–1625 nm	2.0–11.4 ps/nm-km	Mode field diameter: 8.6 ± 0.4 μm at 1550 nm

Source: Copyright © 2005 Furukawa Electric North America, Inc.

available low loss window of the silica fiber into a number of bands whose exploitation will depend on the availability of the relevant technology. A summary of the optical bands and their respective wavelengths is summarized in Table 5.7.

A number of fibers have been introduced to address the use of these optical bands. The aim is to reduce the amount of dispersion in the intended band of operation and also to minimize the impact of fiber nonlinearities. According to the ITU-T G.655 recommendations, the dispersion over the C-band should lie in the range 1 to10 ps/(nm × km). Then the operation can be extended into the L-band because the dispersion will still be low there. Both C- and L-bands are used for long haul applications. ITU-T recommendation G.656 is designed to extend operation into the S-band. With chromatic dispersion in the range of 2 to 14 ps/(nm × km) from 1460 to 1625 nm, this fiber can be used for Coarse WDM (CWDM) and Dense WDM (DWDM) throughout the S-, C-, and L-bands. These fibers are referred to as Non-Zero Dispersion-Shifted Fibers (NZ-DSF).

A summary of the performance of the Furukawa Electric North America, Inc., NZ-DSF fiber is given in Tables 5.8 and 5.9.

The dispersion of the NZ-DSF at 1550 nm is nearly four times less than that of the standard single-mode fiber, reducing the amount of dispersion compensation needed in

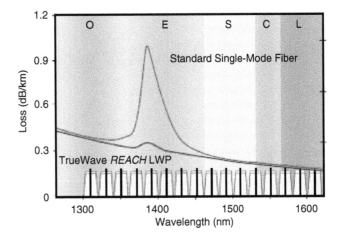

Fig. 5.19 Use of NZ-DSF in a coarse WDM channel allocation scheme.
Source: Copyright © Furukawa Electric North America, Inc.

long-haul systems (see Chapter 6). The reader should be aware that all leading fiber manufacturers market NZ-DSF fiber.

A significant part of the cost of DWDM systems is the cost of the narrowband active and passive optical components required by the specifications. The availability of the NZ-DSF fibers has enabled a substantial reduction of the cost of WDM by spacing the channels at 20 nm. This channel spacing accommodates 16 channels from 1310 to 1625 nm and is supported by ITU-T recommendation G.652C. A schematic of the spectral loss and the CWDM channel scheme is shown in Figure 5.19.

Special Fibers

The discussion so far has been limited to fibers designed for transmission. However, there are a large number of fibers designed with the aim of addressing special requirements. By virtue of necessity we will outline briefly the fibers designed for special needs in the field of optical communications.

Erbium-doped silica fiber. This fiber is used as the gain medium in optical fiber amplifiers for the C-band of the spectrum (1530–1565 nm).

Polarization-preserving fiber. This fiber is used in situations where the polarization of the optical radiation inhibits the use of nonpolarization preserving fibers. Bearing in mind that the normal transmission fiber does not preserve polarization, polarization-preserving fibers are used in laser fiber tails, external modulators, polarization mode dispersion compensators, and in general devices (active and passive) that require control of the polarization of the radiation.

Photosensitive fiber. This fiber is used in the fabrication of Bragg gratings, which are used in dispersion compensators employing the Bragg effect and other optical fiber components.

Bend-insensitive fiber. The aim of this fiber is to minimize the losses due to the tight bending of fibers and finds applications in the packaging of optical devices.

The parameters of the fibers previously enumerated vary from manufacturer to manufacturer, and the reader is encouraged to look at the Web sites of fiber manufacturers.

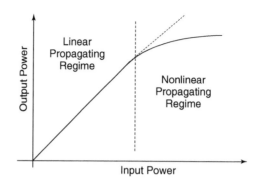

Fig. 5.20 Typical nonlinear characteristic of optical fiber.

Nonlinear Optical Effects in Fibers

The discussion of fiber performance was based implicitly on a linear model. That is, when the input is doubled the output is also doubled. However, the optical fiber is essentially a waveguide loaded with a dielectric material, the glass, and under suitable conditions the behavior of the glass departs from a linear characteristic. This is illustrated in Figure 5.20, where the input to output relationship is no longer linear.

The study of nonlinear optical effects belongs to the field of nonlinear optics, but because the effects on optical fiber communications are serious, a brief outline will be presented here (see also Chraplyvy et al., 1984).

The origin of nonlinear optical effects lies in the response of the atoms and molecules of the glass dielectric in high electric fields. One might be surprised that in spite of the low optical powers used in fiber communications, the nonlinear behavior of the fiber impairs system performance.

The electric field in a medium with effective area A_{eff} and power P is given by

$$|E| = \sqrt{\frac{2 \times 377}{n}} \times \sqrt{\frac{P}{A_{\text{eff}}}} \tag{5.48}$$

With a semiconductor laser of 10 mW and a fiber with an effective area of 53 μm^2, the electric field in the fiber is

$$|E| = \sqrt{\frac{2 \times 377}{n}} \times \sqrt{\frac{P}{A_{\text{eff}}}} = \sqrt{\frac{2 \times 377}{1.5}} \times \frac{10 \times 10^{-3}}{53 \times 10^{-12}} = 3.0 \times 10^5 \text{ V/m}$$

The electric field experienced by the electron in the hydrogen atom is of the order of 5×10^{11} V/m, so one could think that the field generated by a laser beam of 10 mW would not give rise to nonlinear optical effects. Although the optical power is low, the interaction length between the optical radiation and the glass dielectric can be very long when the fiber operates with optical amplifiers. The effect is cumulative, and system performance is impaired.

The channel power used in this simple example refers to one laser, but in a WDM system the number of channels can exceed 100, so the total power can be very high.

In nonlinear effects associated with the propagation of light in fibers there are two fiber parameters that dimension the impact of the effects. The first is the effective area of the radiation mode, A_{eff}. It is defined as shown in equation 5.51:

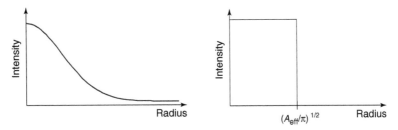

Fig. 5.21   Concept of the effective area of A_{eff}.

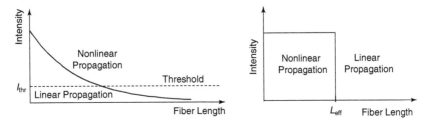

Fig. 5.22   Concept of the effective length, L_{eff}, in nonlinear optical effects.

$$A_{\text{eff}} = \pi r_0^2 \tag{5.49}$$

where r_0 is the mode field radius. The concept is illustrated in Figure 5.21.

The second is the effective length, L_{eff}, and it is defined by

$$L_{\text{eff}} \equiv \int_0^L \exp(-\alpha z)dz = \frac{1-\exp(-\alpha L)}{\alpha} \tag{5.50}$$

where L is the fiber length and α is the attenuation coefficient of the fiber, which here is not in dBs. The attenuation coefficient of the fiber in this equation is derived from the loss in decibels through the equation

$$\alpha = \frac{\alpha_{\text{dB}}}{4.343} \tag{5.51}$$

For low-loss fiber $L_{\text{eff}} \approx L$ and for high loss fiber, $L_{\text{eff}} \approx 1/\alpha$. Because the fiber for optical transmission has low loss, the first approximation is usually used. The physical meaning of L_{eff} is simple. We know that for low power the nonlinear effects are negligible, so there is a "threshold" of power above which the propagation is nonlinear and below which it is linear. This concept is illustrated in Figure 5.22.

For example, consider a fiber with a loss of 1.8 dB/km and length of 25 km. Using the approximation for low loss, $L_{\text{eff}} \approx L$. If we use equation 5.50 the result is 15.5 km. In a system with optical amplifiers, the effective length of the system is given by

$$L_{\text{eff}} = \frac{1-\exp(-\alpha l)}{\alpha} \times \frac{L_{\text{total}}}{\Delta l} \tag{5.52}$$

where L_{total} is now the total system length and Δl the spacing between amplifiers. It is clear that to reduce the system effective length it is better to use as few amplifiers as possible.

The nonlinear effects relevant to fiber communications are

[1] Single-channel systems
    [1a] Self-Phase Modulation (SPM)
    [1b] Stimulated Brillouin Scattering (SBS)
[2] Multichannel systems
    [2a] Cross-Phase Modulation (XPM)
    [2b] Stimulated Raman Scattering (SRS)
    [2c] Four-Wave Mixing (FWM)

[1a] Self-Phase Modulation

To understand the self-phase modulation effect, one has to understand the effect of high intensity on the refractive index of the material. This impact of optical power on the refractive index can be stated mathematically as

$$n(t) = n_1 + \Delta n(t) = n_1 + n_2 E^2(t) \tag{5.53}$$

where n_1 is the conventional refractive index of the silica glass and n_2 is the nonlinear refractive index. The value of n_2 is 3.2×10^{-20} m^2/W and it is therefore very small compared to the linear refractive index, n_1, which for silica glass at 1550 nm is 1.444. The difference is 20 orders of magnitude, but the impact of n_2 is still important because of the long length interaction between material and radiation and the high intensity of the radiation.

Over a fiber length L the total phase shift due to $\Delta n(t)$ is given by

$$\Delta\varphi(t) = \int_0^L \frac{\omega_0}{c} \Delta n(t)dz = \int_0^L \frac{2\pi}{\lambda} \Delta n(t)dz = \frac{\omega_0}{c} n_2 E^2(t)L \tag{5.54}$$

$$\Delta\omega(t) = -\frac{d}{dt}\Delta\varphi(t) = -\frac{d}{dt}\left[\frac{\omega_0}{c} n_2 E^2(t)\right]L \tag{5.55}$$

To proceed, one needs an analytical expression of the electric field. For convenience let us assume that the input light pulse has a Gaussian envelope:

$$E(t) = E_0 \exp\left[-\frac{(t-t_0)^2}{\tau^2}\right] \tag{5.56}$$

where E_0 is the peak pulse amplitude, t_0 is the time position of the pulse, and τ is the half-width at the $1/e$ point of the peak amplitude. Substituting equation 5.56 into equation 5.54, we obtain

$$\Delta\varphi(t) = \Delta\varphi_{max} \exp\left[-\frac{2(t-t_0)^2}{\tau^2}\right] \tag{5.57}$$

$$\Delta\phi_{max} = \frac{\omega_0}{c} n_2 E_0^2 L \tag{5.58}$$

where the length L corresponds to the nonlinear interaction length, L_{eff}. The maximum frequency change is obtained by substituting equation 5.57 into equation 5.55:

$$\Delta\omega(t) = \Delta\varphi_{max}\frac{4(t-t_0)}{\tau^2}\exp\left[-\frac{2(t-t_0)^2}{\tau^2}\right] \quad\quad (5.59)$$

At the center of the pulse the frequency shift is zero. To find the position of the maximum frequency shift we calculate

$$\frac{d}{dt}[\Delta\omega(t)=0)] \quad\quad (5.60)$$

One finds that the maximum frequency changes occur at $(t - t_0) = \pm\tau/2$. For example, consider the Gaussian pulse with $\tau = 1$ psec, $\Delta\varphi_{max} = 8\pi$. The results are summarized in Figure 5.23.

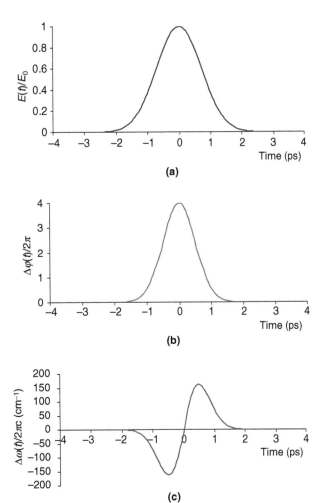

Fig. 5.23 (a) Normalized input pulse amplitude; (b) phase shift due to nonlinearity; (c) frequency chirp.

(a)   (b)

Fig. 5.24 SPM followed by negative and positive dispersion: (a) $D < 0$, additional pulse broadening; (b) $D > 0$, pulse compression.

For the central position of the pulse the frequency chirp is linear between $\pm\tau/2$ and is given by

$$\omega(t) = \omega_0 + \left(\frac{4\pi n_2}{\lambda_0 \tau^2}\mathrm{L}\frac{P}{A_{\mathrm{eff}}}\right) \times t \tag{5.61}$$

The impact of SPM on the performance of an optical system depends on the dispersion. If there were no dispersion in the fiber, SPM would have no impact on performance.

First consider operation at $\lambda > \lambda_0$. Then $D < 0$ and the pulse will broaden because the leading "red" frequencies travel faster and the "blue" frequencies slower. For $\lambda > \lambda_0$ the situation changes substantially. Now, the "red" frequencies in the rising edge of the pulse are delayed and the "blue" frequencies at the falling edge are speeded up; consequently, the pulse width contracts. However, as the blue frequencies catch up with the red ones and overtake them, the pulse starts broadening again.

This feature of the interaction of SPM with anomalous dispersion is used in soliton transmission. Figure 5.24 summarizes the interaction of SPM with fiber dispersion.

[1b] Stimulated Brillouin Scattering

The physical origin of Stimulated Brillouin Scattering (SBS) is the interaction between an incident optical wave and the elastic acoustic wave induced by the radiation. A threshold can be established for the onset of SBS, and it is given by

$$P_{\mathrm{th}}^{\mathrm{Brillouin}} \approx 21 \times p_{\mathrm{Pol}} \times \frac{A_{\mathrm{eff}}}{g_B L_{\mathrm{eff}}} \times \left(1 + \frac{\Delta f_{\mathrm{source}}}{\Delta f_{\mathrm{Br}}}\right) \tag{5.62}$$

where p_{Pol} takes the value 1.5 if the polarization is completely scrambled and 1.0 otherwise; A_{eff} is the effective area of the mode; g_B is the Brillouin (Br) gain, which is around 4×10^{-11} m/W at 1550 nm; L_{eff} is the effective fiber length; $\Delta f_{\mathrm{source}}$ is the bandwidth of the incident radiation; and Δf_{Br} is the bandwidth of the SBS effect, which is approximately 20 MHz.

In the worst case for $\Delta f_{\mathrm{source}} \ll \Delta f_{\mathrm{Br}}$ and $p_{\mathrm{Pol}} = 1$ with $A_{\mathrm{eff}} = 50~\mu\mathrm{m}^2$ and $L_{\mathrm{eff}} = 20$ km the threshold power is 1.3 mW. This is very low and care must be taken to enhance the threshold. For high-speed modulation this is not a problem. In SBS the scattered wave propagates backward and the effect appears in the receiver as a reduction of the power of the wave carrying the information. Figure 5.25 illustrates the basic features of SBS.

The SBS is a narrowband effect and consequently does not interact with other waves unless they fall within its bandwidth. For this reason SBS once suppressed in individual channels is not an issue in WDM systems.

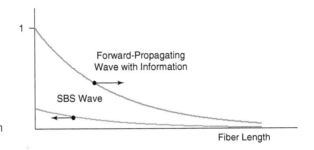

Fig. 5.25 Stimulated Brillouin scattering (SBS) effect.

[2a] *Cross-Phase Modulation*

The SPM discussed earlier operates within each channel. However, if a channel has enough power to induce the effect on the refractive index of the material for another channel propagating through the same material (that is to say, in a WDM system) then impairment arises in this other channel. This effect is referred to as Cross-Phase Modulation (XPM). The total induced frequency chirp, $\Delta\omega_{NL}(t)$, is given by

$$\Delta\omega_{NL}(t) = -\gamma \times L_{eff}^{System} \times \left(\frac{\partial P_n}{\partial t} + 2 \sum_{j\neq n}^{N} \frac{\partial P_j}{\partial t} \right) \tag{5.63}$$

where

$$\gamma = \frac{2\pi n_2}{\lambda A_{eff}} \tag{5.64}$$

Because the XPM term is twice that of a single channel in the first-order design of WDM, only the XPM term is taken into consideration.

[2b] *Stimulated Raman Scattering*

With an very intense incident optical wave, Stimulated Raman Scattering (SRS) can occur in which a forward-propagating wave, referred to as the Stokes wave, grows rapidly in the fiber so that most of the energy of the incident wave is transferred into it. The incident radiation is also referred to as the pump.

In a fiber the forward-scattered wave grows as

$$I_s(L) = I_z(0) \times \exp(g_R I_p L_{eff} - \alpha_S L) \tag{5.65}$$

where g_R is the Raman gain coefficient; I_p is the intensity of the incident radiation; $I_S(0)$ is the initial value of the forward-scattered wave; L_{eff} is the effective length; α_S is the fiber loss at the wavelength of the Stokes wave; and L is the fiber length. The Raman gain, g_R, in silica fiber extends over a large frequency range (up to 40 THz) with a broad peak near 13 THz.

This behavior is due to the noncrystalline nature of silica glass. In amorphous materials such as fused silica, molecular vibrational frequencies spread out into bands that overlap and become a continuum, unlike most materials where Raman gain occurs at specific frequencies. The Raman gain for silica is shown in Figure 5.26 measured with a pump at 1.0 μm.

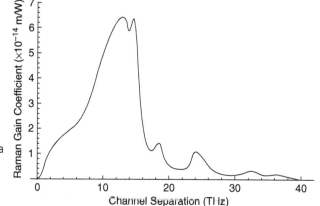

Fig. 5.26   Raman gain of silica fiber.

Source: Based on Ramaswami and Sivarajan (2002) and Stolen (1980).

The impact of the SRS on the performance of a single-channel optical system is negligible because the threshold power required is

$$P_{th}^{Raman} \approx 16 \frac{A_{eff}}{g_R L_{eff}} \tag{5.66}$$

and for a fiber with $A_{eff} = 50 \ \mu m^2$, $g_R = 6 \times 10^{-14}$ m/W and with $L_{eff} = 20$ km, the threshold is reached for 0.666 W.

Considering that optical systems are low-power systems in terms of optical power, single-channel systems are not affected by SBS. With optical amplifiers and single-channel systems one has to use equation 5.50 for L_{eff}. The threshold given by equation 5.66 is defined as the threshold at which the power of the forward-scattered wave, the Stokes wave, equals the incident radiation (pump).

The Raman gain coefficient at 1550 nm is approximately 6×10^{-14} m/W. This is of course much smaller than the gain coefficient of SBS, but the fundamental difference is in the bandwidth of the two processes. From Figure 5.26 the gain extends up to 15 THz (125 nms), and two waves 125 nm apart will still be coupled. Although SRS is of no particular importance in the performance of single-channel systems, in WDM systems it becomes very important.

The reason for this is the wide bandwidth of the SRS. Let us consider a WDM system and one wavelength, say λ_0, that is at 10 THz in the Raman gain profile. All the channels with shorter wavelengths act as pumps and amplify the wavelength at λ_0, losing photons (power) in the process. The SRS process can be used for optical amplification, and the amplifiers using it are referred to as Raman amplifiers. In a WDM system the SRS can lead to performance impairments if care is not taken. Figure 5.27 illustrates the effect of SRS in a four-channel WDM system.

It is clear that SRS cannot be avoided in WDM systems, and the way to handle it is to accept impairment in system performance and design with this as a constraint. If this approach is acceptable, then the loss of power to the channel of the shortest wavelength is the constraint. After some assumptions, such as that (1) the channel spacing is equal, (2) the power in each channel is the same, and (3) all channels fall within the Raman gain

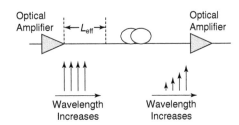

Fig. 5.27 Impacts of SRS in a four-channel WDM system. Notice that the channel of the shortest wavelength is most depleted.

bandwidth, the power loss by the channel with the shortest wavelength, say channel 0, is given by

$$P_0 = \frac{[(N-1)\Delta f][NP_{ch}]}{6 \times 10^4} \times \frac{g_{Rpeak} \times L_{eff}}{A_{eff}} = \frac{B_{total} \times P_{total}}{6 \times 10^4} \times \frac{g_{Rpeak} \times L_{eff}}{A_{eff}} \qquad (5.67)$$

The penalty for channel 0 is now given by

$$\text{Penalty}|_0 = -10\text{Log}[1 - P_0] \qquad (5.68)$$

If a penalty equal to or less than 0.5 dB is desired, then $P_0 \leq 0.1$ and after some changes in the units of parameters,

$$P_{total} \times B_{total} \times L_{eff} < 40 \times 10^3 \quad \text{mW} \times \text{nm} \times \text{km} \qquad (5.69)$$

This equation is conservative because it ignores chromatic dispersion; with chromatic dispersion in the system the right side of equation 5.69 is nearly doubled. Using a channel spacing of 0.8 nm (10 GHz), an amplifier spacing of 80 km, and an L_{eff} of 20 km, the limitations imposed on channel power by SRS are shown in Figure 5.28.

It is clear that for a small number of channels SRS does not impose a strong upper bound on power per channel, but for a large number of channels care should be taken to control the impact of SRS.

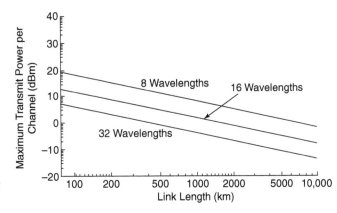

Fig. 5.28 Maximum power limitations imposed by SRS in a WDM system.

Source: Based on Ramaswami and Sivarajan (2002).

[2c] Four-Wave Mixing

The origin of four-wave mixing is the dependency of refractive index on optical power. This dependency not only gives rise to SPM but also induces new frequencies to appear. This effect is called Four-Wave Mixing (FWM) and it is independent of the bitrate but dependent on channel spacing and the chromatic dispersion present in the propagation path. It is a complex effect to analyze, but FWM efficiency is reduced by increasing the channel spacing and the chromatic dispersion between channels. New frequencies are generated by the process

$$\omega_{ijk} = \omega_i + \omega_j - \omega_k \quad \text{with} \quad i, j \neq k \qquad (5.70)$$

where the indices i, j, and k refer to the frequencies present. Assuming that N frequencies are present, the number of new frequencies is given by

$$\text{Number of new frequencies} = N(N-1)^2 \qquad (5.71)$$

As an example, consider a three-channel system. The number of new frequencies expected to be induced by the process is 12. The spectrum of the 12 new frequencies is shown in Figure 5.29.

It clear that the penalty of FWM is twofold: First, there is loss of power to the new frequencies generated but falling outside the bandwidth of the original three-channel system; second, there is interference as some of the new frequencies coincide with the original frequencies. The assessment of the performance is complex, and we can only illustrate the impact of performance for a few cases (see Figure 5.30).

It should be clear that the impact increases as the dispersion between channels decreases (for conventional single-mode fiber and dispersion-shifted fiber) and the channel spacing is reduced.

Closing this section on nonlinear fiber effects, one might wonder what fiber parameters one should expect from a fiber suitable for long-haul multichannel operation. The key to answering this question is equation 5.67. From that equation it is clear that the propensity of the fiber to induce nonlinear effects depends on the effective area. For this reason fiber manufacturers have introduced fibers with larger effective areas. For example, the Large

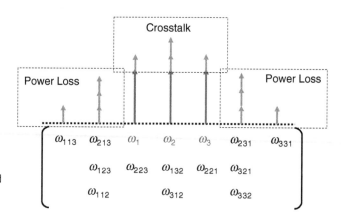

Fig. 5.29 Spectrum generated from a FWM process for a three-channel system.

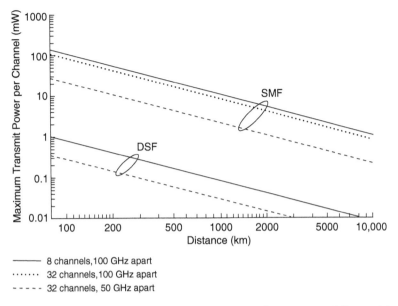

Fig. 5.30 Performance limitation imposed by the FWM process for two types of fiber and two classes of WDM systems.

Effective Area Fiber (LEAF) manufactured by Corning, Inc. has an area of 72 μm^2; the True-wave XL Ocean fiber manufactured by OFS also offers the same effective area. The standard SMF has an effective area around 50 μm^2. The Corning Vascade L1000 fiber offers an effective area of 101 μm^2. Looking at research results, a fiber design using a novel approach claims an effective area of 1470 μm^2 (Wong et al., 2005).

System design with optical nonlinearities present is a complex undertaking and after some preliminary calculations one has to use a simulation package that solves the nonlinear propagation equation (nonlinear Schrödinger equation) under the conditions imposed by the system details. A good introduction is given in Agrawal (2005).

Future Evolution of Optical Fiber Communications

Silica glass proved to be the ideal material in its optical and mechanical properties and also for manufacturing cabling and reliability. The losses have been reduced to such an extent that any further progress in that direction will be met with the law of diminishing returns. There has been work on new glass materials whose intrinsic theoretical losses are less than that of silica. They are collectively referred to as telluride and fluoride glasses. Figure 5.31 shows the theoretical loss spectrum of these glasses.

The interesting aspect of these losses is that, at the wavelengths where optical fiber systems operate now, silica glass is still as good as or better than these new glasses. Therefore, to exploit the theoretical low fluoride losses, the optical window has to shift to longer wavelengths. Here semiconductor optical devices will have difficulties delivering the performance one has come to expect in the C- and L-bands and around 1300 nm. In spite of the theoretical low losses, experimental work with fluoride material has not delivered

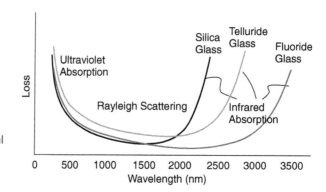

Fig. 5.31 Theoretical performance of some potential new fiber materials.

Source: FibreLabs, Inc. Used with permission.

results that will give confidence for the future. Therefore, from the material perspective, silica glass will continue to dominate the telecommunications market as far as one can see. As for multimode and single-mode fibers, the trend is to use single-mode fiber because it is future proof for bandwidth and its cost is low because it is manufactured in large volume. However, multimode fiber will keep a part of the market because of its special properties.

There was a time (late 1970s to late 1990s) when multimode fibers were considered obsolete on the grounds that operation in the fundamental waveguide mode provides better dispersion performance. However, it subsequently became apparent that there will be a huge market for optical LANs, data communications networks, and access networks. Multimode fibers offer greater ease of handling and splicing and therefore their role is being reevaluated, especially for cost-sensitive, short-span applications.

References

Agrawal, G. P. (2001) *Nonlinear Fiber Optics,* Third Edition, Academic Press.

Agrawal, G. P. (2005) *Lightwave Technology, Telecommunications Systems,* Wiley-Interscience.

Buck, J. A. (2004) *Fundamentals of Optical Fibers,* Second Edition, John Wiley & Sons.

Chigusa, Y., et al. (2005) Low-Loss Pure-Silica-Core Fibers and Their Possible Impact on Transmission Systems, *IEEE J. Lightw. Technol.* 23:3541–3450.

Chraplyvy, A. R., et al. (1984) Optical Power Limits in Multi-Channel Wavelength Division Multiplexing, *Electron. Lett.* 20:58–59.

Epworth, R. E. (1978) The phenomenon of modal noise in analogue and digital optical fibre systems, *Proceedings 4th European Conference on Optical Communications,* 492–501.

Hecht, J. (1999) *City of Light: The Story of Fiber Optics,* Oxford University Press.

Kao, K. C., and Hockman, G. A. (1966) Dielectric Fibre Surface Waveguides for Optical Frequencies, *Proceedings IEE* 113(July):1151–1158.

Miya, T., Teramuna, Y., Hosaka, Y., and Miyashita, T. (1979) Ultra Low Loss Single Mode Fibre at 1.55 μm, *Electron. Lett.* 15:106–108.

Ramaswami, R., and Sivarajan, K. (2002) *Optical Networks—A Practical Perspective,* Second Edition, Morgan Kaufmann.

Senior, J. M. (1992) *Optical Fiber Communications,* Second Edition, Prentice Hall.

Stolen, R. H. (1980) Nonlinerity in Fibre Transmission, *Proceedings IEEE* 68:1232–1236.

Wong, W. S., et al. (2005) Breaking the Limit of Maximum Effective Area for Robust Single-Mode Propagation in Optical Fibers, *Opt. Lett.* 30:2855–2857.

6 Optical Fiber Transmission

Paul Urquhart
Universidad Pública de Navarra

Introduction

Optical fiber communication provides data capacities that are unavailable from any other transmission means, and so it exerts a substantial influence on the information society (Hecht, 1999). Optical cables have been installed in many environments—within buildings, under streets, suspended from overhead pylons, and across the world's deepest oceans (Urquhart, 2004; Keiser, 2006; Lin, 2006; Halabi, 2003; Chesnoy, 2002). The networks they serve carry data for many end users, including domestic consumers, manufacturing, financial services, hospitals, educational establishments, and government departments. The history of the subject is one of record-breaking achievements, especially as measured by distance, total capacity, and number of customers served.

This chapter explains the main physical aspects of fiber transmission systems, without specific reference to network topology. The treatment is system oriented but, when appropriate, points to implications for the main categories of fiber networks. Many physical phenomena influence system design and implementation (Agrawal, 2002). Moreover, the broad spectrum of component technologies that are required, ranging from the transmitter to the receiver, all affect system performance (Liu, 2005). Our main aim is to describe the various operating impairments, explain how they influence the system's bit error rate (BER), and discuss ways to compensate for them.

The topics considered in this chapter can be categorized as several interlocking themes:

- Single- and multiple-channel transmission, which includes Synchronous Digital Hierarchy (SDH) and Wavelength Division Multiplexing (WDM) standards
- Transmission impairments, which represent the influence exerted by amplifiers (especially noise), chromatic dispersion, polarization mode dispersion, and nonlinear optical effects

- The role of the receiver and how it influences BER—the main physical measure of a network's performance
- Overall system design, which includes a number of essential techniques to overcome transmission impairments:
  - Mitigation and compensation of chromatic dispersion and polarization mode dispersion
  - Avoidance of the worst effects of nonlinear crosstalk
  - Use of optimized modulation formats and forward error correction (FEC)

The design of optical systems and networks requires knowledge of many effects, and the design techniques used are constantly evolving. This chapter explains the underlying engineering principles. It does not aim to review the most recent results at the time of writing, which, if history is a reliable guide, will soon be superseded.[1]

Single-Channel Transmission

When optical fibers were initially introduced, they were designed to carry a single stream of optical pulses by simple on-off keying of the signal light. As systems have developed, and fibers have been required to transport ever more information, the transmission techniques have become more refined. Basic properties of light impulses have been exploited and innovative methods have been developed to achieve higher efficiencies for data transmission. This section explains the operation of single-channel systems to lay a foundation for understanding the more recent advances.

Optical Binary Digits: "Bits"

Light guided by an optical fiber is normally a stream of optical pulses to convey a digital data sequence (Agrawal, 2002; Ramaswami & Sivarajan, 2002). The pulse stream is divided into equal, short time intervals known as "bitslots" or "bit intervals." The most common type of data sequence in optical communications is "binary," in which the data is in the form of logical 1s and 0s. If there is a pulse in a bitslot, it is a 1 and if there is none, it is a 0 (Conradi, 2002). Indeed, the word "bit" was derived from "binary digit."

Nonbinary forms of digital communication are also possible; examples include "duobinary" and "multilevel signaling." There can be, for example, three logic levels, such as 0, 1, and 2, supplied by no pulse, a low-power pulse, and a high-power pulse, respectively (Gallion, 2002). However, such formats are less common. Analogue optical modulation is also possible but its use is confined to specialist applications and so is not considered in this chapter. Digital logic is preferred because it is more tolerant to noise and is compatible with the large numbers of computers and other electronic processors that are connected to telecommunications systems.

An optical channel is formed by launching the modulated output from a laser into an optical fiber and then receiving the pulses at the far end of the system to recover the information in the form of electrical 1s and 0s. The laser is an electrical-to-optical energy converter; the detector at the output is the opposite, converting back to electrical signals. The "datarate," or "bitrate," of the channel is simply the number of bits per second.

---

[1] A glossary is provided at the end of this book.

Light in an optical fiber can be characterized by its optical intensity, phase, wavelength, and polarization. The most common type of optical transmission is phase insensitive, which means that at the transmitter there is a simple electrical current-to-light conversion and at the detector there is a reverse conversion (Ho, 2001). Similarly, most transmission systems use unpolarized light, which is, ideally at least, a superposition of all states of polarization. An expression commonly used for binary data sequences that do not make use of the light's phase or polarization properties is "On-Off Keying" (OOK), according to which the light is either on or off.

Nonreturn-to-Zero and Return-to-Zero Signals

The two main types of OOK transmission are known as "Nonreturn to Zero" (NRZ) and "Return to Zero" (RZ), both of which are illustrated in Figure 6.1. In the ideal case, light is transmitted for the entirety of each 1 bit in NRZ modulation. In contrast, it is present for only a fraction of the duration of the 1 bits in RZ modulation. The percentage of the bitslot that is occupied by a pulse is called the "duty cycle," and common values are between 25 and 50 percent. (NRZ modulation can be regarded as RZ with a 100 percent duty cycle.) For most of this chapter we refer to NRZ usage, but a later section will address RZ and other modulation formats. The datastream in Figure 6.1 is shown as perfect rectangles. In reality, the RZ pulses are close to Gaussian in the time domain and the NRZ pulses are often modeled as super-Gaussians (Norimatsu and Yamamoto, 2001), as shown in the figure inserts.

For medium- and high-datarate operation, the output from an unmodulated laser of wavelength λ should be continuous wave (cw), single-longitudinal mode, and narrow bandwidth. When modulated, the wave experiences spectral broadening to frequency bandwidth Δv, and in wavelength space it is given by equation 6.1.

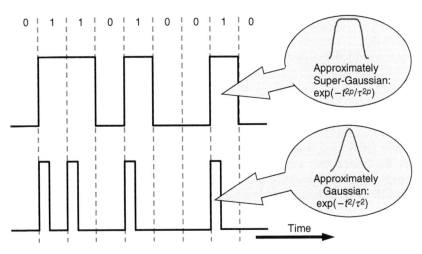

Fig. 6.1 Bitstreams: (*top*) NRZ; (*bottom*) RZ. The actual pulse shapes can often be modeled by the mathematical forms in the inserts (p is a positive integer and τ is a constant with the dimensions of time).

$$\Delta\lambda = (\lambda^2/c) \cdot \Delta v \qquad\qquad (6.1)$$

The magnitude of Δv depends on several influences, which include the bitrate, the temporal profile of the pulses, the nature of the digital data sequence, and the modulation format used. In general, by Fourier transform considerations (Bracewell, 1986), short pulses have broad bandwidths; long pulses, narrow bandwidths. Therefore, the bandwidth increases with the bitrate. Figure 6.1 shows that RZ signaling uses shorter pulses than NRZ signaling and so results in a greater Δv. The influence of the modulation format is considered further in a later section of the chapter.

In RZ and NRZ modulation, the bandwidth of a pulse stream is proportional to its bitrate. This means that, in the absence of any adjustment to the output power from the transmitter, the power spectral density of a channel (W/Hz) varies inversely with its bitrate. However, operation at the same BER requires that the power spectral density be maintained at the receiver. Therefore, for higher-bitrate operation we need a transmitter that provides higher-power pulses and/or we need to increase the inline amplification (Kartalopoulos, 2004). As a concrete example, if all else is equal and we upgrade operation from 2.5 Gbit/s to 10 Gbit/s, the average optical channel power incident on the receiver must be raised by a factor of 4 (i.e., 6 dB). Normally high-datarate systems require both higher launched powers and the maintenance of channel powers through appropriate amplification strategies. Both of these become increasingly demanding at very high bitrates. Later in the chapter we describe some of the problems that can occur when the average optical power in the fiber is too high.

Optical Transmitters

The lasers that provide the optical pulses are semiconductor based (Liu, 2005), and they can be driven by the direct injection of electrical current. Operation in this manner is simple and low cost and is commonly used for short-span and lower-bitrate applications, such as Local-Area Networks (LANs) and access networks. However, the pulses that emerge from directly modulated lasers have a positive "chirp," which means that their wavelength changes during the lifetime of the pulse.

Unfortunately, positively chirped pulses become greatly broadened in the time domain when they are guided by fibers with positive chromatic dispersion (see the later relevant section). After propagating over a sufficient distance they spread into neighboring bitslots, making it impossible for the receiver at the end of the system to interpret the bits reliably (Wilner & Hoanca, 2002). The problem of positively chirped pulses can be overcome by using a cw laser followed by an external modulator, such as a semiconductor electro-absorption modulator (Beck Mason, 2002) or a lithium niobate Mach-Zehnder modulator (Madabhushi, 2002). Unfortunately, such devices do not fully extinguish the light in the 0 bitslots; for this reason they are said to operate with a finite "extinction ratio." (As defined in a later section, an infinite extinction ratio is obtained when there is no optical energy in the 0s.) Nevertheless, in long-span and high-datarate communications, pulses that are not fully extinguished are usually preferable to those with significant positive chirp, which is why such applications mostly use external modulators.

The output from a laser can be nonideal in a number of respects that influence the system's overall performance. Short-haul and lower-datarate communications can be very cost sensitive and therefore operators often use cheap multilongitudinal-mode lasers with direct modulation (Liu, 2005). The resulting transmitted bandwidths can be very large, but because such network categories do not exert significant propagation impairments (as discussed throughout most of the chapter), the quality of the received signals is within acceptable bounds.

A variation on the same theme occurs when the intention is to provide very narrow-band light prior to external modulation but the laser exhibits residual optical side modes that deteriorate the quality of the received signals, especially after traveling many kilometers in a nonzero-dispersion fiber. Another effect is that the output from any laser oscillator is never completely constant in time because of an effect called "Relative Intensity Noise" (RIN), which is in the form of small-scale and random-intensity fluctuations on the output (Agrawal, 2002; Liu, 2005). The consequence for the system is a reduction in the optical signal-to-noise ratio (SNR) at the receiver, as will be discussed in a later section.

Synchronous Digital Hierarchy

Optical communications can transmit data for many end applications. However, it is advantageous to limit the number of datarates and data protocols in order to ensure compatibility of the equipment used and to constrain costs. Nearly all data conveyed by transcontinental, national, and regional networks, as well as a large percentage of the data on metropolitan networks, conforms to a set of standards known as Synchronous Digital Hierarchy (SDH) (Sexton & Reid, 1997). The details of SDH, which is used in most countries, are available from the wide-ranging recommendations of the Telecommunications Sector of the International Telecommunications Union (ITU-T) (ITU-T G.707, 2007; ITU-T G.780, 2004; ITU-T G.691).

A slightly different standard, Synchronous Optical Network (SONET), has been adopted in North America. However, because many of the standardized datarates and other operating characteristics in SDH and SONET are identical, throughout this chapter we refer only to SDH. Interested readers should refer to Shepard (2001) for a more SONET-oriented treatment.

In addition to the benefits of cost and interoperability, optical communications systems based on SDH offer several advantages. These include

- The ease of routing, dropping, and inserting in the time domain at digital cross-connects, thanks to the synchronous nature of the data
- Advanced network management functions, which are enabled via a sophisticated data frame structure
- The ability to interoperate with other data protocols, such as Internet Protocol (IP) and Asynchronous Transfer Mode (ATM)
- The availability of protection switching in the event of fiber or equipment failure

Most of the details of SDH are outside the scope of this chapter, but detailed explanations are provided in Chapter 10 and in Sexton and Reid (1997).

One aspect of SDH that does directly relate to optical fiber transmission is the channel datarates used. As the acronym SDH specifies, there is a hierarchy of datarates. They are all designated "STM-m," which means "Synchronous Transport Module" at level m, where m = 1, 4, 16, 64, or 256. STM-1 operation is at 155.52 Mbit/s and the reason for this value follows from the data frame structure used (Ramaswami & Sivarajan, 2002; Sexton & Reid, 1997). It includes the data payload plus a number of bytes that function as administrative unit pointers and regenerator and multiplex section overheads, which are vital for error monitoring

Table 6.1 Datarates Used for Synchronous Digital Hierarchy Operation

SDH Terminology	Exact Datarate (Mbit/s)	Common Abbreviation
STM-1	155.52	155 Mbit/s
STM-4	622.08	622 Mbit/s
STM-16	2488.32	2.5 Gbit/s
STM-64	9953.28	10 Gbit/s
STM-256	39813.12	40 Gbit/s
Not standardized at the time of writing		160 Gbit/s and above

STM = synchronous transport module.

and network management. The higher SDH datarates are simply factors of four above 155.52 Mbit/s, as specified in Table 6.1.

As an approximate guide, at the time of writing metropolitan area networks operate at all rates up to 10 Gbit/s; wide-area terrestrial networks use 2.5 and 10 Gbit/s and (more recently) 40 Gbit/s; older subsea networks are at 2.5 Gbit/s; and nearly all new subsea networks work at 10 Gbit/s. Although many laboratory experiments have been performed with 160-Gbit/s channels, operation at such high rates has yet to be standardized or commercialized.

The final aspect of SDH communications that we mention relates to the important function of clock recovery in the receiver, performed to ensure that the decision process necessary for low error bit recognition occurs at the correct instant in the bitslots. Clock recovery is much more difficult when there are long strings of 1s, and for this reason systems normally include electronic bit scramblers at the transmitter and descramblers at the receiver. Reducing the probability of many consecutive 1s also reduces certain types of nonlinear crosstalk, as discussed in later sections on nonlinear crosstalk and system design and implementation.

Some networks function at datarates that do not comply with SDH standards. In some instances, they can be mapped onto the SDH frame structure but in others this cannot be performed efficiently. One example is a specific transatlantic subsea network called TAT-12/TAT-13 (Bergano, 1997), which operates at 5 Gbit/s. However, owing to the need for expensive customized terminal equipment, it is unlikely that future cables will be designed to this rate. A more general example is a protocol commonly used for storage area networks, called "Fibre Channel" (this spelling is used even in the USA) (Keiser, 2002; Fibre Channel Industry Association, 2008). Fibre Channel rates include 1.06, 2.12, and 4.24 Gbit/s. A final example is "Gigabit Ethernet," widely used in optical LANs, which operates at 1.25 Gbit/s (Keiser, 2002).

Wavelength Division Multiplexing

Wavelength division multiplexing (WDM) allows us to augment the capacity of a point-to-point optical fiber by launching M wavelengths, each of which is modulated to form a channel (Ramaswami & Sivarajan, 2002). In this way, modern systems have demonstrated up to multi-Tbit/s data transport. This section explains the various categories of WDM and where they are applied.

The Concept

The primary justification for point-to-point WDM is financial: It is usually cheaper to operate with a small number of relatively high-capacity fibers than with many single-channel fibers.

In a WDM system, the fiber's total capacity is the sum of the channels' individual datarates. Laboratory prototypes of WDM systems have been demonstrated with M up to ~1000, but at the time of writing M is usually less than ~100 in commercial systems.

All channels are commonly transmitted at the same rate, but this does not necessarily have to be the case (Ito, Sekiya, & Ono, 2002). Indeed, the channels do not need to use the same data protocols, and some can even be analogue. WDM allows operators to mix various combinations of SDH, Gigabit Ethernet, Fibre Channel, Internet Protocol (IP), Frame Relay, and other possibilities by assigning one to each wavelength according to customer needs. From the point of view of datarate and protocol, WDM is said to be "transparent," which means that *logically* the M channels act entirely independently of each other.

The Operating Spectrum

The most common type of transmission fiber in use in Europe and North America is often called "Standard Single-Mode Fiber" (SMF), and its specifications have been agreed on by the ITU-T, as detailed in ITU-T recommendation G.652 (ITU-T G.652). The fiber operates as a single-mode waveguide at wavelengths beyond about 1250 nm, and it has acceptable micro- and macro-bending losses out to nearly 1650 nm. In recent years it has become possible to fabricate G.652 fibers with low absorptions due to OH⁻ ions and heavy metal impurities. Consequently, their losses are often less than 0.5 dB/km over the whole of the 1250- to 1650-nm range and their best performance is ~0.2 dB/km at 1550 nm. With the aid of WDM, such fibers therefore offer great operating flexibility and potentially vast total bandwidth. For convenience the available spectrum has been divided into bands, as stated in Table 6.2.

WDM can be categorized in various ways, but in terms of carrier wavelengths we can speak of Coarse (CWDM) and Dense (DWDM). CWDM is used for low-cost but relatively low-capacity operation, while DWDM is a higher-cost technology that can provide very high capacities. Normally, both CWDM transmission and DWDM transmission are unidirectional, so that two-way propagation between a pair of nodes requires two fibers. However, some very cost-sensitive applications use a variant of CWDM, which could be termed "wavelength diplexing," where two (or a small number of) channels are propagated in opposite directions in the same fiber.

Wavelength and frequency plans for both CWDM and DWDM have been standardized by the ITU-T (ITU-T G.694.2; ITU-T G.694.1) because it is recognized that there are significant benefits to all affected parties in common operation. The use of agreed wavelengths allows operators to provide "midspan meet," which is the point at the physical or logical boundary

Table 6.2 Optical Fiber Telecommunications Transmission Band Names and Wavelength Ranges

Name	Wavelength Range (nm)
O-band ("original")	1260–1360
E-band ("extended")	1360–1460
S-band ("short")	1460–1530
C-band ("conventional")	1530–1565
L-band ("long")	1565–1625
U-band ("ultra-long")	1625–1675

where two networks must exchange signals. Expensive interfacing equipment would be required in the absence of standardized channel plans. Agreeing on the working wavelengths also helps to reduce equipment costs. Production volumes increase and unit costs decrease if all component manufacturers supply their products to the same set of wavelengths. Overall, the greatest benefit of WDM standardization is to the end customer of the telecommunications services.

Coarse Wavelength Division Multiplexing

The ITU-T standard G.694.2 (ITU-T G.694.2) specifies a grid of wavelengths for CWDM, as illustrated in Figure 6.2. The objective is to use a large portion of the ~400-nm fiber operating range at a low cost by spacing the channels far apart (Thiele & Nebeling, 2007). Up to eighteen channels are available but the ones at 1271 and 1291 nm are less commonly used than the others. The 20-nm channel separation depicted in Figure 6.2 is designed to allow relatively cheap filters and lasers. The emission wavelength of a semiconductor laser varies with temperature and so a Thermo-Electric Cooler (TEC) and its control circuits are normally required to achieve a constant output. However, if the channels are positioned at 20-nm intervals, as shown, it is possible to tolerate temperature-induced drifting, which can be up to ±6.5 nm. Therefore, we can omit the TEC and so reduce costs. Furthermore, with such coarse spacing we can employ wider passband filters, usually based on relatively cheap optical thin-film structures.

An important application for unidirectional CWDM is in metropolitan area networks, especially in "metro-access" rings. In most cases, CWDM operation is unamplified. However, the channels at 1531 and 1551 nm lie within the gain bandwidth of the C-band Erbium-Doped Fiber Amplifier (EDFA), should it be needed. Semiconductor optical amplifiers can also easily provide gain for channels at 1311 nm and other wavelengths, as described by Thiele and Nebeling (2007).

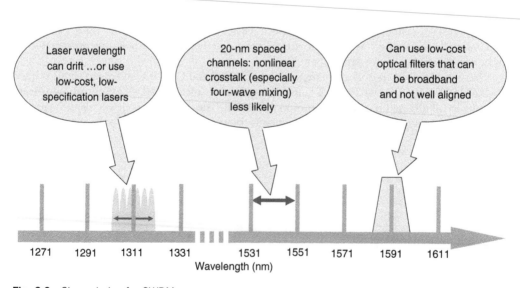

Fig. 6.2 Channel plan for CWDM.

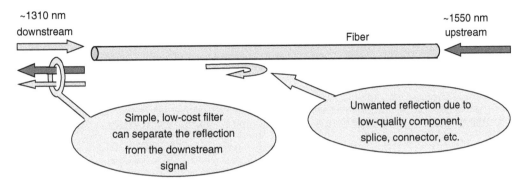

Fig. 6.3 Bidirectional operation with two wavelengths. The wavelengths used are usually from the CWDM grid.

In wavelength diplexing the objective is to propagate light in two directions in the same optical fiber, as shown in Figure 6.3. The main applications are in "Passive Optical Networks" (PONs) to satisfy the needs of access networking, but the strategy might also be of interest to optical fiber-based LANs or other forms of fiber–computer interconnects (see Volume 3). Access technologies are particularly cost sensitive, and so the aim is to use only one fiber to guide the light in two directions. Unfortunately, bidirectional operation is potentially problematic if the transmitters at the opposite ends of the fiber use the same wavelength. Any phenomenon, such as unintended point reflections or Rayleigh backscattering, that provokes the guided waves to counter-propagate causes crosstalk at the receivers. However, the difficulty is easily overcome by sending the counter-propagating signals at different wavelengths because low-cost optical filters in front of the two receivers can easily separate the intended signal from any unwanted reflected light.

The wavelengths used in diplexed operation are commonly on the CWDM grid. For example, the access network categories known as B-PON, G-PON, and E-PON ("Broadband," "Gigabit," and "Ethernet" PONs, as the case may be) use wavelength combinations of ~1490 nm and ~1550 nm downstream and ~1310 nm upstream (Keiser, 2006; ITU-T G.983.1–G.983.5; ITU-T G.984.1–G.984.6; ITU-T G.985).

Dense Wavelength Division Multiplexing

Dense Wavelength Division Multiplexing (DWDM) is used in high-capacity and long-span optical transmission systems. The channel plan, which is illustrated in Figure 6.4, is specified in terms of (absolute) frequencies rather than wavelengths. ITU-T standard G.694.1 (ITU-T, 2006) specifies a central reference frequency of 193.1 THz and an equally spaced grid to higher and lower values. The standards documentation advises the use of an accurate value for the speed of light in a vacuum when converting between frequency and wavelengths: $c = 2.99792458 \times 10^8$ m/s. With this value, the reference frequency is the equivalent of ~1552.52 nm, which is in the lowest loss region of silica fiber.

The grid lines shown at the top of Figure 6.4 are at 100-GHz intervals, but they do not all have to be occupied. For example, we could use every second line or some irregularly spaced combination. An operator might want to use only 4–8 lines approximately equally spaced across the gain profile of an Erbium-Doped Fiber Amplifier (EDFA) in intermediate-

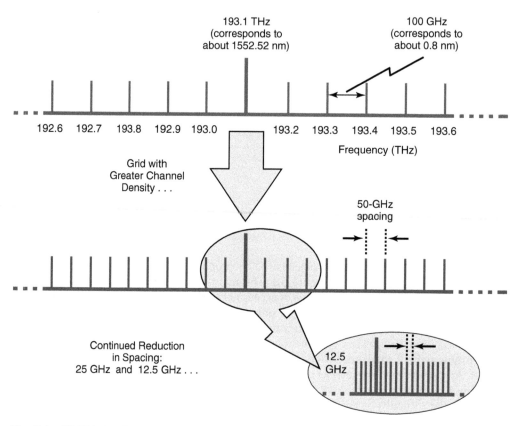

Fig. 6.4 DWDM: the ITU-T frequency grids.

capacity, but acceptable-cost, metropolitan area networks. The DWDM grid also permits very high channel densities. For example, by using 50-GHz spacing, it is possible to fill the C-band gain profile of an EDFA with about 80 channels. Such use is most likely to be found in longer-span national and international systems.

Figure 6.4 indicates the possibility of even denser grid lines: at 25 or 12.5 GHz. Channels spaced at 25 GHz (approximately 0.2 nm) would be for WDM operation with many closely spaced channels to achieve very high spectral efficiencies. However, such operation can be technically demanding. The laser and filter wavelengths have to be tightly specified, which increases their costs.

Very close channel spacing also increases the likelihood of crosstalk due to four-wave mixing and cross-phase modulation (the subject of a later section). Moreover, operation with 25-GHz channels precludes very high datarates (such as 40 Gbit/s) because the modulation bandwidths are so great that they spread over several grid lines. The grid lines at 12.5 GHz are for a special purpose: to permit operators with a fiber type known as Dispersion-Shifted Fiber (DSF) in their networks to mitigate nonlinear crosstalk by launching unequally spaced channels (see the later sections on cross-phase modulation, four-wave mixing, and zero-dispersion wavelength). These 12.5-GHz lines are of much less interest to users of nonzero-dispersion fibers, such as the G.652 category mentioned earlier.

Amplification in Optical Systems

Wavelength division multiplexed systems were initially developed for long-haul applications. In these systems, it is essential to boost the signal at regular intervals to overcome the attenuation of the fiber and components in the transmission path. Electronic regeneration is one option, but the cost of conversion from optical to electrical and back to optical for each channel is high. Moreover, the process becomes even more demanding as channel datarates are increased. Optical amplifiers offer the exclusive benefit that they can boost all channels together, often irrespective of their datarates, without the need for optical-to-electrical conversion.

Optical Amplifier Characteristics

Optical amplifiers are required to overcome loss as a result of transmission fiber, optical components, and power splitters. The main types are the erbium-doped fiber amplifier and other rare-earth variants, such as the Thulium-Doped Fiber Amplifier (TDFA), the Fiber Raman Amplifier (FRA), the Semiconductor Optical Amplifier (SOA), and the Erbium-Doped Waveguide Amplifier (EDWA) (Desurvire, 1994, 2002; Srivastava & Sun, 2002; Dutta, 2003; Kasamatsu, Yano & Ono, 2002; Zimmerman & Spiekman, 2004; Torres & Guzmán, 2007; Aozasa, 2007; Morito et al., 2003).

Although their geometries, gain processes, and pumping requirements differ, optical amplifiers have important common features. They produce noise, demonstrate gain saturation when the signal powers become significant, have wavelength-dependent spectral profiles, show Polarization-Dependent Gain (PDG), and can exhibit transient behavior for short intervals after channels are switched on and off. Because these phenomena have consequences for optical networks within which the amplifiers reside, we address them in the following subsections. However, the issue of noise, which is so important for network design, is covered separately in the section Amplifier Noise in Optical Systems. Furthermore, the signal output powers from optical amplifiers can be high enough to cause nonlinear crosstalk in the transmission fiber. This aspect is described in the final section of the chapter.

Optical Amplifier Locations

Amplifiers can be used in numerous network configurations, as depicted in Figure 6.5. Parts A through E illustrate discrete amplifiers, while Part F shows a distributed amplifier. A power amplifier (otherwise known as a "booster" or "post" amplifier) follows the transmitter and so is often used in medium to heavy saturation. Power amplifiers are commonly EDFAs pumped at 1480 nm to maximize pump power–to–signal power conversion efficiency. They are beneficial to the system when the transmitter cannot provide pulses of the required power (perhaps because it includes a high-loss external modulator) or can only do so with unacceptable chirp. Then a low-chirp, but low-power, transmitter can be followed by a power amplifier.

Line amplifiers are often spaced periodically in a system, especially in subsea operation. However, in terrestrial use geographical features or the logistics of electrical power supplies commonly impose locations at irregular intervals. Line amplifiers normally operate in intermediate saturation, and they need well-equalized gain profiles, as explained in the next two subsections.

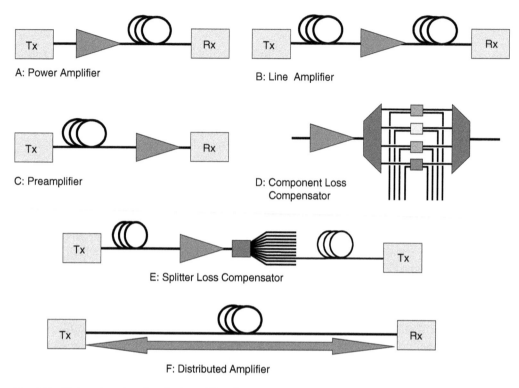

A: Power Amplifier

B: Line Amplifier

C: Preamplifier

D: Component Loss
Compensator

E: Splitter Loss Compensator

F: Distributed Amplifier

Fig. 6.5 Main categories of optical amplifiers in networks.

Preamplifiers are located before the optical receiver and ensure sufficient incident optical power prior to photon-to-electron conversion to enable good bit error rate (BER) performance, as described later in the chapter. They should be very low noise and normally operate in rather light saturation. The required noise performance is usually achieved by an EDFA pumped at 980 nm (because it provides lower noise than 1480-nm pumping). Sometimes a preamplifier is incorporated within the receiver module.

Parts D and E in Figure 6.5 represent a number of possibilities in which the aim of the amplifier is to overcome losses that are due to splitters, passive components, or even complex subsystems, rather than the result of lengthy spans of transmission fiber. There are many examples, but three important ones are

- At optical cross-connects within metropolitan-area and wide-area networks, where the large number of constituent components can create node losses exceeding 20 dB
- At dispersion-compensating fiber modules (the details of which are described in a later section)
- In long-reach PONs, which have target splitting ratios of 1×1024 or more (plus fiber spans of up to ~100 km) (Davey et al., 2006)

A distributed fiber amplifier is shown schematically in Part F of Figure 6.5, and two types have been demonstrated: the EDFA with very low Er^{3+}-doping levels and the FRA. Of these only the FRA has been commercially exploited and this situation is not likely to change in the foreseeable future. Distributing the gain offers the advantage of simultaneously reducing overall optical noise and nonlinear crosstalk. A fuller discussion is to come.

Channel Power Equalization

All optical amplifiers exhibit wavelength-dependent gain and, in the absence of corrective measures, they cause the problem for WDM systems shown in Figure 6.6. Even if the spectral gain variation is only a few decibels, the effect is cumulative and, after passage through only a small number of amplifiers, the channel powers can vary markedly. The most favored channel can be sufficiently powerful to cause nonlinear crosstalk in the transmission fiber or to saturate the detector, while the least amplified channel arrives at the receiver with such low power that it has an unacceptable bit error rate. There are three main strategies for gain equalization described here: pre-emphasis, Gain-Equalizing Filters (GEFs), and the control of pumping conditions.

"Pre-emphasis" (Menif, Rusch, & Karásec, 2001) is the technique of launching unequal, but carefully selected, channel powers in order to obtain spectrally equalized channels at the end of the system, as illustrated in Figure 6.7. The required values are predicted by an optimization procedure along with a detailed model of the amplifiers (such as the Giles–Desurvire theory of EDFAs [Desurvire, 1994, 2002]) and knowledge of the transmission fiber's loss. The pre-emphasis technique is successful and is applied commercially, but the predicted solutions are valid only for one network state. Any changes due to capacity upgrades or revisions in geographical routes require the calculation to be redone. Similarly, the inclusion of reconfigurable optical add-drop multiplexers or cross-connects proves par-

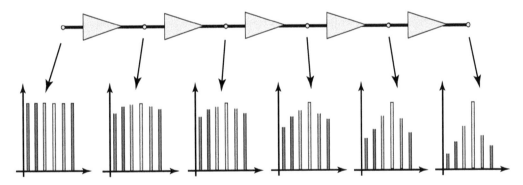

Fig. 6.6 Schematic of the progressive difference in channel powers when optical amplifiers with unequalized gains are used in WDM systems.

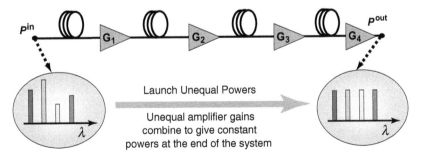

Fig. 6.7 Schematic of pre-emphasis to obtain equal channel powers at the end of a WDM system with carefully selected launch powers.

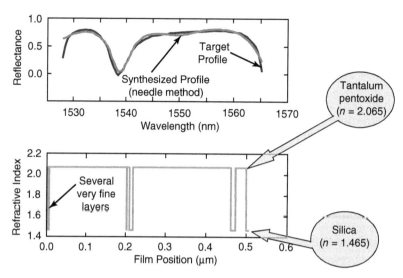

Fig. 6.8 Computed thin-film GEF to be used in transmission with a C-band erbium-doped fiber amplifier. *Top:* the filter reflectivity profile superposed on the gain profile of the EDFA. *Bottom:* the refractive index structure of the thin films. (Results produced by Miguel Zamora [2005].)

ticularly challenging for pre-emphasized networks because the channel powers can differ greatly after reconfiguration.

Gain-Equalizing Filters (GEFs) are normally used to achieve flattened spectra in discrete amplifiers (Othonos et al., 2006). Because of the nature of the semiconductor band gaps, SOAs have parabolic gain profiles (Dutta, 2003; Zimmerman & Spiekman, 2004). However, the semiconductor material has a high refractive index, causing reflections, and so low-finesse Fabry-Pérot modes can be superposed. Even with measures such as buried and angled waveguides and antireflection coatings, the ripple can be ~0.5 dB or more. Nevertheless, the overall parabolic shape can be straightforward to equalize with suitable filters. C-band EDFAs have a more problematic gain profile with a rather narrow peak at ~1532 nm, a minimum at ~1540 nm, and a broader plateau around 1555–1565 nm (Srivistava & Sun, 2002). Consequently, carefully designed GEFs are required. An example based on thin-film technology (Othonos et al., 2006; Miguel Zamorra, 2005) is shown in Figure 6.8.

GEFs should have low excess losses and, for fear of deteriorating overall noise performance, they should not be located at the amplifier's input. (The explanation for this is provided in a later subsection on the Friis cascade formula.) The favored technologies are long-period fiber gratings, thin-film filters (with an isolator), and Arrayed Waveguide Gratings (AWGs) (Othonos et al., 2006). A problem with all of these components is that they are normally static filters. Amplifier gain profiles can change significantly with the launched pump power and the number and power of the channels. Therefore, system upgrades and reconfigurations alter the gain spectrum, which means that static GEFs are often no longer able to achieve the necessary equalized powers. Dynamic GEFs, based on liquid crystals or other means, have been successfully demonstrated (Afonso et al., 2006; Barge, Battarel, & De Bougrenet de la Tocnaye, 2005). Their operation requires additional power-monitoring and control software, which increases the amplifier's cost and complexity but they are necessary to enable wavelength-routed networks.

In fiber Raman amplifiers, gain equalization is usually achieved by multi-wavelength pumping (Ritwitt, 2005; Islam, DeWilde, & Kuditcher, 2004). Each pump creates its own spectrally dependent gain profile that is shifted to lower frequencies with a peak at ~13.2 THz and a full width at half-maximum of ~7 THz in silicate glass. As will be discussed later, an important category is the dispersion-compensating fiber amplifier module, in which up to about a dozen pumps can provide very wide and flat spectral profiles in discrete modules. Distributed Raman gain is also commonly used, with the aim of overcoming loss in the transmission fiber, but then three to six pumps are more common. In either case, as the pumps propagate in the fiber, each one contributes to the amplification of the signals. Moreover, the shorter-wavelength (i.e., higher-frequency) pumps provide gain for pumps with longer wavelengths. Indeed, the conceptual difficulties are increased by the saturating influence of the bidirectional amplified spontaneous scattering, the presence of signal-to-signal Raman power transfer, and Rayleigh backscattering. Consequently, there are many interactions within the fiber and the power propagation equations are normally very complicated.

The pump powers to provide flattened gains can be predicted by appropriate use of Raman amplifier models, together with a suitable optimization procedure. Good agreement with experiment is possible. Thus, spectral profiles of over 100-nm bandwidth with subdecibel gain ripples have been achieved in DCF amplifier modules (Namiki, Emori, & Oguri, 2005), and in this sense the technique is successful. Unfortunately, multi-wavelength pumping creates several problems. As the number of pumps is increased, the gain spectrum becomes ever more sensitive to the launched pump powers and complicated control algorithms are required as a result. Moreover, the pump units are expensive and have demanding heat dissipation requirements. When multi-wavelength pumping is used with distributed amplification, high total powers are launched into the fiber transmission medium, raising potential concerns about the integrity of the fiber and, worse, about eye and skin damage to users. This aspect is considered further in the last section of the chapter, which discusses the benefits of distributed Raman amplification.

Operation in Intermediate Saturation

Line amplifiers are normally operated in intermediate saturation (Desurvire, 1994; Desurvire et al., 2002; Srivastava & Sun, 2002). In WDM systems this is usually unavoidable if the required output powers are to be provided for the M channels. Multichannel operation with sufficient signal powers to cause saturation can be problematical for SOAs but beneficial for other amplifier types.

In SOAs the conduction band has a lifetime of around 100 ps, which is comparable to the bit period of 10-Gbit/s operation; thus the population inversion can vary from one bit to another. (In contrast, EDFAs have ~10-ms metastable state lifetimes and, unless the data-rate is very low, there is an averaging effect.) Where there are two or more WDM channels in an SOA, each one changes the state of saturation and as a result imposes a modulation on all of the others. This phenomenon, called "cross-gain modulation," is an important limitation, especially in systems at channel datarates of 10 Gbit/s or higher. The effect can be reduced somewhat by using one strongly saturating wavelength that does not convey any data, together with the M data channels, or by encouraging laser oscillation outside the range of channel wavelengths (Zimmerman & Spiekman, 2004; Gutiérrez-Castrejón & Filios, 2006; Michie et al., 2007). Cross-gain modulation remains a constraint, however. For this reason SOAs are most likely to be applied where the number and datarate of the channels

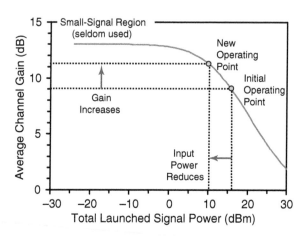

Fig. 6.9 Self-regulating behavior of a C-band EDFA in intermediate saturation: 40 channels with 100-GHz spacing. The vertical axis is defined as $G^{(dB)} = 10\log_{10}[\{\Sigma P_{out}(\lambda_i)\}/MP_{in}]$, where $M = 40$, fiber length is 6 m, launched pump power is 900 mW, and pump wavelength is 1480 nm. The amplifier is heavily pumped and has a suboptimal length to increase saturation output power. (Simulation by García López, 2008.)

are low, such as in optical access networks, lower-capacity metropolitan networks, and fiber-based LANs. (This issue is discussed in detail in Volume 3.)

In EDFAs and FRAs, operation with several decibels of gain compression below the small signal limit can be beneficial for the whole system. This situation is illustrated in Figure 6.9, which is a graph of the gain variation of an EDFA with launched signal power summed over all channels (García López, 2008). It shows a self-regulating mechanism. If the input channel power decreases by a few decibels during the system's operating life, the gain increases and so the output power remains close to its original value. The opposite occurs in the event of an increase in the input signal power. Thus, when used in intermediate saturation, total output power varies slowly with total input power.

Input signal powers can change for a number of reasons. One possibility is a "soft failure" (i.e., gradual power reduction) of an amplifier pump in a previous part of an amplifier chain. Another is extra loss after performing a repair in a subsea system, where it is necessary to add a cable span that is at least twice the depth of the sea at the point of cable damage (Horne, 1989).

The process shown in Figure 6.9 can be regarded as a form of passive automatic gain control. When there is a chain of amplifiers in intermediate saturation, the signal output powers tend rapidly to the original value as they progress along the chain. In a WDM system, self-regulation refers to the total input power of all of the channels (as in Figure 6.9). Sometimes the power variation affects only one channel—for example, owing to the reconfiguration of an OADM somewhere in the network. The power of the affected channel then differs from the others throughout the remainder of the amplifier chain, even if each amplifier has a well-flattened gain profile. In such circumstances, other measures, such as dynamic GEFs (Afonso et al., 2006; Barge, Battarel, & De Bougrenet de la Tocnaye, 2005) are required to reestablish power equalization.

Amplifier Noise in Optical Systems

All of the amplifiers in an optical system produce noise, which interacts in often subtle ways as it is passed through successive amplifiers toward the receiver. The noise, superposed on the signals at the detector, influences the assignment of the detected 1s and 0s to their correct values. This section explains how we account for noise accumulation in an amplified fiber system.

When an amplifier's gain does not depend on the phase of the incident signals, it is said to be "phase insensitive." The majority of amplifiers—such as rare-earth–doped fiber and waveguide amplifiers, FRAs, and SOAs—fall into this category (Srivastava & Sun, 2002; Dutta, 2003; Kasamatsu, Yano, & Ono, 2002; Zimmerman & Spiekman, 2004; Torres & Guzmán, 2007; Aozasa, 2007; Morito et al., 2003). Arguments based on the Heisenberg uncertainty principle show that phase-insensitive amplifiers always add noise to a signal (Desurvire, 1994). Therefore, when a noisy signal enters such an amplifier, three main things happen:

- The incident signal experiences gain, G.
- The incident noise also experiences gain, G.
- The amplifier adds some noise of its own.

Consequently, the amplifier reduces the signal-to-noise ratio (SNR) of any signals that pass through it. The noise produced by the main optical amplifier types is continuous wave and broadband (and therefore incoherent). In EDFAs, EDWAs, TDFAs, and SOAs this noise is due to amplified spontaneous emission, while in fiber Raman amplifiers its cause is amplified spontaneous scattering. To simplify the terminology, we use the expression "Amplified Spontaneous Emission" (ASE) in all instances.

To a good approximation, the ASE produced by all fiber amplifiers is unpolarized, but there can be over 1 dB of polarization dependence from SOAs unless preventive measures are taken. The ASE is present during both the logical 1 and logical 0 bits and is carried to the detector along with the signal (Olsson, 1989). Normally, there is an optical filter at the detector (such as the de-multiplexer in a WDM system), and during the 1s it allows only the modulated signal plus a small spectral slice of the ASE to enter. In such circumstances, there is a coherent superposition of the signal with the ASE at the (square-law) detector and the dominant noise component is due to what is called signal-ASE beat noise. Should an optical filter not be present, there will be additional noise components because of the coherent superposition of every infinitesimal spectral slice of the ASE with every other one—the resulting term is ASE-ASE beat noise. However, in many practical situations, such as DWDM operations, narrow band width filtering is used and thus ASE-ASE beating occurs to a small extent only during the 0 bits.

The Optical Noise Figure

The figure of merit to quantify an amplifier's noise performance is called its optical *noise figure*, and it should be as *low* as possible. The noise figure allows one amplifier to be compared with another. It is dimensionless and defined as the ratio of the optical signal-to-noise ratios at its input and output:

$$F = SNR_{\text{in}} / SNR_{\text{out}} \qquad (6.2)$$

Given that the amplifier adds its own noise, F must be greater than unity. Indeed, it follows from quantum mechanical considerations (Desurvire, 1994; Bristiel, 2006; Haus, 1998) that the lowest noise figure is 2 (i.e., 3 dB). Such performance is an ideal case, known as the "quantum limit," and amplifier manufacturers always strive to design devices that come as close as possible to it.

A semiclassical treatment of an optical amplifier, such as an EDFA, allows the optical noise figure to be quantified in terms of its gain, $G = P_{out}/P_{in}$. Assuming that the signal is narrowband, an optical filter of bandwidth B_O (Hz) is placed before the detector to eliminate ASE-ASE beat noise; assuming that all of the photon numbers are reasonably large (Desurvire, 1994),

$$F = (1/G) \cdot (1 + P_{ASE}/h \nu_s B_O) \tag{6.3}$$

P_{ASE} is the exiting amplified spontaneous emission power (watts) that is forward-propagating with respect to the signal of frequency, ν_s, and it is measured within B_O, as defined by the optical filter; h is Planck's constant (6.63×10^{-34} J $\cdot$ Hz^{-1}). The first term on the right of equation 6.3 of magnitude $1/G$ is due to shot noise, which arises because of the corpuscular nature of light. The number of photon arrivals fluctuates from one time interval to the next, and the effect is more pronounced as the signal power decreases (i.e., at lower photon numbers). The quantities G and P_{ASE} can be provided from amplifier models or measured in a laboratory. In all practical amplifiers G and P_{ASE} are wavelength dependent and so F varies as we tune the signal across the amplifier's gain profile.

Equation 6.3 can aid our understanding of optical system design. We start with the apparently trivial case of a passive element, such as an optical fiber of length L (m) and loss coefficient α_s (m^{-1}), and calculate its noise figure. Being passive, the element generates no ASE and its "gain" is in fact a loss, having the value $G = P_{out}/P_{in} = \exp(-\alpha_s L)$. Substitution into equation 6.3 then shows that the noise figure is $F = \exp(+\alpha_s L) = 1/G$. Therefore, for the purpose of comparing one network with another, we can assign noise figures to the fibers or passive components. A brief explanation is that, as the signal power decreases in magnitude, the noise due to the random arrival times of the signal photons (i.e., the shot noise) becomes ever more important. Even though no spontaneous emission photons are generated, the ratio of signal to shot noise is reduced merely through signal attenuation. For this reason, it is sometimes said that the signal "sinks into the noise." (Further details of why we assign a noise figure to a loss element are presented in Bristiel [2006].)

The Friis Cascade Formula

Equation 6.3 can be used to calculate the noise figure of two elements in series. From a quantum mechanical viewpoint, the following derivation is nonrigorous, but fuller treatments are presented in Haus (1998, 2000) and Desurvire (1994, 1999).

Let there be two amplifiers (or loss elements) of gains G_1 and G_2 and noise figures F_1 and F_2. The total gain for two amplifiers in series is $G_{tot} = G_1 \cdot G_2$. If we can assume that the ASE power is not sufficient to cause saturation, and therefore reduce the gain, the total forward-propagating ASE output power is $P_{ASE\text{-}tot} = G_2 P_{ASE\text{-}1} + P_{ASE\text{-}2}$. This equation applies because the second amplifier amplifies the noise of the first one and adds some noise of its own. The overall noise figure of the two amplifiers in series is $F_{tot} = (1/G_{tot}) \cdot (1 + P_{ASE\text{-}tot}/h \nu_s B_O)$. By making the appropriate substitutions, it follows that

$$F_{tot} = F_1 + (F_2 - 1)/G_1 \tag{6.4}$$

Equation 6.4 is the noise figure cascade formula, which is attributed to the Danish radio engineer Harald Friis (Brittain, 1995). It should be remembered that linear units, not decibels, are to be used when substituting into it. Moreover, the formula should *not* be used in this form when there are very low photon numbers (Haus, 1998, 2000; Desurvire, 1999).

The Friis formula enables some simple but important deductions. First, the cascade formula for amplifier gains is commutative: $G_{tot} = G_1 \cdot G_2 = G_2 \cdot G_1$, which means that the order in which we place the amplifiers in a system does not affect the power budget. This is *not* the case with noise figures—equation 6.4 gives a different answer if we swap the order of the two amplifiers. (A simple example illustrates the point: If $G_1 = 10$, $G_2 = 20$, $F_1 = 5$, and $F_2 = 4$, the total noise figure is either 5.3 or 4.2, depending on amplifier order.) The process of amplifying the input signal plus ASE and then adding more ASE is necessarily order dependent.

As a second deduction, it can easily be proved by mathematical induction that the total noise figure of a chain of N amplifiers is given by

$$F_{tot} = F_1 + (F_2 - 1)/G_1 + (F_3 - 1)/(G_1 G_2) + \ldots + (F_N - 1)/(G_1 G_2 \ldots G_{N-1}) \qquad \textbf{(6.5)}$$

A noise figure can therefore be assigned to an entire system, in which some of the elements with gains G_i and noise figures F_i are true amplifiers (such as EDFAs) while the others are passive components and transmission fibers. Unfortunately, such a treatment does not provide a complete picture of a network's performance because so many factors, such as nonlinear effects, chromatic and polarization mode dispersion, and time interval errors also contribute. In nearly every practical case, what matters in digital communications is a system's BER (the topic of the next section). Equation 6.5 nevertheless allows us to understand which of several network or subsystem designs provides better noise performance, and it is in this spirit that it is applied here.

Our third deduction is that, when the noise figures, F_i, are similar in value and the gains, G_i, are greater than unity, equation 6.5 predicts that the first term, F_1, makes the dominant contribution and the influence of the other elements decreases along the chain. The consequence for network design is that, where possible, the first amplifier in a chain should have the lowest noise figure. It is another implication of the noncommutative nature of the Friis formula.

A fourth deduction arises by applying equation 6.4 to a loss element, such as a fiber span, plus an amplifier. When the loss $\exp(-\alpha L)$ is between the transmitter and the amplifier, the total noise figure is

$$F_{tot} = F_{amp} \exp(+\alpha L) \qquad \textbf{(6.6)}$$

Alternatively, if the amplifier precedes the loss,

$$F_{tot} = F_{amp} + [\exp(+\alpha L) - 1]/G_{amp} \qquad \textbf{(6.7)}$$

In all cases of practical interest, where G_{amp} is greater than unity, equation 6.6 yields a significantly higher value than equation 6.7. This is another important consequence of the noncommutative nature of the Friis cascade formula. When converted to decibels, equation 6.6 states that $F_{tot}^{dB} = F_{amp}^{dB} + A^{dB}$; and one way in which this is often expressed is to say that any loss, A^{dB}, that comes before an amplifier's gain medium adds to the global noise figure "decibel for decibel."

Equations 6.6 and 6.7 have several practical implications. In terms of amplifier devices or subsystems, it is essential to minimize the losses that precede the gain element. For example, any components such as filters, couplers, and isolators that are required in an EDFA should ideally be located after the doped fiber. (Refer to the discussion of GEFs in an earlier section on channel power equalization and the discussion of amplification in dispersion-compensating modules in the section Optical Dispersion Compensation.) Another example of the importance of loss in front of the gain is in the SOA, where the gain medium is a waveguide of rectangular cross-section and a much higher refractive index than glass. Thus, there is a large semiconductor–to–fiber-coupling loss at the input, which is difficult to reduce. For this reason (and not because of the internal gain processes [Zimmerman & Spiekman, 2004]), SOAs usually have higher noise figures than fiber-based amplifiers.

The Noise Figure of Optical Amplifier Chains

An additional useful insight provided by equation 6.5 relates to amplifier chains, such as in transoceanic systems. We could use a small number of high-gain amplifiers spaced far apart or a large number of low-gain amplifiers. This choice is of great economic importance because of the high cost of remote amplifiers and their electrical power supplies. Figure 6.10 illustrates the choice when the objective is to provide net end-to-end power transparency. The total system length, L, consists of N equal segments of fiber with loss α, each followed by an amplifier of gain $\exp(+\alpha L/N)$.

We assume for simplicity that all amplifiers have a noise figure, F, regardless of how they are used in the system. (This is an approximation because amplifier noise figures normally vary with gain, but it becomes increasingly acceptable as the number of amplifiers in the chain increases.) Substitution in equation 6.5 allows us to predict that the composite noise figure of the entire system is

$$F_{comp} = 1 + N[F \exp(+\alpha L/N) - 1] \qquad (6.8)$$

F_{comp} is plotted as a function of N for a 500-km network in Figure 6.11 for four different amplifier noise figures, F. In accordance with intuition, it is beneficial to use amplifiers with a low noise figure, but Figure 6.11 also shows that overall performance improves with the number of amplifiers, which might at first appear surprising because each one produces noise.

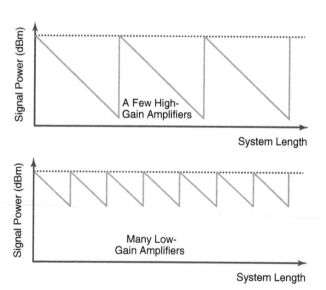

Fig. 6.10 In a transmission system the choice is a small number of high-gain amplifiers or a large number of low-gain amplifiers. The objective in either case is to ensure that the input and output powers are the same.

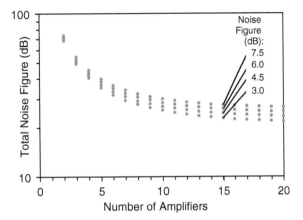

Fig. 6.11 500-km system with a loss of 0.2 dB/km and equally spaced amplifiers: How the total noise figure varies with the number of amplifiers used. See Figure 6.10. Note that nonlinear crosstalk and dispersion are ignored.

There are two ways of interpreting the situation: mathematical and physical. From a mathematical viewpoint, the use of many low-gain amplifiers allows us to place our first amplifier earlier in the system and so suffer less from the adverse effect of a large loss immediately after the transmitter, which the Friis formula predicts as having a particularly detrimental effect. In terms of physics, when there are many low-gain amplifiers, the signal power cannot fall to levels where the shot noise is as influential. In other words, if the signal power is always maintained at a high value, the noise component associated with the random photon arrivals makes a less important contribution to the overall SNR (Bristiel, 2006).

Figure 6.11 allows us to conclude that if noise is the dominant concern, we must use the (expensive) strategy of many low-gain amplifiers. However, the fiber nonlinear effects also influence our decision. (See the later section Nonlinear Crosstalk in Amplifier Chains.)

Rayleigh Backscattering

The amplifier noise that we have considered so far is due to ASE, but there is another contribution that can be important in amplified systems: Rayleigh Backscattering (RBS) (Jiang et al., 2007; Bromage, Winzer, & Essiambre, 2004). Even in fiber that has been fabricated to a high quality, all waves experience a weak distributed reflection without any shift in frequency that is due to microscopic perturbations in the refractive index on the scale of one-quarter wavelength. RBS acts on the pump, the signals, and the bidirectional ASE. Moreover, the retro-propagating waves also experience RBS, resulting in what is known as double RBS. Normally the powers, especially for double backscattering, are very low, but the presence of amplifier gain can allow them to grow.

When there is a chain of discrete amplifiers (e.g., EDFAs), the signals experience multiple RBS as they pass through the transmission fiber and the amplifiers provide periodically spaced gain elements. When the gain is high or there are many amplifiers, the amplified backscattered power can become sufficient to contribute to the system's total noise. The scattered waves arrive at the detector, having made multiple transits through the chain. Single and double backscattering of the bidirectional ASE from the amplifiers enhances the noise power, while double backscattering of the signal causes time-delayed replicas to be superposed at the detector. In both cases, the BER is increased.

RBS can also be substantial within the fabric of fiber-based optical amplifiers. It grows most when the amplifier medium is long, the effective area is small, and the germania content of the glass is high. These circumstances can sometimes apply in L-band EDFAs, where the doped fibers can be ~200 m, but they are most marked in FRAs.

In Raman amplifiers RBS is nearly always a significant contributor to the overall noise figure because the Raman pumps provide sufficient gain to enable the backscattered waves to grow and contribute additional noise. When RBS is included, equation 6.3 must be modified to give

$$F_{\text{int}} = (1/G_{\text{net}}) \bullet (1 + P_{\text{ASE}}/hv_s B_O + P_{\text{DRB}}/2hv_s B_e) \tag{6.9}$$

in which F_{int} is the "intrinsic" noise figure and P_{DRB} is the forward-propagating RBS power as measured within an electrical filter of bandwidth B_e (Hz). The "net" (or "input-output") gain, G_{net} of an FRA of length L is calculated by numerically solving complicated propagation equations, but a simple approximation for single-channel operation can be obtained by assuming that the signal power is low:

$$G_{\text{net}} = P_{\text{out}}/P_{\text{in}} = \exp(\Gamma P_{\text{pump}} L_{\text{eff}} - \alpha_s L) = G_{\text{on-off}} \bullet \exp(-\alpha_s L) \tag{6.10}$$

P_{pump} is the launched pump power, which can be either co- or contra-directional with respect to the signal, Γ is the fiber's gain coefficient, and α_s is the loss at the signal wavelength. L_{eff} is the fiber's effective length, given by $L_{\text{eff}} = [1 - \exp(-\alpha_p L)]/\alpha_p$, in which $\alpha_p = \alpha (\lambda_{\text{pump}})$. Equation 6.10 identifies an alternative definition of gain, known as "on-off gain," $G_{\text{on-off}}$. It is the ratio of output signal power with and without the pump turned on. $G_{\text{on-off}}$ is commonly used, especially for distributed amplifiers.

The Noise Figure of Distributed Raman Amplifiers

To understand how distributing the gain influences noise performance, we can compare two strategies that provide the same end-to-end power budget. In the first we use a distributed FRA of gain G_{net}; in the second, there is a passive fiber of the same length, L, and signal loss, α_s, followed by a discrete amplifier. In order to achieve the same output power, the discrete amplifier must have a gain, $G_{\text{on-off}}$. The noise figure of the distributed FRA is given by equation 6.9. In contrast, by equation 6.6, the total noise figure when using the discrete amplifier must be $F_{\text{discrete}} \exp(+\alpha_s L)$.

We now pose the question: What must the noise figure of the discrete amplifier be to give the same value as the distributed Raman amplifier? The answer is provided by equation 6.10, equating

$$F_{\text{discrete}} \exp(+\alpha_s L) = (1/G_{\text{net}}) \bullet [1 + P_{\text{ASE}}/hv_s B_O + P_{\text{DRB}}/2hv_s B_e] \tag{6.11}$$

When this applies, F_{discrete} is termed the "effective noise figure" of the distributed FRA and, by using equation 6.10, it is given by

$$F_{\text{eff}} = F_{\text{discrete}} = (1/G_{\text{on-off}}) \bullet [1 + P_{\text{ASE}}/hv_s B_O + P_{\text{DRB}}/2hv_s B_e] \tag{6.12}$$

which shows the importance of on-off gain. F_{eff} is smaller than F_{int} and in realistic circumstances it is commonly less than unity. (In other words, when expressed in decibels, it is often a negative number.)

The possibility of a negative (decibel) value of an effective noise figure may be a surprise, but it is important to realize what it means. It does *not* imply that the Raman amplifier reduces the system's noise in any way, but it tells us how much better the distributed gain is than a passive lossy fiber followed by a discrete amplifier. This is consistent with our general observation from the Friis formula that the best overall noise performance can be obtained by bringing the gain element as close as possible to the transmitter end of the system. The argument presented here partly explains why, despite the presence of RBS, Raman amplifiers have received intense attention in optical systems research laboratories and have been extensively commercialized.

Bit Error Rate

The fundamental purpose of an optical transmission system is to carry data from one location to another with the best possible accuracy. In practice, the presence of noise and other propagation impairments cause errors to occur in the received datastream and an aim of systems design is to ensure that the number of errors is kept to an acceptable level and at an acceptable cost. The factors that affect the bit error rate are therefore discussed in this section.

The Receiver

In this section we assume the use of "direct detection," in which there is no retention of phase information in the detected optical signal. In this case, the role of an optical receiver is to convert a photonic datastream into an electric current and interpret it to decide whether the individual bits are logical 1s or 0s. The detector in the front end of the receiver is a semiconductor diode and, for operation in the 1250- to 1650-nm range, is normally based on the InGaAs system. The two main types are p-i-n ("positive–intrinsic–negative" layers) (Taguchi, 2002) and Avalanche Photodiodes (APDs) (Kobayashi and Mikawa, 2002).

The p-i-n detector is normally lower cost and relatively low noise, but the APD offers an internal current gain, M_{apd}, through the process of impact ionization, leading to greater sensitivities. Unfortunately, the APD imposes an additional noise term. Optical receivers also include electronic amplifiers and filters, together with clock recovery and decision circuits. The clock recovery unit (Ellis, Smith, and Patrick, 1993; Philips et al., 1998) isolates a frequency component at $f = B$, the optical bitrate, from the received signal in order to synchronize the decision process.

When using direct-detection optical systems the transmitted pulse stream is converted into a time-varying electric current, which is proportional to the incident optical power, P_{inc}, via the relation $I_P(t) = RP_{inc}(t)$. R is the detector's "responsivity," given by $R = \eta q/h\nu_s$, in which q is the charge of an electron and η is the (semiconductor device-dependent) photon-to-electron conversion efficiency, which can often be close to unity. Typical values of R are around 1 W/A (Agrawal, 2002; Liu, 2005). The decisions that take place in a receiver are of special importance in all forms of digital communications, and the role of the decision circuit is to interpret the individual bits, deciding whether each one is a 1 or a 0.

Bitstream Corruption

In a real optical system, the arriving bitstream is corrupted as a result of the transmitter laser, the fiber transmission medium, the optical amplifiers, and (on conversion to a photocurrent) the receiver itself. As explained in an early section of the chapter, the emitted pulse stream from the transmitter is less than ideal in three main respects:

- The pulses exhibit high-frequency fluctuations called "Relative-Intensity Noise" (RIN).
- The pulses can have unintended chirp.
- In cases where the laser is biased slightly above threshold, the logical 0s are not normally at zero optical power.

The fiber medium imposes chromatic and polarization mode dispersion and a number of effects that lead to nonlinear crosstalk, as described in later sections. The optical amplifiers always add noise, as described in an earlier section. Temperature drifting of the fiber, mistiming at any digital cross-connects through which the signals pass, and interactions between amplifier noise and nonlinear optical effects in the fiber can cause jitter, which is a randomization of the pulse arrival times. Furthermore, the receiver circuitry imposes thermal noise on the signal and the semiconductor detector adds a "dark current" to the signal, even during the 0 bits. Collectively, these influences combine to ensure that the electrical bitstream at the decision circuits differs significantly from ideal 1s and 0s.

Before discussing in greater detail BERs and the factors that influence them, we describe a useful means of visualizing bitstreams with various levels of corruption (Kartalopoulos, 2004). Figure 6.12 shows three representative "eye diagrams," as would be displayed on an oscilloscope screen for NRZ bits. (Recall from the section Single-Channel Transmission that in NRZ signaling the transmitter laser is turned off for the full duration of the 0s and on for the full duration of the 1s.) An eye diagram is formed by repeatedly sampling and superposing all of the possible bits that could form a short sequence of the signal in the time domain. For example, for three bits the combinations are 000, 001, 010, ..., 111. If the bits are uncorrupted (such as at the top of Figure 6.12) the pattern that is formed resembles an open eye (hence the name). However, if there is much noise, timing errors, pulse spreading due to dispersion, and so forth, the bit degradation is clearly displayed through a closure of the eye.

The characteristics of an eye diagram give a useful insight into the effects that are likely to cause bit errors. Although the precise details can differ in many ways, if we were to sample one of the channels at (1) the transmitter output, (2) around the midway point in the system, and (3) at the receiver end, we would often expect to obtain eye diagrams that resemble those in Figure 6.12.

Definition of BER

When a corrupted bitstream arrives at the decision circuit in the receiver it is in the form of a photocurrent, as shown schematically in Figure 6.13. The decision circuit sets a threshold value, I_D, and measures the current at an appropriate instant within the bitslot, as indicated by the arrows. If the current is above I_D, the symbol is interpreted as a 1; if below I_D, as a 0. The threshold current could be arbitrarily chosen to be at the 50 percent level but, as will be argued, other values can improve the number of correct decisions.

The decision instant is chosen to be at a point in the bitslot where there is the clearest differentiation between 1 and 0. For example, in NRZ signaling the most obvious choice is in the center and, in terms of Figure 6.12, this is when the eyes are wide open. The vast majority of the bits are correctly assigned to their proper values. However, once in a while, as shown on the right of Figure 6.13, the signal can lie on the wrong side of the threshold and in that case we have a bit error. A bit error can be *either* a 1 that should have been a 0 *or* a 0 that should have been a 1.

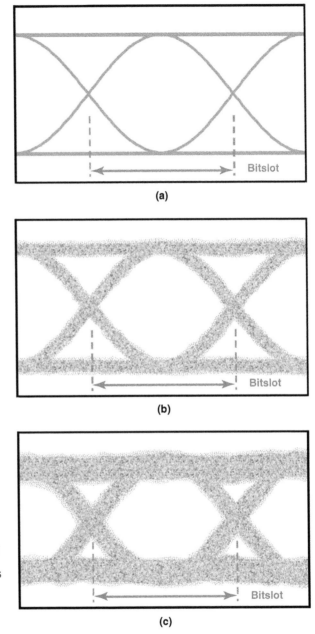

(a)

(b)

(c)

Fig. 6.12 Three eye diagrams for NRZ transmission showing progressive deterioration of the bits because of noise. The diagrams become more complicated in the presence of dispersion, jitter, and nonlinear effects.

The bit error rate is defined as the mean number of errored bits divided by the total number received in a time interval that is sufficiently long to provide realistic averaging. Optical transmission systems are commonly designed to provide BER values that are better than 10^{-9} or 10^{-10}. Indeed, recent standards documents recommend that performance be no worse than 10^{-12} (ITU-T, G.959.1), in some cases without the use of Forward Error Correction. Operation to achieve such values can be demanding. The point is illustrated by

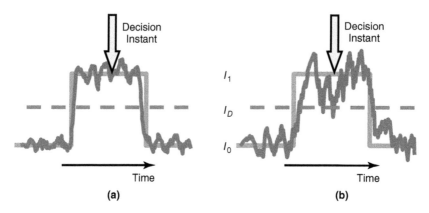

Fig. 6.13 Schematic of individual corrupted bits, showing the photocurrents at the logical 1, 0, and decision thresholds (I_1, I_0, and I_D, respectively). (a) Slightly corrupted bitstream: Most decisions are correct. (b) Badly corrupted bitstream: More decisions are wrong.

comparing with transmission by wireless or (metallic) coaxial cable, where BER = 10^{-6} is considered to be good.

Another perspective on optical systems is that when operating at 10 Gbit/s, a 10^{-12} BER implies that on average there will be one error every 1.7 minutes whereas systems that perform at $BER = 10^{-15}$ have mean times between errors of 11.6 days. One consequence of such performance is that there is a practical limit to the BER values that can realistically be measured in experimental systems; indirect methods (Bergano, Kerfoot, and Davidson, 1993) or computer modeling often must be used to predict system performance limits.

Formulating BER

A straightforward formulation of BER can be provided if we are able to assume that all of the noise processes that affect the 1 and 0 bits are Gaussian, as illustrated schematically in Figure 6.14. In that case, the probability density functions (Kartalopoulos, 2004) of the photocurrents are given by

$$p(I, I_{1,0}) = \left(\frac{1}{\sigma_{1,0}\sqrt{2\pi}} \right) \exp\left\{ -\frac{(I - I_{1,0})^2}{2\sigma_{1,0}^2} \right\} \tag{6.13}$$

In equation 6.13, $I_{1,0}$ are the mean photocurrents of the 1s or 0s, according to the subscript, and $\sigma_{1,0}^2$ are the corresponding variances (i.e., the squares of the standard deviations) due to all of the noise sources that act on them. It is important to stress that the equation does not apply in all cases. For example, shot noise is a Poissonian process that can only be approximated by Gaussian statistics (Kartalopoulos, 2004) and the inaccuracy can be significant in APD-based receivers (Kobayashi & Mikawa, 2002). Furthermore, nonlinear transmission effects can impose bit distortions that are not appropriately modeled by Gaussian probability distribution functions. In some amplified systems it is preferable to replace equation 6.13 with a χ^2 function (Chan & Conradi, 1997) and in others accurate results may require numerical models together with more demanding computation techniques.

BER is calculated using the following equation:

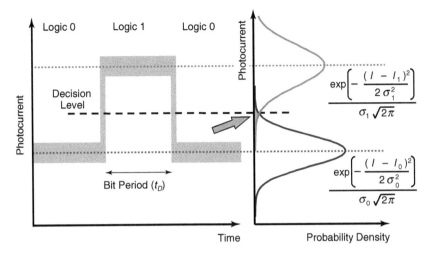

Fig. 6.14 Schematic of the probability density functions for the detected photocurrents of 1 and 0 bits. It is assumed that Gaussian functions apply. The arrow indicates the overlap region corresponding to bit errors.

$$BER = \frac{1}{2}P(0|1) + \frac{1}{2}P(1|0) \qquad (6.14)$$

in which $P(0|1)$ is the probability of deciding 0 when the bit is a 1 and $P(1|0)$ is the opposite. The two factors of ½ result from the fact that in most transmission schemes there are equal numbers of 1s and 0s in the datastream and therefore they are equally probable. (Refer to the description of bit scrambling in the section Synchronous Digital Hierarchy.)

The functions $P(0|1)$ and $P(1|0)$ are shown in Figure 6.14 as the small shaded area (indicated by the arrow) where the probability density functions cross to the "wrong" side of the decision threshold. They are obtained by summing (i.e., integrating) all of the values of the probability density function in the regions of mistaken interpretation

$$P(1|0) = \int_{-\infty}^{I_D} p(I_1) \cdot dI = \frac{1}{2} erfc\left(\frac{I_1 - I_D}{\sigma_1 \sqrt{2}}\right) \qquad (6.15)$$

$$P(0|1) = \int_{I_D}^{+\infty} p(I_0) \cdot dI = \frac{1}{2} erfc\left(\frac{I_D - I_0}{\sigma_0 \sqrt{2}}\right) \qquad (6.16)$$

The probability density functions are from equation 6.13. *erfc* is known as the "complementary error function," which is tabulated in Abramowitz and Stegan (1972) together with a comprehensive list of its properties. It results from the definite integral of a Gaussian function

$$erfc(u) = \frac{2}{\sqrt{\pi}} \int_{u}^{+\infty} \exp(-t^2) \cdot dt \qquad (6.17)$$

By the use of equations 6.14, 6.15, and 6.16, we have

$$BER = \frac{1}{4}\left\{ erfc\left(\frac{I_1 - I_D}{\sigma_1 \sqrt{2}}\right) + erfc\left(\frac{I_D - I_0}{\sigma_0 \sqrt{2}}\right)\right\} \qquad (6.18)$$

A key parameter in equation 6.18 is the decision threshold current, I_D, because we are at liberty to set it to any value between I_0 and I_1. The best procedure is to select a value that makes the BER as low as possible. In this case, the necessary value is found by differentiating and equating to zero: $d(BER)/d(I_D) = 0$. If we make use of the definition of *erfc(u)* in equation 6.17, we find that

$$\frac{(I_D-I_0)^2}{2\sigma_0^2} = \frac{(I_1-I_D)^2}{2\sigma_1^2} + ln\left[\frac{\sigma_1}{\sigma_0}\right] \tag{6.19}$$

In most circumstances, σ_0 and σ_1 are sufficiently close that the logarithmic term in equation 6.19 can be ignored. Then we obtain

$$\frac{I_D-I_0}{\sigma_0} - \frac{I_1-I_D}{\sigma_1} = Q \tag{6.20}$$

equation 6.20 defines Q, which we identify as a quality factor. Large values of Q are obtained when the standard deviations of the signal currents, σ_0 and σ_1, are small. The optimal value of the decision threshold, I_D, is found from equation 6.20 to be $I_D = (\sigma_0 I_1 + \sigma_1 I_0)/(\sigma_0 + \sigma_1)$. Equation 6.20 also leads to an expression for Q in terms of the means and standard deviations of the photocurrents:

$$Q = (I_1-I_0)/(\sigma_0+\sigma_1). \tag{6.21}$$

In the special case where there are equal noise currents on the 1s and 0s, $\sigma_0 = \sigma_1$, and so $I_D = (I_1 + I_0)/2$. In other words, only in these particular circumstances is the best threshold current at the 50 percent level.

It is straightforward to substitute into equation 6.18 to obtain an expression for the BER in terms of the Q value:

$$BER = \frac{1}{2}erfc\left(\frac{Q}{\sqrt{2}}\right) \tag{6.22}$$

Sometimes the complementary error function is inconvenient to compute or we wish to use equation 6.18 to provide further analytical results. Fortunately, it can be approximated when $Q > 3$ to $BER = \{exp[-Q^2/2]\}/\{Q(2\pi)^{1/2}\}$ by taking the first term in the series expansion (Gallion, 2002).

Figure 6.15 shows how the BER depends on Q, from which it is obvious that the variation becomes very rapid as Q increases. Two useful values to remember are $BER \approx 10^{-9}$ when $Q = 6$ and $BER \approx 10^{-12}$ when $Q = 7$. However, these and other values in Figure 6.15 apply only so long as the assumption of Gaussian noise statistics is valid.

The Laser Extinction Ratio

The quality factor, Q (and hence BER), depends on the photocurrents of the 0 and 1 bits, I_0 and I_1, and their corresponding variances, σ_0^2 and σ_1^2. Many approximations result from assuming that the transmitter laser plus the external modulator provide full extinction and therefore $I_0 = 0$. In that case,

$$Q = I_1/(\sigma_0+\sigma_1) = RP_{inc-1}/(\sigma_0+\sigma_1)$$

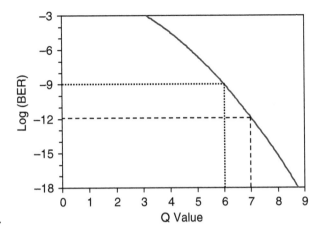

Fig. 6.15 Variation of BER with the Q value, showing two useful values: $Q = 6$ and $Q = 7$.

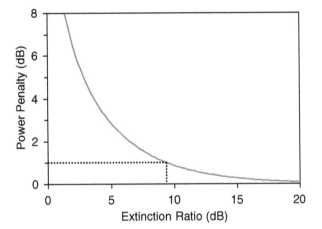

Fig. 6.16 Variation of the optical power penalty with respect to the extinction ratio. Marked an extinction ratio of 9.4 dB corresponds to a 1-dB power penalty.

where R is the responsivity, as previously defined. Alternatively, we can define an extinction ratio of the 1 and 0 bits, $r_{ex} = P_{inc\text{-}1}/P_{inc\text{-}0}$, together with an average optical power arriving at the receiver, $P_{av} = R(I_1 + I_2)/2$ (see the earlier section on optical transmitters). Total extinction gives an infinite value of r_{ex}. By substituting into the expression for Q, we have

$$\delta = 10\log_{10}\left[\frac{P_{av}(r_{ex})}{P_{av}(0)}\right] = 10\log_{10}\left[\frac{r_{ex}+1}{r_{ex}-1}\right] \tag{6.23}$$

δ, called the "power penalty," tells us by how much (in decibels) we must increase the power arriving at the detector in order to obtain the same Q value (i.e., the same BER). To put it another way, when all else is equal, a given receiver requires an increased average incident optical power in order to provide the same BER when the transmitter cannot provide full extinction. The power penalty is a quantification of that additional power.

The power penalty is of interest to optical transmission designers, who need to set an upper limit on the performance degradation they are willing to tolerate. Equation 6.23 is plotted in Figure 6.16 and shows how δ varies with the extinction ratio. For example, the power penalty is less than 1 dB when r_{ex} is greater than 8.7. Because extinction ratios are often expressed in decibels, the equivalent value is 9.4 dB. In comparison, ITU-T standards G.691 and G.959.1 for SDH operation at rates from STM-4 to STM-256 recommend minimum extinction ratio values of 6 dB, 8.2 dB, 9 dB, and 10 dB, depending on the system datarate, wavelength range, and transmission span (ITU-T G.959.1).

Other Influences on BER

Many more results have been derived from the theory of bit error rates that leads to equation 6.23, but it is not possible to reproduce them here. Instead, we comment on some of the BER's main features. The simple analysis that leads to equation 6.21 makes no reference to the effects that cause the statistical variation of the noise currents. There are many contributors, but here we mention five terms with their variances:

- Shot noise, σ_{sh}^2
- Thermal noise, σ_{th}^2
- Relative-intensity noise (RIN), σ_{rin}^2
- Signal-ASE beat noise, $\sigma_{s\text{-}ase}^2$
- ASE-ASE beat noise, $\sigma_{ase\text{-}ase}^2$

As was explained in an earlier section on amplifier noise, the last two of these apply only if optical amplifiers are used.

One of the useful consequences of random independent processes as described by Gaussian statistics is that the total variance of the combined probability density function is the sum of the individual variances (Riley, Hobson, & Bence, 2004). Therefore, in general we have

$$\sigma_{tot}^2 = \sigma_{sh}^2 + \sigma_{th}^2 + \sigma_{rin}^2 + \sigma_{s\text{-}ase}^2 + \sigma_{ase\text{-}ase}^2 \qquad (6.24)$$

in which σ_{tot}^2 can be the total noise variance of a 1 or a 0 bit, as appropriate. Expressions for the terms in equation 6.24 are presented in Steele, Walker, and Walker (1991).

As was stated in an earlier subsection on the optical noise figure, shot noise results from the random arrival times of the photons in a stream. The fluctuating photon number is translated into shot noise on the photocurrent, which depends on the incident optical power. Thermal noise results from random motion of the electrons in the conductors that constitute the electronic circuitry in the receiver. It varies inversely with the load resistance on the electronic amplifier and so can be reduced (but not eliminated) with appropriate receiver circuit design. Both shot noise and thermal noise are proportional to the bandwidth, B_e, of the electrical filter in the receiver, but there is limited scope to reduce either by use of a very narrow passband, for fear of eliminating important spectral components from the signal.

The RIN results from the superposition in the semiconductor laser of the random field of the amplified spontaneous emission with the coherent field of the stimulated emission. We saw in the subsection on optical transmitters that it is manifest as high-frequency intensity fluctuations on the signal, even when the injected electrical current is highly stabilized and constant in time.

An earlier section on amplifier noise explained that $\sigma_{s\text{-}ase}^2$ results from the beating (i.e., the coherent superposition) of the signal with all of the spectral components of the ASE from the amplifiers in the system, while $\sigma_{ase\text{-}ase}^2$ is due to the beating of each spectral component of the ASE with every other one. The ASE-ASE beat noise can be greatly reduced with the aid of an optical filter, and the wavelength de-multiplexer in a WDM system commonly satisfies this need for each channel. However, signal-ASE beat noise is always present in amplified systems. The best that can be done to minimize it is to use optical amplifiers with low noise figures and to follow the system design guidelines that were presented in an earlier section and that will be further discussed in the final section of the chapter.

Another complication is the type of detector chosen: APD or p-i-n. The avalanche process in APDs provides a gain, M_{apd}, which enhances the receiver's sensitivity but unfortunately imposes additional noise that is normally accounted for by including an APD noise figure, F_{apd}, within the expression for the variance of the shot noise, σ_{sh}^2 (Steele, Walker, & Walker, 1991). Receiver manufacturers optimize the value of M_{apd} to achieve a balance between the sensitivity enhancement that an APD can offer and the additional noise that results from the impact ionization process, on which its operation depends.

Inspection of equation 6.21 reveals that, when there is a good extinction ratio, it is possible to achieve a high Q value by minimizing all of the noise sources (so that σ_1 and σ_0 are low) and/or by increasing the power of the 1s (so that I_1 is high). Another way of saying this is that, once we have done everything to minimize the noise, the only way of achieving acceptably high Q values (and hence lower BERs) is to augment the incident optical power. Consequently, even when the transmitter offers a high extinction ratio, all of the noise sources in equation 6.24 impose a power penalty, δ_{noise}.

Equation 6.24 can be used in various ways to substitute into equation 6.21 and therefore to deduce the BER. In an unamplified system the noise of the 0 bits is given by $\sigma_0^2 = \sigma_{th}^2$, owing to the fact that there is neither RIN nor shot noise at the instants when the transmitter is turned off. However, this does not apply for the 1s, so we must use $\sigma_1^2 = \sigma_{sh}^2 + \sigma_{th}^2 + \sigma_{rin}^2$. The noise on the 1s and 0s can be visualized in terms of Figure 6.14, in which the two Gaussian curves have different widths.

When we use equation 6.21, in which the transmitter has a high extinction ratio, $Q \approx R\,P_{inc\text{-}1}/(\sigma_1 + \sigma_0)$. It follows that the 1s must have a higher incident power, $P_{inc\text{-}1}$ to achieve the same BER in the presence of RIN than they would require in an ideal RIN-free transmitter. Therefore, we can specify a RIN penalty, δ_{rin}. Fortunately, the RIN can be a relatively small contributor to the BER in many unamplified systems with high-quality transmitters.

Optical Preamplifiers

Optical preamplifiers, as shown schematically in Part C of Figure 6.5, are key elements in high-capacity and long-span optical communications systems. Sometimes the amplifier, such as an EDFA, is constructed as a constituent element within the receiver unit. The preamplifier provides gain, G, so that the 1 bits arrive at the detector with power $GP_{inc\text{-}1}$ and are then converted into a mean photocurrent, $I_{P1} = RGP_{inc\text{-}1}$. However, a preamplifier adds signal-ASE beat noise and (small levels of) ASE-ASE beat noise to the 1s and ASE-ASE beat noise to the 0s. In most situations the two beat noise terms dominate over both the thermal and shot noise components. If we can also neglect RIN to a first approximation and say that the transmitter has a high extinction ratio, equations 6.21 and 6.24 give

$$Q = \frac{RGP_{inc\text{-}1}}{(\sigma_{s\text{-}ase}^2 + \sigma_{ase\text{-}ase}^2)^{1/2} + \sigma_{ase\text{-}ase}} \qquad (6.25)$$

Equation 6.25 can be used to demonstrate that the receiver sensitivity improves substantially when a preamplifier is used (Agrawal, 2002; Steele, Walker, & Walker, 1991) and this result applies despite the addition of amplified spontaneous emission noise on both the 1 and 0 bits. The optical power incident on the detector is greater by a factor of G, and it normally more than compensates for the additional noise terms.

The best results for an optically preamplified system depend on the preamplifier having a low noise figure. For this reason, when EDFAs are incorporated into the receiver node, they tend to be pumped at 980 nm (Becker, Olsson, & Simpson, 1999). In an earlier section it was argued that distributing the gain with the aid of Raman amplification provides a lower effective noise figure than can be achieved by an equivalent discrete preamplifier. Even though Raman amplifiers are subject to an additional noise component due to Rayleigh backscatter, this argument remains valid when calculating BER, and it has been shown experimentally that distributed amplifiers can provide impressive overall system performance (Morita et al., 2000; Masuda, Kawai, & Suzuki, 1999).

It would appear from the arguments presented in this section that all we need to do to achieve good BER performance is to ensure that a sufficiently high signal power is incident on the detector (with or without the aid of an optical preamplifier). Unfortunately, as will be explained when we discuss nonlinear optical transmission, high propagating powers increase the likelihood of unacceptable nonlinear cross-talk in the optical fiber. The nonlinear optical effects can sometimes cause even greater bit corruption than the various noise terms. In real systems what we observe is a "BER floor." Increasing the incident optical power reduces BER to an extent, after which nonlinear optical effects become dominant and improvement is no longer possible. A key challenge for optical designers is to optimize their systems to provide acceptable BER within a design window in which the transmitted powers are sufficiently high to ensure good SNRs but sufficiently low to avoid nonlinear crosstalk. Some of the issues that designers must consider are explained in the final section of the chapter.

Chromatic Dispersion: System Implications

Chromatic dispersion causes pulses to spread in time, and this can limit the capacity of a system. This section provides an overview of the origin and implications of chromatic dispersion and discusses the tolerance of systems toward it. This section refers exclusively to single-mode fibers. Multimode operation is explained in Chapter 5.

Origin of Chromatic Dispersion

The most important (and most easily understood) design of optical fiber is the step index structure. This type has only two refractive indices: a relatively high value for the core and a lower value for the cladding. Light propagates in optical fibers as guided waves. High-capacity optical systems nearly always operate in the fundamental waveguide mode, which has an intensity distribution that is a rotationally symmetrical bell-shaped function (Mynbaef & Scheiner, 2001).

The main constituent of optical transmission fibers is vitreous silica (SiO_2) because when purified it is highly transparent. The core region is created by adding a few percent germania (GeO_2) or other dopants to raise the refractive index. Alternatively, fluorine (F) used as a dopant can lower the refractive index slightly. The refractive index of both pure and lightly doped SiO_2 decreases slowly towards longer wavelengths. This variation of refractive index with wavelength is called "material dispersion," or, more fully, "the material component of chromatic dispersion."

In a dielectric fiber the optical power travels in both the core and the cladding. Continuity of electromagnetic field is required at the core–cladding interface, which means that the overall waveguide mode must propagate at a velocity jointly influenced by the two refractive indices. Light travels faster in a glass with the refractive index of the cladding than in one with the refractive index of the core. Optical waveguide theory accounts for this with the concept of an effective refractive index, n_{eff}, to determine the speed of light of the whole guided mode. n_{eff} is governed by the ratio of power in the core to power in the cladding. For a mode to be guided, n_{eff} must be greater than the power of the cladding but less than that of the core.

In a step index fiber two factors dictate the core–cladding ratio of guided power. The first is the difference between the two refractive indices. When the indices are similar, light spreads into the cladding to a greater extent, but when they are different the guided wave is more tightly confined to the core. The second factor is the ratio of core diameter to wavelength of light. The mode spreads further into the cladding when this ratio is relatively small.

We now have a simple chain of logic. The fiber's refractive index profile, its core diameter, and the wavelength of light determine the extent to which the guided mode spreads into the cladding. The ratio of power in the core to that in the cladding dictates the effective index of the guided mode. However, the effective mode index governs the speed of light of the whole guided wave. Thus, for a fixed waveguide structure with a given core diameter and refractive indices, n_{eff} varies with the wavelength of the propagating wave. The variation is small and is known either as the "waveguide dispersion," or formally, "the waveguide component of chromatic dispersion."

The Chromatic Dispersion Coefficient of Single-Mode Fibers

Let us now turn to digital optical communications, which consist of pulses to constitute the 1s and (ideally) no light during the 0s. Whether the required pulses are created by directly modulating the current injected into a semiconductor laser or by following a cw laser with an external modulator, the presence of the modulation broadens the bandwidth of the emitted light, as explained in the earlier section on RZ and NRZ signals. Therefore, even a single-channel optical transmission system has a spread of wavelengths. Moreover, a high-bitrate channel requires shorter pulses, which have greater bandwidths.

We have seen that there are two mechanisms by which the refractive indices can vary with the wavelength of light: the material dispersion and the waveguide dispersion. However, a pulse code–modulated signal has a spread of wavelengths and so experiences a corresponding spread of each of the core, cladding, and effective refractive indices. This means that each pulse experiences a range of speeds as it passes through the optical fiber. The variation is small but accumulates over the many kilometers of transmission and so becomes significant. Sometimes the shorter-wavelength components travel faster than the long-wavelength components; sometimes the opposite occurs. However, in both cases the pulses are broadened in the time domain as they travel in the fiber. High-datarate channels have shorter pulses and therefore wider spectral bandwidths. The result is greater broadening.

Dispersion is often explained in terms of group velocity. A pulse can be regarded as consisting of a "group" or a "packet" of wavelengths. The group velocity quantifies how fast the entire group travels, as distinct from the wavefronts within the group. As the different wavelength components of the group encounter different refractive indices, they

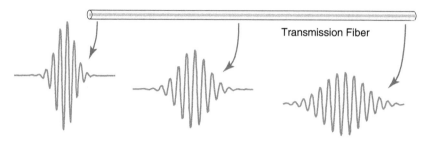

Fig. 6.17   Schematic of pulse spreading due to chromatic dispersion, as represented by packets of wavelengths.

travel at fractionally different speeds and therefore the pulse spreads in time, as illustrated in Figure 6.17. For this reason we often speak of "Group Velocity Dispersion" (GVD), which means the same as "chromatic dispersion."

When a pulse propagates a distance, L, in a fiber, two of its spectral components with a wavelength separation, $\Delta\lambda$, experience transit times that differ by ΔT, known as the "group delay," given by Wilner and Hoanca (2002):

$$\Delta T = DL\Delta\lambda \qquad (6.26)$$

in which D is called the "dispersion coefficient" of the fiber. D can be specified in wavelength space by

$$D(\lambda) = -\frac{\lambda}{c}\frac{d^2 n_{eff}}{d\lambda^2} \qquad (6.27)$$

n_{eff} is the fiber's effective index and λ is the average wavelength of the spectral components. If we were to substitute expressions for n_{eff} from fiber waveguide theory (Ramaswami & Sivarajan, 2002), we could identify the two main constituents of D:

$$D = D_{mat} + D_{WG}, \qquad (6.28)$$

They are the material and waveguide components, the physical significance of which is as previously discussed. (In fact, there is a third term, the "profile dispersion," but it is small and can often be neglected [Mynbaef & Scheiner, 2001].)

Equation 6.26 describes a linear increase of the group delay with the fiber length. The parameter D is measured in ps/(km·nm), which can be interpreted as picoseconds of group delay per kilometer traveled in the fiber per nanometer of bandwidth. D_{mat} and D_{WG} are both wavelength dependent and can be positive or negative. This means that D can be positive, negative, or zero. When D is negative, the fiber is said to be in the "normal dispersion regime"; when D is positive, the fiber is in the "anomalous dispersion regime." It is very important to stress that propagating pulses broaden for all nonzero values of chromatic dispersion. When D is negative, the long-wavelength components travel fastest but when it is positive, the short-wavelength components are the fastest.

The chromatic dispersion coefficient is wavelength dependent. However, there is no analytical expression for the wavelength variation of refractive index of the various fiber glasses and so it is convenient to fit empirical equations to experimental data. Manufacturers often use equation 6.29 for "standard single-mode fiber" (Mynbaef & Scheiner, 2001):

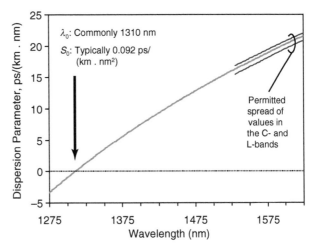

Fig. 6.18 Typical wavelength variation of chromatic dispersion for a standard single-mode fiber, showing the approximate range of values in the C- and L-bands that conform to ITU-T recommendation G.652.

$$D(\lambda) = \frac{S_0}{4}\left[\lambda - \left(\frac{\lambda_0^4}{\lambda^3}\right)\right]$$ (6.29)

The constant λ_0 is the wavelength of zero dispersion. $S_0 = S(\lambda_0)$ is the gradient, otherwise known as the "dispersion slope," $dD(\lambda)/d\lambda$, evaluated at $\lambda = \lambda_0$. For fibers that comply with the ITU-T G.652 standard (ITU-T G.652), λ_0 should lie in the range 1300 to 1324 nm, which can be achieved with appropriate selection of refractive indices and core diameter. Equation 6.29 is plotted in Figure 6.18, which indicates the range of values in the C- and L-bands. Typically D is about +17 ps/(km·nm) at 1550 nm.

System Tolerance of Chromatic Dispersion

Let us now consider the time domain. Dispersive pulse broadening causes the logical 1s to spread into one or more adjacent bitslots in a process known as "Intersymbol Interference" (ISI). When a 1 is neighbored by 0s the optical energy within the 1 bitslot is reduced and transferred to the 0s. However, consecutive 1s interact in a more complicated manner. Interference occurs between them and can cause peak powers that can be much greater than twice the time-averaged channel power. By considering equation 6.21, we can see that the photoelectric currents that constitute the Q value at the receiver are affected because the values of I_0 can be increased while the values of I_1 can be either raised or lowered, depending on their position in the bit pattern. The overall consequence is a greater probability of the 1s and 0s crossing to the "wrong" side of the threshold current at the instant of decision, as illustrated in Figure 6.13.

In common with the noise phenomena described in an earlier section on optical system noise, we can compensate for the lower Q values due to chromatic dispersion by increasing the power of the pulses incident on the detector; when expressed in dB, the increase is the dispersion power penalty. The maximum chromatic dispersion power penalty for acceptable system performance depends on the datarate, the nonlinear optical effects, the modulation format, whether forward error correction is used, and various other factors. However, common values are 1 to 2 dB. If we can assume the use of NRZ transmission, high extinction at the transmitter, and negligible transmitted pulse chirp, the power penalty can be expressed as a function of ε, the fraction of a bit period in which pulse spreading occurs. ε is 0.306 and 0.491 for power penalties of 1 dB and 2 dB, respectively (Ramaswami & Sivarajan, 2002).

A simple relationship between transmission span, L, bitrate, B, and signal bandwidth at the transmitter, $\Delta\lambda$, follows from equation 6.26 and allows us to estimate the operating limits. If we interpret ΔT as the inverse of the bit period, $1/B$, we have $|D| \bullet L \bullet B \bullet \Delta\lambda \leq \varepsilon$. The frequency bandwidth, Δv (in Hz), is normally higher than the bitrate (Gallion, 2002), and we can include a factor, b, which depends on the modulation format being used so that $\Delta v = bB$. Substitution for $\Delta\lambda$ from equation 6.1 yields

$$|D| \bullet L \bullet B^2 \bullet \lambda^2 \leq \varepsilon c/b \qquad (6.30)$$

where λ is the central wavelength of the pulses. The modulus of D results from the fact that pulse broadening occurs for both positive and negative dispersion coefficients.

Equation 6.30 shows that the maximum propagation length varies inversely with the square of the bitrate, and there are two contributions to this dependence. First, by Fourier transform considerations, the pulse bandwidth varies as the inverse of the pulse duration and therefore in proportion to the bitrate. Second, the duration of the neighboring bitslots into which the pulses spread is the inverse of the bitrate.

Equation 6.30 is plotted in Figure 6.19 to give approximate values for 1- and 2-dB power penalties and for when the pulses expand to one whole bit period ($\varepsilon = 1$). We assume for simplicity that $b = 1$. The transmission distance reduces greatly with bitrate. For example, transmission at 2.5 Gbit/s permits operation up to 1160 km when $\varepsilon = 1$. However, at 10 and 40 Gbit/s, the span is reduced to 74 and 4.6 km, respectively, showing that uncompensated chromatic dispersion can render all but the shortest systems inoperative.

It should be noted that the values given in Figure 6.19 are merely indicative. They assume that the linewidth of the unmodulated laser is narrow, that the pulses that constitute the 1s are unchirped, that there are negligible spectral sidelobes, and that nonlinear optical effects in the transmission fiber can be neglected. Still, it is clear that finding the means to overcome the limits imposed by chromatic dispersion is essential, especially as the datarate is increased. This is the subject of the following section.

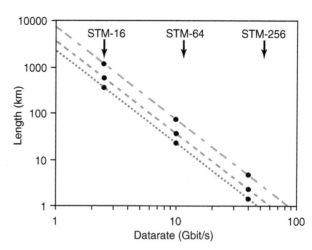

Fig. 6.19 Maximum transmission distance for NRZ modulation limited by chromatic dispersion in a G.652 fiber for which $D = +17$ ps/(km · nm) at 1550 nm. The values of ε in equation 6.30 are 1.0 (*top*), 0.491 (*center*), and 0.306 (*bottom*).

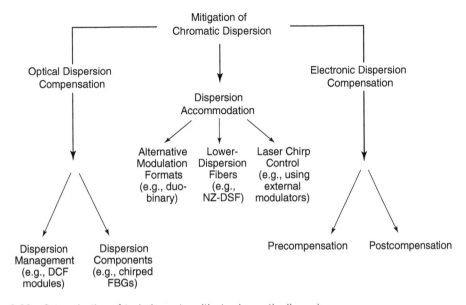

Fig. 6.20   Categorization of techniques to mitigate chromatic dispersion.

Chromatic Dispersion: Mitigation

The following subsections contain an overview of the main techniques for mitigating chromatic dispersion.

Techniques

Figure 6.20 categorizes the techniques we will consider. The methods are not mutually exclusive, and high-performance systems normally use more than one.

"Dispersion accommodation" refers to a collection of techniques to make chromatic dispersion less detrimental. "Dispersion compensation" is a collection of techniques that counteract the pulse broadening and that can be categorized as optical or electronic. At the time of writing, most installed systems use optical means, but operators and system vendors are increasingly devoting their efforts to Electronic Dispersion Compensation (EDC). Because electronic methods offer potential cost savings and sometimes overall performance improvements, there is a migration away from optical techniques. However, even if EDC eventually dominates, legacy optical compensators will remain in use for many years.

Dispersion Accommodation

Chromatic dispersion can be mitigated by setting the terms in equation 6.30 to values that reduce it. We designate such methods as examples of "dispersion accommodation." Unfortunately, despite the λ^2 dependence, there is limited scope to lessen the difficulties merely by appropriate selection of wavelength. With the exception of short-span and low-datarate applications, such as unamplified PONs (Gilfedder, 2006), where chromatic dispersion is seldom a constraint, fiber loss dictates that we favor the C- and L-bands. Nevertheless, with reference to equation 6.30, we can obtain useful improvement in two other ways: through alternative modulation schemes and with fibers that are designed for reduced dispersion coefficients.

Much effort has been devoted to developing low-bandwidth pulse-coding formats (as described in the later section Modulation Formats) and in this way we can reduce the value of b. Concerning the transmission fibers, their refractive index profiles can be designed to differ from being step index and so alter the waveguide dispersion in equation 6.28 to more negative values. As a result, the total chromatic dispersion, D, can be adjusted to the point where λ_0 is about 1550 nm, resulting in what are known as Dispersion-Shifted Fibers (DSF) (ITU-T G.653).

Unfortunately, as will be explained in later sections (Cross-Phase Modulation and Four-Wave Mixing), operation in the region of zero dispersion causes increased nonlinear cross-talk and so the strategy is not favored by most network operators. However, we can compromise and design fibers for dispersions in the C-band of about 3 to 8 ps/(km·nm); such types, called "nonzero-dispersion-shifted fibers" (NZ-DSF) (ITU-T G.655), have lower dispersion without incurring unacceptable nonlinear crosstalk.

A third dispersion accommodation technique, which can be regarded as a type of modulation format, is to provide careful control of transmitted chirp from what is called a "Chirp-Managed Laser" (CML) (Chandrasekhar et al., 2006). The earlier section Single-Channel Transmission, explained that positively chirped pulses from a directly modulated laser experience considerable broadening in positive dispersion fiber. However, when negatively chirped pulses are launched, they initially narrow and then broaden again. With careful optimization of both phase and amplitude of the emitted pulses, we can operate over longer distances before pulse broadening becomes unacceptable.

By using a combination of alternative modulation formats (which includes CMLs) and lower-dispersion fibers, system spans can be improved by up to one order of magnitude. Thus, with reference to Figure 6.19, the maximum transmission spans for 10- and 40-Gbit/s operation could be around 500 km and 50 km, respectively. While there are applications for such performance, it does not serve the needs of many national and continental trunk routes. Therefore, additional measures are required to overcome dispersive pulse spreading.

Optical Dispersion Compensation: Negative-Dispersion Fiber

The main optical compensation technique is called "dispersion management" (Wilner & Hoanca, 2002). As stated previously, if we use Positive-Dispersion Fiber (PDF), the short-wavelength components of a pulse travel fastest, but in Negative-Dispersion Fiber (NDF) the long-wavelength components travel fastest and arrive first. We can therefore alternate the two types of fiber in our system, with the effect on the pulses illustrated in Figure 6.21, which shows that they expand and then contract again. In order to provide ideal dispersion compensation in this way, the fiber lengths and dispersion coefficients must obey

$$|D_{PDF}|\, L_{PDF} = |D_{NDF}|\, L_{NDF} \qquad\qquad (6.31)$$

High-capacity systems have many periods that each conform to equation 6.31, and the overall variation is called a "dispersion map," an example of which is shown in Figure 6.22. The objective is to ensure that the accumulated dispersion is within acceptable bounds by the end of the system. Consequently, the pulse duration at the receiver is close to that at the transmitter and so the power penalty is tolerable.

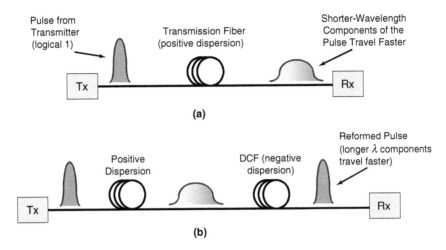

Fig. 6.21   Schematic of the role of dispersion-compensating fiber in a simple point-to-point system. (a) Uncompensated; (b) compensated.

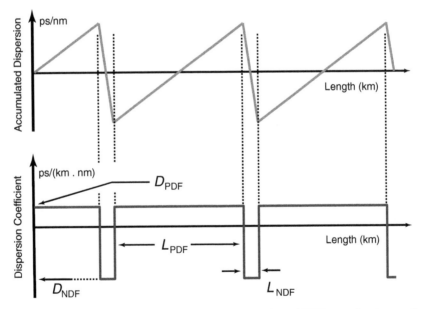

Fig. 6.22   Dispersion map consisting of alternating spans of PDF and NDF according to equation 6.31 (not to scale).

Operating with high-datarate channels and/or WDM is a little more complicated. As Figure 6.18 shows, a fiber's dispersion coefficient, D, varies with wavelength. High-speed systems have very broad bandwidth pulses, which means that D varies slightly across the pulse spectrum. Similarly, in WDM operation each channel experiences a different value of D. In both cases, the consequence is that equation 6.31 applies only at one particular wavelength; for all others it is merely an approximation. Figure 6.23 shows the outcome in

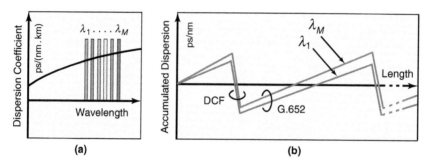

Fig. 6.23   (a) Variation of the dispersion coefficient of the PDF over *M* channels. (b) Consequences for the channels with the most extreme wavelengths. (Effects are exaggerated for illustration.)

WDM operation if some channels are not well compensated because equation 6.31 does not apply sufficiently closely. Therefore, we must take the dispersion slopes of the two fibers into account and an additional equation, equation 6.32, must be satisfied (Wandel et al., 2001, 2002):

$$|S_{PDF}| L_{PDF} = |S_{NDF}| L_{NDF} \qquad (6.32)$$

in which $S = dD/d\lambda$. Normally the values of D_{PDF} and S_{PDF} are fixed by the type of transmission fiber in use and, if it is G.652 compliant, the values can be deduced from Figure 6.18. However, the negative-dispersion fiber has to be designed to ensure the closest possible agreement with equations 6.31 and 6.32 over the range of operating wavelengths.

There are two strategies by which alternating PDF and NDF can provide dispersion management: Dispersion-Compensating Modules (DCMs) and incorporating the NDF within the cables. On the whole, operators prefer to use DCMs in terrestrial systems and cabled NDF in subsea operation. When network usage changes (for example, by incorporating additional optical add-drop multiplexers or WDM channels), the dispersion map of the whole system may have to be revised. The use of DCMs to compensate for dispersion enables simple upgrading by module replacement. In contrast, transmission loss in subsea communications is a critical concern and is minimized by including the negative-dispersion fiber within the overall system span. Moreover, there is seldom, if ever, sufficient space within submerged repeaters for bobbins of negative-dispersion fibers.

The simplest version of a DCM is a packaged bobbin of fiber having a negative dispersion, known as "Dispersion-Compensating Fiber" (DCF). Its dispersion coefficient, D_{DCF}, should have a high absolute value, with 100 to 200 ps/(km·nm) a common range. The use of these values in equation 6.31, together with data from Figure 6.18, shows that if the spacing between the DCMs is 50 to 80 km, 5 to 10 km of DCF is needed on the bobbin.

There are many proprietary designs of DCF, normally having refractive index profiles that differ significantly from step index. Most of them achieve the required values of D and S by the use of relatively high refractive index but narrow diameter cores. Such fibers have small effective areas and therefore are more susceptible to nonlinear optical crosstalk, as described in the later section Nonlinear Optical Transmission.

Unfortunately, the waveguide designs necessary for DCF normally require high levels of GeO_2 doping (Mendez & Morse, 2007), which leads to increased RBS and therefore greater loss. RBS causes increased noise in amplified systems, as discussed in the subsection on the noise figure of optical amplifier chains. The overall loss of DCF is also a concern. Whereas the (positive-dispersion) transmission fiber has loss coefficients in the range $\alpha = 0.20$ to 0.25 dB/km in the C-band, we can anticipate values that are two to four times as high from the DCF. For this reason, DCF manufacturers strive to design for the largest possible $|D_{DCF}|$, simultaneously with the lowest value of loss; therefore, they quote a figure of merit, $FOM = |D_{DCF}|/\alpha_{DCF}$, measured in ps/(nm·dB), which should be as high as possible (Wandel et al. 2001, 2002).

The small-core waveguide structure of the DCF causes another problem: splicing loss (Yablon, 2005). The DCF has to be spliced at both ends to the relatively large-core transmission fiber and, even with the use of special mode field adaptation techniques, we can anticipate a typical loss of 0.5 dB at each end.

DCMs often incorporate optical amplification to compensate for their losses. When EDFAs are used, the DCF can be included between two spans of the erbium-doped fiber and each one can be pumped forward, backward, or bidirectionally as necessary to optimize overall performance. Such "midstage insertion" within the amplifier's gain medium is a compromise configuration that provides a reasonably low overall noise figure without unacceptably sacrificing pumping efficiency.

Another approach, as explained in the earlier section Amplification in Optical Systems, is to use the DCF as a Raman gain medium and thus the DCM itself becomes a discrete fiber Raman amplifier (Namiki, Emori, & Oguri, 2005). The small effective core area of the DCF facilitates relatively high-efficiency pumping, and the pump lasers and their power supplies can be included in the DCM package. DCMs are normally located in equipment racks in node buildings. Whether Raman or erbium gain is used, it is necessary to provide electrical powering to the pump lasers. While they are not a problem in node buildings, electrical cables can be highly inconvenient and costly if geographical constraints impose placement of the DCM in the field. Therefore, network planners always try to avoid such situations.

Subsea transmission requires carefully optimized dispersion compensation (Bickham & Cain, 2002). Because the systems can be very long, loss minimization is a key issue. Moreover, many of the system specifications are constrained by the accumulation of nonlinear crosstalk. Typically, the optimal dispersion map consists of about two or three parts PDF to one part NDF, in which the periodicity coincides with the 40- to 50-km spacing between repeaters. The term "Inverse-Dispersion Fiber" (IDF) or "Reverse-Dispersion Fiber" (RDF) is commonly used for the portions that have negative values of D. The PDF in subsea systems is normally a type of NZ-DSF with $|D_{PDF}| = 4\text{--}8$ ps/(km·nm). Thus to satisfy equation 6.31 we expect $|D_{NDF}|$ to be about 8–24 ps/(km·nm). In contrast to the DCF used in modules, it is possible to achieve the required refractive index profiles without either very small core sizes or very high GeO_2 doping. This brings four benefits:

- The loss coefficient tends to be only slightly higher than that of the PDF.
- Noise due to the interaction of RBS with the amplifiers is less of an issue.
- The PDF–NDF splice losses can be acceptably small.
- The fiber's effective area is larger than the DCF in modules, which leads to lower nonlinear crosstalk.

Optical Dispersion Compensation: Chirped Fiber Bragg Gratings

Several component technologies have been developed to provide discrete optical dispersion compensation (Othonos et al., 2006). Figure 6.24 illustrates the one that is the most favored: the chirped Fiber Bragg Grating (FBG) (Kashyap, 1999; Ennser, Laming, & Zervas, 1998).

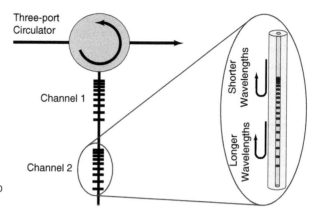

Fig. 6.24 Discrete dispersion compensators based on chirped FBGs and an optical circulator. Where necessary, more than two gratings can be used.

The FBG is a segment of optical fiber, typically a few centimeters long, with a low-magnitude, longitudinal perturbation to its core refractive index. If the perturbation is periodic, such as a sinusoid with respect to the length, there will be a very narrow bandwidth reflection of any incident light with a peak wavelength in the glass that is twice the periodicity of the perturbations. However, the structure used in a dispersion compensator is a chirped grating, which is one that has a superposed variation so that the period of the refractive index changes slowly over the length of the grating. Therefore, as illustrated in Figure 6.24, the refractive index perturbations are more closely spaced at one end of the grating than the other. In this way, the grating reflects a greater range of wavelengths, as illustrated by the arrows in the enlarged insert in the figure.

When a pulse enters the chirped grating, its long- and short-wavelength components propagate into the grating to different extents and therefore experience different phase delays. It is thus possible to a good approximation to use the chirped grating to cancel the group delay that the pulse has accumulated during its progress over a span of a transmission fiber. Figure 6.24 shows how the chirped FBG is normally used with a circulator to ensure that the reflected pulse can continue on its path through the system.

FBG dispersion compensators offer the advantage over DCMs of being low form factor components that do not contribute significant nonlinear crosstalk. Their loss is dominated by the circulator and is typically 1.5 to 2.5 dB for the double passage of the light. In contrast to DCMs, there is no need to include dedicated amplification. Moreover, chirped FBGs can be incorporated into optical add-drop multiplexers to perform simultaneous channel selection and dispersion compensation (Ellis et al., 1997).

Unfortunately, chirped FBGs operate over a much narrower bandwidth than NDF-based solutions and so can only compensate for one or a small number of channels. Three challenges for grating manufacturers are

- To ensure that the group delay is spectrally uniform (i.e., free from high-frequency ripples) (Kashyap, 1999)
- To provide a means, such as athermal packaging (Lo & Kuo, 2002, 2003), to cancel any wavelength drifting of the reflectance profile
- To fabricate the grating with low polarization mode dispersion and polarization-dependent loss. (See the later discussion of PMD due to optical components.)

Electronic Dispersion Compensation

Electronic (Chromatic) Dispersion Compensation (EDC) has recently become a topic of active research (Chandrasekhar et al., 2006; Yu & Shanbhag, 2006; McGhan et al., 2006; McGhan, 2006). Its underlying principles have been understood for a number of years, but the High-speed Application-Specific Integrated Circuits (ASICs) to carry out the necessary processing were not available for application at 10 Gbit/s and 40 Gbit/s until recently. There are two categories of EDC, according to whether the operation is performed at the transmitter or at the receiver end of the system (McGhan, 2006). Indeed, it is possible to use both in a system, and EDC can be combined with other (optical) dispersion accommodation and compensation techniques.

Transmitter EDC, depicted in Figure 6.25, uses an electro-optic modulator that controllably exerts both amplitude and phase modulation on the signal in response to commands from Digital Signal Processing (DSP) electronics. Ideally, it should be able to achieve any phase variation, but this is not always the case in some device designs. The aim is to provide simultaneous pre-emphasis and prechirping of the signal in anticipation of the chromatic dispersion that it is going to encounter during its transit in the system fiber. In this way, the arriving pulses should be confined as closely as possible to their bitslots.

The modulator shown in Figure 6.25 is a dual-parallel Mach-Zehnder design, which can be based on lithium niobate technology. It can access all phase states but is an expensive device with an insertion loss of 10 to 15 dB. An alternative configuration, called the "dual-drive Mach-Zehnder modulator" is lower cost and imposes a loss of only 3 to 6 dB, but it

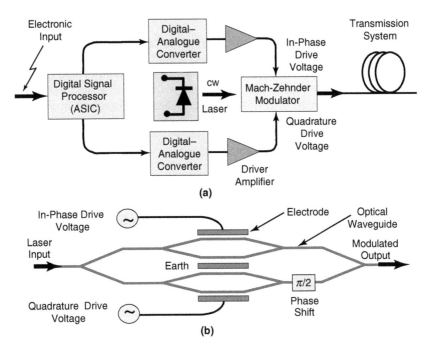

Fig. 6.25 Transmitter-end electronic dispersion compensation. (a) Block diagram of the main components. (b) Detail of one design of the electro-optic modulator configuration, known as the dual-parallel Mach-Zehnder modulator.

cannot explore the full combination of phase and amplitude states (McGhan, 2006). Nevertheless, it is a good compromise, especially for use with NRZ signals.

The DSP electronics shown in Figure 6.25 include the key chromatic-dispersion–compensating filter, together with coding, linear pulse shaping, and nonlinear filtering. Until recently, the design of digital-to-analogue converters was especially difficult for high-speed operation. The EDC filter within the DSP electronics must be configured to match the optical fiber link.

Transmitter-end EDC can be carried out via user input with a knowledge of the anticipated dispersion in the link, by completely automated optimization, or by an intermediate approach, where the user provides a target value around which an optimum can be found. Where the optimization is completely automated, there must be a feedback channel from the receiver end. It can be provided if the communication system consists of two fibers guiding the light unidirectionally in opposite directions, which is normally the case in medium- and long-haul operations.

When the EDC is at the receiver, it is performed after the detector. The main method employed, called "Maximum-Likelihood Sequence Estimation" (MLSE) (McGhan, 2006; Downie, Sauer, & Hurley, 2006), uses a fast analogue-to-digital converter to digitize raw received data samples; from them it deduces the most likely bit sequence. Receiver-end EDC can adapt dynamically to changing network conditions, such as after the reconfiguration of an optical add-drop multiplexer. Moreover, it has the advantage of not requiring expensive and lossy opto-electronic modulators. However, in 10-Gbit/s applications, receiver-end EDC tends to be limited to system spans of a few hundred kilometers because the required processing grows exponentially with the dispersion that has been accumulated in the system. Therefore, it may be most useful for cost-sensitive applications, such as long-reach PONs and metropolitan networks.

In contrast, transmitter-end EDC has provided chromatic dispersion compensation for 10-Gbit/s systems of up to 5200 km that consisted entirely of G.652 fiber and use no optical dispersion compensation (McGhan et al., 2006). Furthermore, with the aid of lookup tables to provide the required phase and amplitude variations, transmitter-end EDC has also been able to compensate for a nonlinear optical effect, which is described in the later section Nonlinear Optical Transmission and known as "self-phase modulation" (Killey et al., 2006; Agrawal, 2001).

The key advantage of EDC over optical techniques is cost, for several reasons:

- The cost of electronic integrated circuits is commonly lower than that of optical components, especially with volume production. (However, this argument is weakened somewhat if advanced electro-optic modulators, such as in Figure 6.25, are used, because they are expensive in their own right, one is required for each channel, and they may impose the need for additional amplification.)
- When receiver-end EDC is used, it is possible to integrate the required processors with the receiver circuitry, which leads to space reductions in addition to any possible financial benefits.
- When some or all of the DCM modules in a system can be eliminated, marked savings result from reducing the additional associated amplification and its electrical power feeds. Belanger and Cavallari (2006) and Jones and colleagues (2006) are techno-economic studies that use these arguments to report potentially significant financial benefits of EDC.

EDC confers benefits on the whole system of a more technical nature. Negative-dispersion fiber (especially DCF for use in modules) has a small effective core area and a relatively high GeO_2 content. Therefore, it is more susceptible to nonlinear crosstalk and RBS, the details of which were explained earlier and will be considered again in the section on nonlinear optical transmission. Indeed, amplified DCMs can be complicated subsystems, especially those using many pumps to provide wide bandwidth and spectrally equalized Raman gain. Their total launched pump powers can be very high, which requires special precautions to ensure eye and fire safety, and their complexity raises a slight concern regarding long-term reliability. One would also hope for reduced network management complexity through the smaller number of line amplifiers in the transmission path.

At the time of writing it is not clear to what extent EDC will displace optical compensation techniques in WDM systems at high channel datarates. We tentatively suggest that, for national and transoceanic applications, EDC combined with FEC and alternative modulation formats will reduce but not eliminate the dependence on NDF and other optical methods. However, the greatest scope for displacing all forms of optical dispersion management is in shorter-span systems, especially high-speed optical LANs, long-reach PONs, and metropolitan networks.

Polarization Mode Dispersion

Polarization mode dispersion is another form of dispersion that particularly affects high-speed signals in longer-range systems. This section describes the factors that give rise to polarization mode dispersion and its characteristics. Methods to minimize it and compensate for it in systems are also discussed.

Nonideal Optical Fibers

Light that is launched into an optical transmission system is preferably unpolarized, which means that it can be regarded as a combination of all possible states of polarization (Damask, 2005). The fiber through which it passes is normally designed for use in the fundamental HE_{11} waveguide mode, but the propagation is in reality the superposition of two orthogonally polarized guided waves. In an ideal fiber the polarized guided waves are "degenerate," which means that they travel with the same group velocity and so arrive at the detector at the same time. However, the real optical fibers and components that the signals encounter differ from the ideal in a number of minute but important respects, all of which influence the signal's State of Polarization (SoP) by creating two group velocities and so destroying the degeneracy.

Transmission and dispersion-compensating fibers and passive optical components exhibit Polarization Mode Dispersion (PMD) and Polarization-Dependent Loss (PDL), and the amplifiers commonly have at least a small Polarization-Dependent Gain (PDG). These effects may be of little concern over short transmission lengths and/or low datarates, but they have a considerable impact on high-speed systems (Neslon & Jopson, 2005). In terms of SDH operation, upgrading from STM-16 to STM-64 often imposes an important transition. Polarization effects rarely limit transmission lengths when working at up to about 2.5 Gbit/s, but they can seldom be ignored at higher rates. At 40 Gbit/s and above, they are of crucial importance.

We start with optical fibers and the cables within which they lie. There are two categories of imperfection that influence the PMD, intrinsic and extrinsic, as illustrated in Figure 6.26. The intrinsic factors, such as stressed glass, core nonconcentricity, and fiber

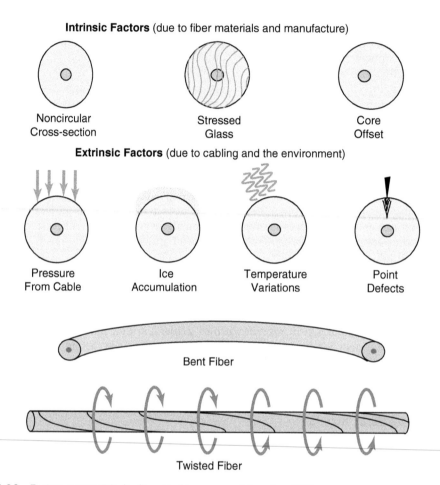

Fig. 6.26   Factors responsible for fiber birefringence and therefore PMD.

noncircularity, are consequences of the quality of fiber fabrication. In contrast, the extrinsic factors, such as bending and twisting of the fiber in the cable, seasonal and daily temperature variations, localized pressure points, and the like, are due to the way in which the fiber and its cable are deployed. Both the intrinsic and extrinsic factors cause "birefringence" in the fiber, which means the existence of a refractive index that is not rotationally symmetric.

There are several issues to note about the effects depicted in Figure 6.26:

- It is normal for the various influences shown to exist simultaneously.
- Even well-fabricated fibers and cables can vary along their length because of the effects illustrated.
- The intrinsic factors are worse in older optical fibers because the seriousness of PMD was not recognized in the 1980s.

- Many of the extrinsic effects (especially those due to temperature, twisting, bending, and localized pressure) vary over time.
- Nearly all of the effects that are shown are probabilistic.

Short and Long Birefringent Fibers

A *short* section of uniformly birefringent optical fiber has a fast and a slow optical axis with different effective refractive indices, n_f and n_s, such that $n_s > n_f$. There are two corresponding propagation constants, β_f and β_s, and the difference between them is given by $\Delta\beta = 2\pi\,(n_s - n_f)/\lambda = \omega\,\Delta n/c$, in which ω is the angular frequency. For the phenomena that cause birefringence in real fibers, Δn is typically around 10^{-7} to 10^{-6}, which is about three orders of magnitude less than the core–cladding refractive index difference in a transmission fiber. We can define the "Differential Group Delay" (DGD), designated $\Delta\tau$, which is the difference in propagation times over a fiber of length L:

$$\Delta\tau = d/d\omega\,(\Delta\beta\,L) = \Delta n L/c + (\omega L/c) \cdot d\Delta n/d\omega \qquad (6.33)$$

Equation 6.33 indicates a linear variation of DGD with length. It consists of a term that does not have explicit frequency dependence and a dispersive term that does. If we ignore the second (dispersive) term, we can define a length, L_B, called the beat length, with the value $\lambda/\Delta n$. If $L = L_B$, $\Delta\tau = \lambda/c = 1/v$. In other words, the beat length is the fiber length that causes the fast mode to advance one optical cycle (2π radians) with respect to the slow mode. Typically, L_B is around 1 to 50 m.

A single section of fiber with constant birefringence does not usefully represent the numerous influences that are present in real transmission. This follows from the statistical nature of and the many contributors to the PMD. Instead, we conceptualize and mathematically model PMD in the manner shown in Figure 6.27. The fiber link is assumed to be composed of many short birefringent fibers, which have randomly oriented axes (Neslon & Jopson, 2005). The coupling between the short sections is caused by twists, bends, splices, localized imperfections, and the like, and it normally varies randomly in time.

A pulse of arbitrary polarization launched into the first section splits into two orthogonally polarized pulses that travel with different group velocities, as described earlier. There is then polarization mode coupling, and the second section decomposes the fast and slow modes from the first section according to the orientation of its axes. In this way, the pulses are passed from one short birefringent section to the next, experiencing randomly oriented fast and slow axes.

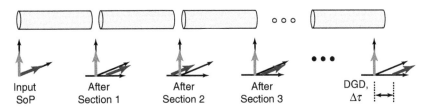

| Input SoP | After Section 1 | After Section 2 | After Section 3 | DGD, $\Delta\tau$ |

Fig. 6.27 Accumulation of DGD over a long fiber link assumed to be the equivalent of many shorter birefringent fibers of randomly oriented axes.

Equation 6.33 specifies a linear increase of DGD with length over one section. However, different behavior applies when considering many such sections because the changing birefringence can cause either addition or subtraction of the DGD. We must therefore account for the overall DGD accumulation in a stochastic manner by using the statistics of a "random walk." If this is done over a long link, it can be shown that $\Delta\tau$ grows as the square root of length. For this reason, PMD is specified in units of picoseconds of differential group delay per square root kilometer distance traveled: $ps/(km)^{1/2}$.

When considering an optical link, we can assign a "correlation length," L_c, which describes whether the DGD grows as the fiber length or as its square root. When $L \ll L_c$, the variation is linear and when $L \gg L_c$, it is as the square root and there is a gradual transition between the two regimes. The correlation length depends on both the intrinsic and extrinsic factors that govern the PMD, but for a modern fiber it is typically 1 km if the fiber is cabled and as little as 1 meter if it is on a bobbin. Two consequences arise:

- Laboratory bench measurements do not necessarily represent the behavior of installed systems.
- For most installed systems, we can nearly always assume the square root behavior.

The Statistical Nature of PMD

Understanding the statistical character of PMD is essential for understanding the performance of a real system. If the DGD of a long fiber link is measured many times, it can be expected to exhibit a range of values. Therefore, we must deal with probability distributions. The probability density function for DGDs in a long-span optical fiber is known as a "Maxwellian distribution," and its validity has been confirmed by many experimental studies (Poole & Nagel, 1997; Bülow & Lanne, 2005). The function is plotted in Figure 6.28, which shows that it is asymmetric with a value of zero when $\Delta\tau = 0$. The mean value of the Maxwellian is $<\Delta\tau>$ and, because of the asymmetry, it is slightly higher than the value corresponding to the peak. It is to the mean that we refer when specifying a fiber's PMD (in $ps/km^{1/2}$). The long "tail" of the Maxwellian distribution to higher values of $\Delta\tau$ is important because it opens the possibility of rare events, in which the DGD can be unusually large and so cause substantial pulse separation (Neslon & Jopson, 2005). The lower graph in Figure 6.28 shows the Maxwellian probability distribution function integrated from $\Delta\tau$ to infinity.

The statistical nature of PMD means that it exerts different influences compared with other more deterministic phenomena, such as chromatic dispersion and nonlinear crosstalk. A key feature is that DGD varies over time. Some effects, such as day-to-night temperature variations, are slow; others, however, such as vibrations caused by human activities, occur on a much shorter timescale. Some environments are relatively stable, such as cables on the sea bed. Others, such as "aerial fibers," where the cables are suspended overhead from poles, are subject to weather variations.

In a given environment, the consequence of random fluctuations is that the DGD can explore a wide range of values. The top graph of Figure 6.28 shows that low-probability events can occur where $\Delta\tau$ is very much greater than the mean; these events are the ones that cause concern. The bottom graph in Figure 6.28 tells us how likely it is for the DGD to exceed a given value, $\Delta\tau$.

From the point of view of optical receivers what matters is the relationship between the pulse separation and the bit period. BER can become unacceptable when the two pulses

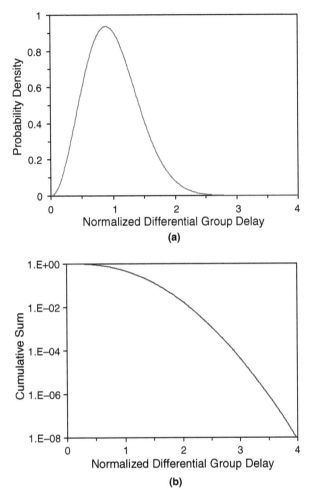

Fig. 6.28 (a) Maxwellian probability distribution for DGD, $\Delta\tau$; (b) cumulative sum from $\Delta\tau$ to infinity, as required in calculating outage probabilities. The horizontal coordinates are defined so that the mean value, $\langle\Delta\tau\rangle$, is unity.

arrive at the detector with times that are a sizeable fraction of the bit period. It is difficult to generalize about permissible performance because it depends on the modulation format used and other more subtle factors, such as interactions with amplifier noise and nonlinear optical effects in the fiber. However, a common rule of thumb is that the DGD becomes problematic when it exceeds one-tenth of the bit duration. (The criterion of 0.1 bit periods is recommended in ITU-T standard G.691 [ITU-T G.691]).

The consequence of PMD for systems is that it is not possible to design a link to guarantee continually available operation. Designers have to accept system "outages" on those rare occasions when the DGD exceeds a predefined BER criterion. Network operators must therefore specify a tolerable outage probability, P_{outage}, for which commonly used values range between 30 minutes ($= 5.7 \times 10^{-5}$) and 1 second per year ($= 3.2 \times 10^{-8}$). Other useful defined parameters are the mean outage rate, P_{outage} (seconds^{-1}), and the mean outage duration, T_{outage} (seconds). The three quantities are related by $T_{\text{outage}} = P_{\text{outage}}/R_{\text{outage}}$.

First- and Higher-Order PMD

If the bitrate is relatively low, the launched pulses are long and therefore have narrow spectral bandwidths. In that case we can assume that the birefringence they encounter is wavelength independent. The regime is called the "first-order PMD approximation," and its effect is seen in equation 6.33 for a short-span fiber in which we can ignore the dispersive term, $d(\Delta n)/d\omega$.

In longer fiber spans we must perform statistical averaging and take care to account for the correct evolution of the state of polarization through a matrix-vector formulation (Haus & Phua, 2005). By this it can be shown that in the first-order approximation we can identify two orthogonal launch states of polarization that provide an undistorted output pulse. These are known as the fiber link's "Principal States of Polarization" (PSPs). Much of the extensive theoretical literature on polarization effects in optical fiber has been formulated in terms of PSPs.

The first-order description of PMD becomes increasingly unrealistic in high-datarate systems, where the pulses are shorter and so have broader bandwidths. In these cases the theory must be generalized to account for the dispersive effects that, as equation 6.33 shows, result from terms in $d(\Delta n)/d\omega$. In a long fiber, as illustrated in Figure 6.27, Δn leads to terms that can be expanded as a Taylor series within the matrix-vector formulation. We can thus identify second- and higher-orders of the PMD.

Second-order PMD has two components. The first, and normally the smaller of the two, is Polarization-Dependent Chromatic Dispersion (PCD), in which the DGD varies with wavelength, leading to polarization-dependent pulse compression or broadening. PCD causes the fiber's effective dispersion to change rapidly with wavelength. The other component of second-order PMD is called PSP depolarization. Commonly larger than PCD, it is a rotation of the PSPs with frequency. PSP depolarization causes spiking and the generation of satellite pulses.

Spun Fibers for Low PMD

The optical fibers that were manufactured and deployed in the 1980s and early 1990s had numerous asymmetries that caused the intrinsic PMD to be as high as 1 ps/km$^{1/2}$, which could create severe problems at high datarates. More recently, manufacturers have invented methods of achieving values of less than 0.05 ps/km$^{1/2}$. The full details of their processes are not necessarily disclosed, but the main technique is called "spun fiber" (Damask, 2005).

In the spun-fiber technique, after being pulled from the perform rod the fiber is still at a high temperature and therefore in a viscous state. At this point it is passed through a device called a spinner, the role of which is to create a fiber that consists of short alternating spans that act as concatenated birefringent fibers with their principal axes aligned at 90° with respect to each other.

As the spinning is performed when the glass is still viscous, the result is different from that when the fiber is twisted at room temperature, because (ideally, at least) it does not impose stress in the glass. There are two mechanisms by which such a fiber reduces the PMD (Galtarossa et al., 2005). First, the correlation length, L_c, is significantly shortened with respect to the beat length, L_b, and this reduces the mean DGD. Second (and more important), any unintended accumulated birefringence in one of the spun half-periods tends to be canceled as the guided wave passes to the next half-period with its axis aligned perpendicularly.

PMD Caused by Optical Components

Optical components also contribute to the overall PMD of a system, and their influence is gaining in relative importance as operating bitrates increase. Sachiano (2005) argues that three potentially significant contributors are

- Dispersion-compensating fiber
- L-band erbium-doped fiber amplifiers
- Chirped fiber Bragg gratings

DCF has waveguide designs and fabrication requirements that differ greatly from those of standardized transmission fibers, and manufacturers strive to ensure that these do not result in additional intrinsic contributions. In C-band EDFAs the rare-earth–doped fiber is typically 10 to 50 m in length and has been found not to impose significant additional contributions to the PMD. However, because L-band amplifiers require longer spans, PMD can become an issue in some circumstances. Fortunately, as reported in Sachiano (2005), the effect can be alleviated by winding the doped fiber on a small-radius bobbin.

As described in the earlier discussion of chromatic dispersion mitigation, chirped fiber Bragg gratings can be used for dispersion compensation of one or a small number of channels by achieving large but negative group velocity dispersions over centimeter lengths. Unfortunately, the combination of possible asymmetries in the writing process, the high chromatic dispersion, and the presence of group delay spectral ripples can lead to high localized PMDs. It is therefore necessary to impose strict standards on the fabrication process and to control the curvature of the gratings within their packages.

PMD Compensation

PMD can be compensated for optically or electronically; in a WDM system there is one compensator per channel (Bülow & Lanne, 2005; Kudou et al., 2000). Whether optical or electronic, PMD compensation is preferably performed at the receiver end of the system. The objective is always to ensure that the accumulated DGD resulting from passage through the system plus the DGD exerted by the compensator sum to a value that is as close as possible to zero bit periods. Unfortunately, the tracking of high-speed variations (especially those caused by mechanical vibrations due to weather or human activities) is very demanding. Therefore, the best that any compensation technique can achieve is to reduce the outage probability to a tolerable value.

Optical compensation is performed *prior to* the detection process. A first-order optical PMD compensator consists of a polarization controller, such as a liquid crystal device, followed by a highly birefringent element, such as a span of high-birefringence fiber. The polarization controller responds to a feedback signal that originates from a measurement of signal quality. The three most widely used quality parameters are derived from

- The degree of polarization, as measured by an optical polarimeter in front of the detector
- An analysis of the electrical spectrum of the detected signal at the output of the compensator
- Electronic monitoring of the eye diagram after the detector

Of these options it should be noted that the polarimeter technique is applicable only to optical PMD compensation.

Electronic PMD compensation is always performed *after* the detection process and is possible at 10 Gbit/s and 40 Gbit/s, thanks to the availability of high-speed ASICs. As with electronic chromatic dispersion compensation, discussed in the previous section, the benefit of the electronic approach is in integrating the compensator functionality with the receiver's own electronics, which leads to lower costs. The main categories, as reviewed in Bülow and Lanne (2005), are the Feed-Forward Equalizer (FFE) and the Decision Feedback Equalizer (DFE). However, because of the square-law nature of direct detection, there is a loss of phase and polarization information once the signal has been converted from the optical to the electrical domain. The result, unfortunately, is a nonzero power penalty. For this reason, electronic equalizers do not always provide the low residual optical SNR penalties that can be achieved with their optical counterparts.

Nonlinear Optical Transmission

When low-power light is guided by an optical fiber over a reasonably short distance, L, it conforms to the requirements of linear optics. The word "linear" refers to the guided-wave equation, which predicts that the accumulated phase change is proportional to the distance traveled according to $\phi_{out} = \beta L$, in which β is the fiber's propagation constant, $2\pi n_{eff} / \lambda$. Similarly, the input and output powers obey the proportionality $P_{out}(\lambda) = P_{in}(\lambda) \exp(-\alpha L)$. In reality, these equations are only approximations because they are based on the assumption that the optical fiber's properties are independent of the propagating intensities. This section examines a number of ways in which optical fiber transmission systems conform to rather more complicated behavior, designated "nonlinear optics."

Nonlinear Refractive Index

Transmission in single-mode optical fibers can differ significantly from the simple operating principles of linear optics, because of a number of factors (Agrawal, 2001):

- The single-mode fiber maintains a high optical intensity over possibly many kilometers.
- Because of the need to provide good power spectral density at the detector, the launched powers in a high-bitrate system might often have to be several milliwatts per channel.
- DWDM systems can carry many channels, often with close wavelength separations.
- Longer-span systems require the use of optical amplifiers, which periodically increase the optical powers.
- The required pulses are shorter as the bitrate is increased and so the instantaneous powers at their peaks can be very high.

Collectively, these factors cause the propagation in the fiber to differ markedly from the linear regime via a number of effects. In this section, we describe the factors that are based on electron oscillations, commonly called "electronic nonlinearities."

The phenomenon of (linear) refractive index results from the excitation of oscillations of bound electrons within the optical fiber's glass by the traveling wave. However, when the optical intensity is high, the oscillations are anharmonic and so must be modeled by nonlinear equations (Butcher & Cotter, 1990). (The term "nonlinear optics" comes from these wave equations.) The anharmonic electronic oscillations produce several effects depending on the optical intensities, the wavelengths present, the states of polarization, the nature of the fiber's glass, and the temporal characteristics of the pulses.

An underlying phenomenon is the intensity-dependent refractive index, which is also commonly called the "nonlinear Kerr effect." An additional component is added to the refractive index to give

$$n_{tot} = n_{lin} + n_{NL}I, \tag{6.34}$$

in which n_{tot} is the total refractive index, n_{lin} is the familiar refractive index from linear optics, and n_{NL} is known as the nonlinear refractive index. I is the optical intensity (W/m^2).

We Remark on two properties of the nonlinear refractive index. The first is that it acts on a sub-picosecond timescale and so, even with channels at or slightly above 160 Gbit/s, any effects resulting from it can be regarded as instantaneous. The second is that the value of n_{NL} depends on the composition of the optical fiber's glass (especially its GeO_2 content), for which various values have been reported according to the measurement technique used. Agrawal (2001) recommends that for standard single-mode fibers we use 2.6×10^{-20} m^2/W for pulses in excess of 1 ns but 2.2×10^{-20} m^2/W for pulses between 1 ps and 1 ns. Slightly greater values are required for germania-rich fibers, such as DCF.

Self-Phase Modulation

The dependence of the refractive index on intensity exerts an additional phase change on a propagating wave according to

$$\phi_{tot}(z = L) = \beta_{lin}L + \beta_{NL}L_{eff} = \frac{2\pi n_{lin}L}{\lambda} + \frac{2\pi n_{NL}}{\lambda}\frac{P_{in}}{A_{eff}}L_{eff} \tag{6.35}$$

$\beta_{lin}L$ is the phase shift used in linear optics, and it acts together with the second group of terms, which include the launched pump power, P_{in}. A_{eff} is the fiber's effective area to account for the radial distribution of the guided mode; and L_{eff} is the "effective length" to account for the progressive reduction of signal power due to the fiber's loss. By integrating $P(z) = P_{in} \exp[-\alpha z]$ over the fiber's length, we obtain

$$L_{eff} = [1 - \exp(-\alpha L)]/\alpha \tag{6.36}$$

L_{eff} can be interpreted as the length over which the same nonlinear phase shift would occur if the fiber were of zero loss. (Recall from equation 6.10 that the small signal theory of fiber Raman amplifiers also uses an effective length but its loss term is at the pump wavelength.)

The logical 1s in an RZ or NRZ transmission system are in the form of pulses that can usually be approximated as Gaussians or super-Gaussians in the time domain. In either case, the intensity at the peak of a pulse is higher than at its extremities, and so by equation 6.35 the peak experiences greater nonlinear phase delays. In other words, each channel undergoes a phase modulation that is induced by its own intensity (Bayvel & Killey, 2002). For this reason, the effect is called "Self-Phase Modulation" (SPM). In terms of the wavelength domain, SPM creates chirped pulses. It occurs whether operation is single channel or WDM, and it follows from equation 6.35 that it is greatest when the launched signal powers are high and in fibers with a small effective area.

An influence that is not apparent from equation 6.35 is that of chromatic dispersion. When both the dispersion and its slope are negligible (e.g., when operating at λ_0 in a low-dispersion–slope fiber) and we use *direct detection*, SPM has no influence on the system's BER. The phase-insensitive nature of the optical-to-electrical bit conversion is not affected by the chirped pulses. However, there is a more complicated interaction between the phase changes resulting from SPM and nonzero values of fiber dispersion. The overall influence depends on the sign of the dispersion and causes pulse broadening when D is negative (the normal dispersion region) but causes pulse narrowing when D is positive (the anomalous dispersion region).

When the pulses broaden beyond their own bitslots there is an increase in the BER. The overall behavior is described mathematically by the "nonlinear Schrödinger equation" (NLSE), which is a partial differential equation in the length and time domains that is formulated in terms of complex electric fields to account for both phase and amplitude changes. The details of the NLSE are outside the scope of this chapter, but they can be found in Agrawal (2001) and Butcher and Cotter (1990), for example.

Cross-Phase Modulation

Other nonlinear phenomena occur in WDM operation because the channels can interact in many ways. An important contributor is called "Cross-Phase Modulation" (XPM), and its influence on the pulse streams can be mathematically modeled by writing an NLSE for each channel and including the cross-modulation terms that are characteristic of the effect. XPM occurs because the intensity dependence of the refractive index applies to every propagating wave in the fiber. Consequently, each channel modulates the phase of the others, without transferring power between them. XPM is always accompanied by SPM, but its influence via the intensity-dependent refractive index is twice as great in the ideal case when a pair of interacting channels are copolarized.

Owing to its influence via the intensity-dependent refractive index, XPM is most pronounced in fibers with small effective areas and when the channel powers are high. These features can be seen in equation 6.35. Moreover, the phase shift imposed by XPM is directly proportional to the effective fiber length, as given by equation 6.36. XPM also depends on the spatial overlap of the pulses of the different channels as they travel along the fiber. Therefore, assuming that there is good bit sequence scrambling to eliminate long strings of 1s, the channels "walk off" from each other when there is nonzero chromatic dispersion (of either positive or negative sign). For this reason, (1) the nonlinear phase shift due to XPM is lower in nonzero-dispersion fibers and (2) the extent of the walk-off depends on the channel spacing. Thus, XPM is more marked in very dense WDM operation.

Four-Wave Mixing

A crucial nonlinear optical effect in WDM systems is Four-Wave Mixing (FWM). We saw that XPM causes each channel to modulate the others without power exchanges. In contrast, FWM does cause power transfer—from each channel to some of the others and/or from each channel to frequencies that are not occupied by channels. In common with the other nonlinear effects, FWM results from the nonlinear response of bound electrons in the fiber's glass, but it differs because it is a parametric process that requires the presence of phase matching. In general, two photons of energy, $h\nu_1$ and $h\nu_2$, create two new photons of energy, $h\nu_3$ and $h\nu_4$.

The FWM process can be regarded in a number of equivalent ways. One of them is as a nonlinear scattering process in which there is momentum conservation that can be expressed in terms of the fiber's propagation constants at the four participating frequencies:

$$\beta_1 \pm \beta_2 = \beta_3 \pm \beta_4 \tag{6.37}$$

By recalling that the phase change experienced by a guided wave is the product of the propagation constant and the fiber length, L, it is clear that equation 6.37 is the phase-matching condition for the FWM process. The participating frequencies are related by the energy conservation condition:

$$h\nu_1 \pm h\nu_2 = h\nu_3 \pm h\nu_4 \tag{6.38}$$

When many channel frequencies are present, large numbers of "mixing products" are generated that satisfy equation 6.38. The process is complicated by the fact that some of these products lie on top of the channel frequencies and others do not. Moreover, the frequencies that create the FWM ($\nu_1, \ldots, \nu_4$ in equation 6.38) do not all need to be distinct. When two of them are identical, we have "degenerate FWM," in which only two new frequencies are generated.

A central requirement for FWM is the condition necessary for the phase matching. FWM can only occur efficiently if the fiber's chromatic dispersion allows equation 6.37 to be satisfied (at least approximately). Therefore, FWM is most efficient (and thus most problematic) when operating at or near the wavelength of zero dispersion. When there is nonzero dispersion, the channels "walk off" from each other but their spatial overlap increases as $|D|$ approaches zero. For the same reason, when the frequency separations of the channels are small they travel at similar velocities in the fiber, which increases the interactions between them.

FWM depends on the presence of at least two channels for the degenerate interaction or three for the nondegenerate case; in all circumstances it has a power-cubed dependence and consequently is sensitive to the launched channel powers. The magnitude of the mixing products increases with the fiber's effective length, and the products have an inverse variation with its effective area.

FWM is at its worst when the channels are equally spaced in terms of frequencies and have small separations. The number of mixing products varies approximately as the cube of the channel number, N, which means that many products can be generated. The difficulty for WDM optical communications is when one or more of the products is passed to the detectors, which happens to the ones that lie within the passband of the de-multiplexer at the receiver end of the system. The result is increased BER.

Other Nonlinear Effects

Other nonlinear effects can occur in optical fiber transmission. We draw attention to two that are based on the principles that have been described in this section. When operating at very high datarates (e.g., 40 Gbit/s and above), the pulses are easily broadened beyond their bitslots by the fiber's chromatic dispersion. When the intensities are sufficiently high, neighboring pulses (logical 1s) can then overlap and interact via the intensity-dependent refractive index. The spectral components of the pulses can thereby cause what are termed "intrachannel XPM" and "intrachannel FWM" (Bayvel & Killey, 2002; Shake et al., 1998; Essiambre, Mikkelsen, & Raybon, 1999).

In common with the other effects described here, intrachannel XPM and FWM depend on the effective length, the fiber's dispersion and effective area, and the launched channel powers. We therefore have the means to reduce their impact on BER by appropriate selection of fiber type(s) and design of the system's power budget. The issue is considered further in the chapter's final section.

Modulation Formats

In the section on single-channel transmission, it was shown that optical data is most commonly transmitted by applying the On-Off Keying (OOK) modulation formats illustrated in Figure 6.1, known as Nonreturn to Zero (NRZ) and Return to Zero (RZ). More advanced modulation formats are chosen to make best use of the available bandwidth and noise capabilities of the channel. This section provides more details on NRZ and RZ transmission and an overview of the most commonly used alternative techniques.

NRZ and RZ Transmission: Power Spectral Densities

We start by assuming for simplicity that the laser at the transmitter has an infinitesimal linewidth and that the pulses are generated by an ideal external modulator that can provide perfect rectangles in the time domain. In RZ operation we define the duty cycle, ξ, as being the ratio of the pulse width to the bitslot and in the special case of NRZ operation, $\xi = 1$. Both NRZ and RZ signaling then provide power spectra that can be represented in frequency space as having continuous and discrete components (Conradi, 2002; Gallion, 2002). In each case, the continuous components are "sinc-squared" functions. In NRZ the discrete component is a solitary frequency tone at the carrier frequency, f_c, while in RZ there is a tone at f_c plus an infinite number both above and below f_c, which is weighted according to a sinc-squared distribution. The normalized spectra are illustrated in Figure 6.29, which follows Gallion (2002) in showing the continuous and discrete components separately.

In both NRZ and RZ signaling the first 0s of the continuous components of the spectra are at $f_c \pm (B/\xi)$, where B is the bitrate of the channel. The bandwidth is commonly defined as being the distance between the first two 0s and so it is $(2B/\xi)$ in general and $2B$ in the case of NRZ modulation. Given that ξ is less than 1 in RZ operation, the channels have broader bandwidths.

As stated in the section on single-channel transmission at the beginning of the chapter, the pulses in both NRZ and RZ modulation tend to differ from perfect rectangles. There are two main reasons for this. First, it is normal to apply filtering to remove the spectral components outside the range $f_c \pm (B/\xi)$, and when this is done the resulting pulses have nonzero rise and fall times. Second, the practical modulators used in optical communications systems can have nonlinear responses to their drive currents and so impose some form of pulse shaping. The elimination of the spectral components beyond the first 0s allows closer channel spacing in WDM transmission without interchannel interference.

NRZ and RZ Transmission: Relative Merits

NRZ channels are normally preferred for low- and intermediate-datarate transmission because they are easily generated by both direct and external modulation. As an example,

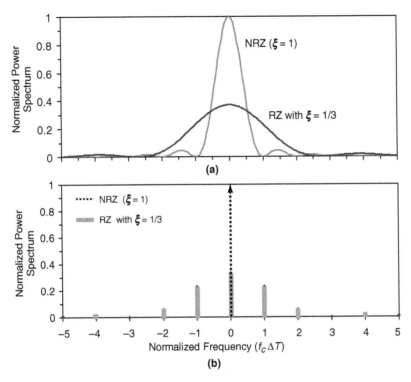

Fig. 6.29 Normalized power spectral densities for NRZ and RZ (ξ = 0.33) modulated signals showing (a) continuous constituents and (b) discrete constituents in which the NRZ and RZ pulses have the same energy. ΔT = bit duration.

NRZ is recommended for use by the various PON standards (Keiser, 2006; ITU-T G.983.1–G.983.5; ITU-T G.984.1–G.984.6; ITU-T G.985). Equation 6.26 shows that group delay due to chromatic-dispersion broadening is directly proportional to bandwidth. Therefore, NRZ pulses experience less dispersive broadening than RZ pulses. The narrower spectral bandwidths of NRZ channels provide better power spectral densities at the detector, and so they experience lower noise penalties than RZ pulses at the same datarate. Furthermore, owing to their relatively narrow spectra, NRZ channels can be placed closer together in a WDM transmission scheme before interference between neighboring channels contributes significantly to BER.

One concern with NRZ operation is if there should be long strings of consecutive 1s or 0s. There is then a loss of information on the digital period and phase, which can lead in turn to a loss of clock synchronization at the receiver. As a result, the bits are not sampled at the appropriate instants. To minimize the problem, bit scramblers and descramblers are used at the transmitter and receiver, respectively, as noted in the earlier discussion of synchronous digital hierarchy.

RZ modulation also provides advantages. The pulses are a fraction of the width of the bitslots and, despite their higher dispersive broadening, they can expand to a greater extent in the time domain before they overlap significantly with adjacent slots. By careful optimization of the duty cycle, the increased flexibility for temporal broadening before unacceptable intersymbol interference can often more than compensate for the pulses' sensitivity to chromatic dispersion. Furthermore, the greater dispersive broadening that the pulses experience in RZ modulation can reduce their peak powers and thus the extent to which they are affected by SPM.

The robustness to nonlinear crosstalk tends to favor RZ over NRZ channels when operating at higher bitrates. High-datarate communications (such as at 160 Gbit/s and above) are often achieved by the technique of "optical time division multiplexing," in which (normally) four trains of short pulses are interleaved in the time domain (Urquhart, 2004; Ludwig et al., 2002). The interleaving process is more tolerant of time interval errors when the pulses are shorter than the bitslots and so RZ modulation is used.

We have seen that both NRZ and RZ offer advantages. On the whole, RZ is preferred above 10 Gbit/s, while NRZ tends to be preferable below that value. Both formats are used in 10-Gbit/s systems, where the arguments are more finely balanced. The decision on which to use can often depend on transmission span and the number and density of the WDM channels.

Alternative Modulation Formats

Other modulation formats have been extensively explored in laboratories (Ho, 2001; Kobayashi et al., 2000); the two main reasons for this are to increase the spectral efficiencies in DWDM systems and to reduce nonlinear crosstalk. Operation on the more closely spaced channels, shown in Figure 6.4, at high bitrates confers significant advantages if it can be done without penalties due to nonlinear effects.

One clear benefit of very dense channel spacing is to make the best use of the C-band, where the fiber loss is lowest. The higher losses in the L-band can be detrimental in very long systems. Moreover, it is normally cheaper to employ 80 channels at 50-GHz spacing within the gain bandwidth of one C-band EDFA than to use a C-band EDFA for one group of 40 channels and an L-band device for another. However, this argument must be stated with caution because the modulators for alternative modulation formats are more complicated and the lasers and filters must operate to more exacting standards when the channel spacing is low, all of which increase costs.

There has been much research on multilevel coding, in which the objective has been to define more than two logic levels and therefore achieve narrower spectral bandwidths. Figure 6.29 shows that, if we take the difference between the first two 0s as our definition, the bandwidth of an NRZ channel is $2f_c\Delta T$. However, it follows from the Fourier transform that the value can be reduced to $f_c\Delta T$ if we use four logic levels. Although we can achieve greater channel spectral densities in WDM systems in this way, the approach requires more complex electronics at both the transmitter and the receiver. Furthermore, the required channel powers are high, which can worsen nonlinear crosstalk. A potentially useful compromise is a format called "duobinary transmission," in which there are three logic levels and a bandwidth of $f_c\Delta T$. In this way, it is possible to extend transmission distances at 10 Gbit/s beyond the values shown in Figure 6.19 (Gu et al., 1994; Walklin & Conradi, 1999).

Another approach to reducing the bandwidth is to recognize that the spectra shown in Figure 6.29 have two symmetrical sets of sidebands and therefore carry twice the information necessary. It is possible to use very sharp filtering to eliminate one of the sides of the spectra and therefore reduce the overall bandwidth. The technique is called "Single-Sideband" (SSB) modulation (Kobayashi et al, 2000). When the suppression is partial, we refer to "Vestigial-Sideband" (VSB) modulation (Bigo et al., 2000).

NRZ signaling and RZ signaling operate using intensity modulation only. The phase attributes of optical waves can also be used. We can achieve phase modulation by using more advanced modulators, which can be dual-drive Mach-Zehnder based, or by using the devices in tandem. One possibility is RZ pulses with a reduced-amplitude carrier wave (Zhu & Hadjifotiou, 2004; Zhu et al., 2002). This is called "Carrier-Suppressed Return to Zero" (CSRZ) modulation, and it suffers less from nonlinear crosstalk than does straightforward RZ transmission. Another technique is to transmit pulses in all of the bitslots but to change their phases between 0 and π, according to whether a 1 or a 0 is being transmitted (Ho, 2001; Zhu et al., 2002). This technique, known as "Differential Phase-Shift Keying" (DPSK), provides increased receiver sensitivies and a greater immunity to nonlinear crosstalk.

Forward Error Correction

The need for error-correcting codes was discussed in Chapter 4. This section considers, in particular, how forward error correction is applied in optical transmission systems.

Error-Correcting Codes

As has been argued, optical transmission is subject to numerous factors that combine to worsen the measured BER. With good system and component design it is possible to ameliorate many of the greatest problems by using the techniques described in the foregoing sections, along with the design principles from the last section of the chapter. However, high-capacity network operation is increasingly being required with BER values of 10^{-12} to 10^{-15} (ITU-T G.691), which is demanding. Even with the aid of all of the physical methodologies described in this chapter, we ultimately reach system capacities and lengths that cannot be exceeded by obvious physical means.

In this section we briefly describe a method, called Forward Error Correction (FEC), in which the signal is encoded to enable bit errors to be corrected by electronic circuits after the receiver (Ait Sab, 2001). Our description of error-correcting codes in this section is introductory. The principles have been known for many years because of their application in nonoptical transmission technologies. The original innovations were often for radio and microwave communications because they suffer numerous adverse influences on signal quality. Whatever physical transport medium is used, the objective is for it to transmit each channel at a higher bitrate than needed for the data that it carries. This is because an additional number of error control bits are added as an overhead to the data, according to a predetermined mathematical algorithm (Kumar et al., 2002). After the receiver we have an electronic bitstream consisting of 1s and 0s, some of which have been badly corrupted during transmission and therefore erroneously assigned at the decision circuit.

Dedicated electronic processors (in the form of ASICs) are located after the receiver circuitry. They perform mathematical operations on the data and the overhead bits and so correct for many of these errors. There are numerous types of FEC that are different in various respects, one of which is how many errors they are able to correct. In a WDM system, FEC is implemented on each channel individually. The sections on chromatic dispersion mitigation and polarization mode dispersion described how electronic compensation for chromatic or polarization mode dispersion can be installed at the receiver end of the system. If this is done, the FEC processors are placed *after* the EDC and the decision circuit, as shown in Figure 6.30.

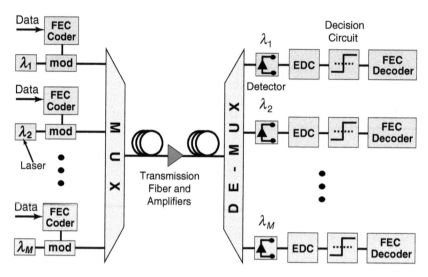

Fig. 6.30 Placement of FEC circuits in a WDM system. EDC = electronic dispersion compensation (if used at the receiver end); mod = electro-optic modulator; MUX and DE-MUX = wavelength multiplexer and de-multiplexer, respectively.

Useful Definitions

Some definitions are useful to know in order to understand the literature on FEC. Let us assume that in a given time interval exactly m bits are transmitted and of these k are information-bearing bits. Consequently, the number of "redundant" bits, which are the ones used for error correction coding, is $(m - k)$. Our first definition is the "rate" of the error correction code being used: $R_{FEC} = k/m$, which is the ratio of the number of information-bearing bits to the total number transmitted. Another defined quantity is the "overhead" of a code that has a rate R_{FEC}: $w = (1 - R_{FEC})/R_{FEC}$. In terms of our integers m and k, the overhead can be re-expressed as $w = (m - k)/k$. Coding overheads are often written as percentages, with typical values in the range of 5 to 15 percent. A third definition is the "redundancy" of the code: $\rho = (1 - R_{FEC}) = w/(1 + w) = (m - k)/m$. Clearly, all of these quantities are related and so are equivalent ways of specifying some of the properties of a code.

The majority of error-correcting codes currently used in optical communications are most effective when the dominant cause of bit errors is noise. This can be a combination of thermal and shot noise, as well as the important contributions from the amplifiers in the system. We saw in the section on bit error rate that when we can ignore the nonlinear effects in the fiber, the BER of a noisy channel can be improved by increasing the incident power on the receiver. An equivalent way of saying the same thing is that we must raise the optical energy of the 1s in the bitslots. The effect of an error-correcting code is rather similar. In the absence of nonlinear crosstalk, we can improve the bit error rate from a high value, BER_{high}, to a low value, BER_{low}, either by increasing the average energy of the pulses incident on the receiver or by applying an error-correcting code.

We are now able to explain another concept that appears in the FEC literature, known as "coding gain" (sometimes referred to as "net effective coding gain" in optical communications papers). We assume OOK modulation, as defined in the early section on optical binary digits, and direct detection, as used in our discussion of bit error rate. The coding gain tells us by how much we have to increase the energy of the pulses incident on the detector in order to achieve the same bit error rate improvement that results from applying the code. Therefore, it is calculated as a ratio of pulse energies and is normally expressed in decibels.

The coding gain can also be expressed in terms of the received Q value, which is specified in equation 6.21 and related to the BER in equation 6.22. If we define $Q_{uncoded}$ as the Q value before application of FEC, and Q_{coded} as the value afterwards, the coding gain is given in decibels as follows (Gallion, 2002; Kumar et al., 2002):

$$\eta_{coding} = 10\log_{10}[R_{FEC} Q_{coded}^2 / Q_{uncoded}^2]. \qquad (6.39)$$

The achievable gain is dependent on the details of the particular FEC that is used. It is difficult to generalize but typical values lie in the range of 2 to 8 dB.

FEC Types

A key challenge for specialists in FEC is to achieve the highest coding gain with the lowest overhead. They have two choices of error-correcting codes: block and convolutional. Block codes are more suitable for use in high-bitrate optical communications systems that require postcoding BER values below 10^{-12} because their overheads are lower. In a block code $(m - k)$ bytes (in which 1 byte = 8 bits) are added to k information-bearing bytes; m is the "codeword length."

The most widely used class of FEC in optical communications is the type of block code called known as "Reed-Solomon" (RS). This is what the FEC literature terms a "nonbinary" subclass of "Bose-Chaudhuri-Hocquenghem" (BCH) codes. (The details are not important in the present context, and the reader is referred elsewhere for an explanation [e.g., Wicker & Bhargava, 1994; Guruswami & Sadan, 1999].) RS codes have been very successfully applied in subsea optical communications and are becoming widespread in terrestrial systems. They have also recently been incorporated into the G-PON standard for fiber access networks (ITU-T G.984.6).

The most common RS codes are designated RS[m, k], in which we use our previous definitions to show that the overhead is $(m - k)/k$. Resulting from its adoption in the ITU-T G.975 standard, the most widely used specific code is RS[255, 239], which has a rate of 239/255, an overhead of 6.7 percent, and a redundancy of 16/255. If, as an example, this code is applied to an STM-16 channel, which, according to Table 6.1, is normally at 2.48832 Gbit/s, the datarate in the optical fiber is raised to ~2.66 Gbit/s. RS[255, 239] can correct up to 8 bit errors from every 255. If the system designer's aim is to achieve a BER of 10^{-13} *after* the application of RS[255, 239], the coding gain is 5.8 dB.

So far this discussion might seem rather abstract. Therefore, we provide a concrete illustration. Let us assume that the system in Figure 6.30 operates with STM-16 channels and is subject to bad amplifier noise. We require the BER after error correction to be no worse than 10^{-13}, and we employ an RS[255, 239] code to achieve this. We can use equation 6.39 to specify the coding gain, together with equation 6.21 to convert between Q value and BER. By substituting $R_{FEC} = 239/255$ and $\eta_{coding} = 5.8$ dB, the required performance can be obtained at the system's output on the far right of Figure 6.30, provided that the BER of the signals incident on the detectors is no worse than 1.4×10^{-4}. Given that operation is at ~2.5 Gbit/s per channel, the mean time between errors without FEC is only 3 μs. However, it is extended to 1 hour and 7 minutes after FEC is applied.

The RS[255, 239] code has served optical communications well, but higher-datarate and longer-span systems, such as national and subsea operations at 10 Gbit/s and above, are very demanding. Systems vendors and operators are therefore exploring the use of more powerful FEC. The most common approach uses "concatenated codes," where two or more shorter codes together provide superior coding gains. If this is done, the overhead can be in the range of 10 to 20 percent.

Another development is the application in optical transmission of a class of FEC known as "turbo codes," which perform very efficiently in the wireless domain and are more recently being explored in optical fiber systems (Berrou, Glavieux, & Thitimajshima, 1993; Heegard & Wicker, 1998).

System Design and Implementation

This section explains the many concepts that system engineers must take into account in whole-system design. It uses all of the techniques presented up to this point.

Design Methodologies

The design of a whole system is multifaceted. It demands knowledge of all transmission impairments (loss, amplifier noise, chromatic and polarization mode dispersion, nonlinear crosstalk, etc.), subsystem and component technologies (transmitters, amplifiers, optical cross-connects, add-drop multiplexers, filters, isolators, passive splitters, detectors, etc.), modulation formats, and FEC coding.

We now have sophisticated software with advanced visualization tools to carry out the modeling and allow developers to understand how design modifications influence BER. System and network suppliers normally perform as much simulation as possible with this software to constrain their need for expensive laboratory testing, especially when developing very-high-capacity systems. In all stages the work has to be undertaken as a major coordinated effort. It is outside the scope of the chapter to describe the planning, modeling, testing, and implementation processes. Instead, this section provides insight into some of the design compromises that must be made.

All of the nonlinear effects that were discussed in the section Nonlinear Optical Transmission cause crosstalk and thereby increase BER. We have seen that the problem becomes greater with high-launched powers. When operating at low datarates we can use relatively low-power transmitters and therefore often avoid the difficulties altogether. However, as we increase the datarate we need ever greater incident powers on the detector to retain the spectral power densities necessary for good BER. In contrast with other influences on BER, such as chromatic dispersion, PMD, amplifier noise, and finite transmitter extinction ratios, it is not possible to improve nonlinear crosstalk merely by increasing the mean received optical power. Such an action would make matters worse. For these reasons, system optimization for high-capacity WDM operation can impose very tight design margins.

Fiber Effective Area

One measure that we can use to lower nonlinear crosstalk relates to the effective area in the denominator of the nonlinear term in equation 6.35. Fiber designs with a large value of A_{eff} are therefore desirable. For the G.652 fiber, to which we referred in previous sections, A_{eff} is around 80 to 85 μm^2 in the C-band, which is rather higher than many other categories, such as DSF and NZ-DSF.

Other specialized types of transmission fiber exist with effective areas in excess of $150\,\mu m^2$—notably, the large effective area designs, which often have a pure silica core. However, they can be expensive and sometimes have relatively high bending loss, which tends to limit them to niche applications, such as repeaterless subsea communications (ITU-T G.654). For these reasons, operators commonly consider the G.652 fiber to be a good multi-application compromise for all terrestrial needs.

As explained in the earlier section on chromatic dispersion mitigation, DCF for use within modules often has a very small effective core area, with values in the 15 to 25 μm^2 range being typical. Consequently, it can contribute significantly to the system's accumulated nonlinear crosstalk. Manufacturers devote much effort to novel DCF waveguide designs to achieve the highest values of A_{eff} and $|D_{\text{DCF}}|$ simultaneously, with the aim of minimizing the required spans. Where it is possible to do so, it is best to minimize the use of DCF through the deployment of other compensating techniques, such as chirped FBGs and (especially) electronic methods.

Nonlinear Crosstalk in Amplifier Chains

We have seen that SPM, XPM, and FWM all depend on the effective length of the transmission fiber, as given by equation 6.36. However, equation 6.35 applies when one or more signals are launched into a single span of length L. In the section on the Friis cascade formula (and with reference to Figure 6.10) we addressed the question of whether the noise figure of a point-to-point system is better with a large number of low-gain amplifiers or a small number of high-gain amplifiers. By using the effective length, we can do the same for accumulated nonlinear crosstalk. We assume that there are no interactions between the amplifier noise and the nonlinear effects in the fiber. (In particular, we ignore a process called "modulation instability" [Agrawal, 2001]). Equation 6.36 can then be adapted for a chain of N amplified fiber spans, in which the objective is to obtain net end-to-end power transparency, as shown in Figure 6.10. L_{eff} becomes

$$L_{\text{eff}} = \frac{N}{\alpha}[1 - \exp(-\alpha L/N)] \qquad (6.40)$$

Equation 6.40 assumes for simplicity that all of the system's fiber has the same loss coefficient, α. It is plotted in Figure 6.31, in which we use the same values of L and α as in

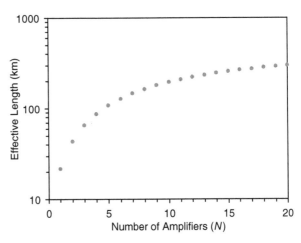

Fig. 6.31 Effective length for nonlinear optical effects as a function of the number of amplified segments. System physical length = 500 km; fiber loss = 0.2 dB/km. Each point represents a system that conforms to the requirements illustrated in Figure 6.10.

Figure 6.11. We see that L_{eff} (and hence the crosstalk) increases with the number of amplified spans. The straightforward reason is that the amplifiers maintain the signal power at a relatively high value throughout the system. Therefore, best performance requires a relatively small number of high-gain amplifiers spaced far apart.

The messages from Figures 6.11 and 6.31 are in conflict. The largest optical SNR is provided by using an amplifier strategy that maintains the signal power at a high level. In contrast, the lowest nonlinear crosstalk is achieved by allowing the signal power to fall to low values. The compromise that system engineers reach is influenced by other factors, such as the logistics of amplifier placement and their associated electrical powering; the desire to make the periodicities of the amplifiers and the dispersion maps coincide; and, crucially, the cost of using many amplifiers. When using discrete amplifiers, such as EDFAs, the optimized solutions tend to favor a 60- to 80-km spacing for terrestrial systems and 40 to 50 km between submerged transoceanic repeaters.

Benefits of Distributed Raman Amplification

As argued earlier, the use of Raman amplification is very effective in providing a compromise between the opposing requirements of maintaining low overall noise figures and minimizing nonlinear crosstalk. Distributing the gain prevents the signal powers from descending to very low values, and so we have the flexibility to launch lower channel powers and/or to extend the length of the amplifier stages.

A simple simulated example that illustrates the longitudinal power variation in a 75-km link is shown in Figure 6.32. It is for a backward-pumped small-signal fiber Raman amplifier based on G.652 fiber. Three pump powers are selected to provide net gains of −3 dB, 0 dB (i.e., the transparency condition), and +3 dB; they are compared with an unpumped fiber. Operators sometimes prefer to have a small net gain (e.g., +3 dB, as shown in Figure 6.32) to compensate for any additional components such as isolators and filters. Various experimental studies have demonstrated good performance by placing the Raman pump lasers up to 100 km apart, even in very-long-haul communications (Ritwitt, 2005; Islam, DeWilde, & Kuditcher, 2004).

Unfortunately, distributed FRAs cause problems that result from their low pumping efficiencies (in terms of dB gain per km of fiber). The necessary total launched powers can be 1 to 3 watts in some cases, which raises concern regarding harm to the eye and skin

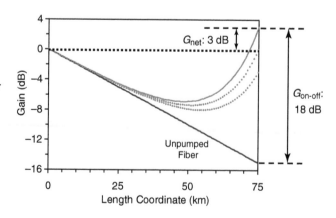

Fig. 6.32 Longitudinal gain dependence of a small-signal fiber Raman amplifier. Pump powers (contradirectional): 823 mW (*top*), 686 mW (*center*), and 548 mW (*bottom*). Raman gain coefficient is $\Gamma = 0.35 \text{W}^{-1} \text{ km}^{-1}$; pump and signal wavelength losses are 0.3 and 0.2 dB/km, respectively; $G_{on\text{-}off}$ and G_{net} are defined in equation 6.10.

should a fiber end be accidentally exposed. An increasing worry is fiber damage through what has come to be known as the "fiber fuse effect" (Sikora, 2006), in which many kilometers of transmission fiber could be rendered irreparable. Given these issues, there is a need for rapid automatic laser shutdown in the event of unforeseen fiber damage.

Although the relatively large effective area of G.652 fiber is beneficial in lowering the potential for nonlinear crosstalk, by equation 6.35 it also reduces Raman amplifier pump efficiency. (The high pump powers reported in Figure 6.32 are evidence for this.) We could resort to other kinds of transmission fibers, such as NZ-DSF, which have sufficient chromatic dispersion to frustrate XPM and FWM, together with values of A_{eff} in the 50- to 70-μm^2 range. Unfortunately, such fibers often have their λ_0 in the S-band. When we use multi-wavelength Raman pumping to obtain spectral equalization, it is possible to obtain four-wave mixing between the *pump* wavelengths (Neuhauser et al., 2001). To overcome this problem, the ITU-T has proposed a relatively new fiber standard, G.656 (ITU-T G.656), according to which D is nonzero in the C- and L-bands and λ_0 no longer lies in the range of wavelengths used for pumping.

Fiber Dispersion Maps

Normally the single most important measure to frustrate nonlinear crosstalk is the appropriate management of the system's chromatic dispersion, as described in the relevant earlier section. Even though the dispersion causes unwanted pulse broadening, it can be beneficial because it is very effective in reducing FWM and XPM via the pulse walk-off effects explained previously. Most operators are therefore reluctant to operate intermediate- and long-haul WDM systems where $|D| \approx 0$. Fortunately, G.652 fiber is very good from this point of view because, as shown in Figure 6.18, it has values of D between +15 and +22 ps/(km·nm) over the C- and L-bands. The dispersion management schemes shown in Figures 6.21 and 6.22 cause successive pulse expansion and contraction, but the PDF and NDF both have nonzero values of $|D|$ and so multichannel nonlinear effects can often be well suppressed.

Many studies have been performed on optimal dispersion compensation strategies in the presence of nonlinear transmission. The dispersion map chosen depends on how the required total capacity is to be provided (e.g., $4M$ channels at 10 Gbit/s or M channels at 40 Gbit/s). Different strategies apply for point-to-point systems and networks that include optical add-drop multiplexers. Logistical factors also influence the choice of dispersion map, such as whether existing fiber is to be reused for higher-capacity operation (for financial reasons) or if new fiber can be justified. Moreover, different maps are optimal for subsea and terrestrial applications.

We cannot summarize all of the options for chromatic dispersion compensation but we do note one feature. The Nonlinear Optical Transmission section described how SPM in the presence of a positive dispersion fiber can lead to pulse compression. We can take advantage of this fact to mitigate BER deterioration by what is known as "optimal under-compensation" (Bayvel & Killey, 2002). In terms of equation 6.32, we have, for standard single-mode fiber,

$$|D_{\text{G.652}}| L_{\text{G.652}} = (1/x) |D_{\text{DCF}}| L_{\text{DCF}} \qquad (6.41)$$

where x is a factor with a typical value of ~0.9, which is found from simulations using the NLSE. An example of an optimally undercompensated dispersion map for use in the presence of nonnegligible SPM is shown in Figure 6.33.

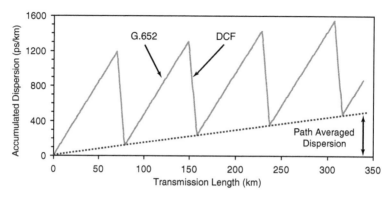

Fig. 6.33 Dispersion map illustrating optimally undercompensated dispersion in which PDF is +17 ps/(km·nm) and NDF is −119 ps/(km·nm). The spans are 70 km (PDF) and 9 km (NDF).

Operation with High Channel Densities

We saw in the earlier section on wavelength division multiplexing that one of the main methods of augmenting system capacities is via channel separations of 50 GHz or less. However, crosstalk due to FWM and XPM is greatest with closely spaced channels. The problem can be reduced by launching the channels with alternating orthogonal states of polarization. If this is done and polarization is preserved throughout the system, the FWM is eliminated and the XPM is reduced by two-thirds. Unfortunately, PMD in the fiber and components causes progressive randomization and so it lowers the effectiveness of the technique. It is also possible to alternate the polarization states between pulses in the same channel with the aim of ameliorating intrachannel nonlinear effects (Bayvel & Killey, 2002).

One of the most promising means of enabling the use of closely spaced channels is via alternative low-spectral-bandwidth modulation formats (Matera et al., 1999). Examples are single-sideband generation, multilevel amplitude shift keying, and duobinary transmission. Furthermore, operation with carrier-suppressed return-to-zero modulation tends to suffer from less nonlinear crosstalk than do many other formats (Zhu & Hadjifotiou, 2004; Zhu et al., 2002). The choice of modulation format is complicated by the interactions between chromatic dispersion, nonlinear effects, and the phases and power spectral densities of the pulses. It is outside the scope of this chapter, but the reader is referred to Ho (2001) for further details.

Operation Close to the Zero-Dispersion Wavelength

Working at or near λ_0, the wavelength of zero dispersion, is sometimes unavoidable. Two such circumstances are in short-reach applications (such as in PONs or some of the smaller metropolitan networks) and when legacy dispersion-shifted fiber is installed. Operators then have to take great care to avoid nonlinear crosstalk. We briefly consider the issues here.

G-PON fiber access network standards recommend that upstream traffic (from the customer to the local exchange) be in the 1260- to 1360-nm range and, as shown in Figure

6.18, it can lie close to λ_0 for G.652 fiber (Keiser, 2006; ITU-T, G.652). The maximum upstream datarate for which full specifications are published at the time of writing is 1.25 Gbit/s (ITU-T, G.984.1–G.984.6). The maximum recommended physical reach is 20 km, the transmitted powers are less than +7 dBm, and the downstream channels are at ~1490 and ~1550 nm on the CWDM grid. All of these values have been selected to ensure that nonlinear crosstalk is not an issue, even though the upstream wavelength can be close to λ_0 in some circumstances.

The earlier discussion of CWDM in the section on WDM, especially with regard to Figure 6.2, raises the question of whether ~20-nm interchannel spacing is sufficient to prevent FWM when at least one of the channels is near λ_0. Okuno et al. (2002) show that for fiber spans of 25 km and launched powers of +4 dBm per channel at 10 Gbit/s, this is not the case, even when two of the laser wavelengths have drifted with temperature so that the most critical channels are 17 nm apart. Therefore, CWDM can be a viable solution across the whole of the fiber's available spectrum for applications such as small-circumference metropolitan rings.

The operation of longer-span DWDM systems in the vicinity of λ_0 is potentially more problematic, and it is an issue for operators in countries (such as Japan and Brazil) where there is a large amount of installed DSF. The usual approach is to use the L-band as much as possible. However, the fiber losses are higher beyond 1565 nm and L-band EDFAs normally require longer spans of doped fiber and higher pump powers than their C-band counterparts, which can make them more expensive.

For the reasons just given, it can be beneficial to place at least a few channels in the C-band. FWM-based crosstalk can be ameliorated by using unequally spaced channels in order to frustrate the necessary phase matching for the effect (Agrawal, 2001). With the aid of optimization algorithms based on the NLSE, it is possible to predict a set of channel frequencies that can minimize the problem. This is why, as explained in the earlier section on dense wavelength division multiplexing, the DWDM grid lines have been specified down to 12.5-GHz spacing. The lasers are selected to coincide with the grid lines that lie closest to the required frequencies.

FEC: Method of Last Resort or Design Requirement?

We have seen that many techniques can be used to improve the physical performance of optical networks and so provide low measured BERs. However, some of the required means are costly, and operators' needs for greater spans and capacities or for wavelength routing can often be so demanding that physical techniques are no longer viable. In particular, a previous subsection explained that PMD is inherently probabilistic and that improvements to fiber fabrication and component quality, or the use of compensation techniques, cannot *guarantee* to prevent outages. In that case, network designers have turned to FEC as a method of last resort.

The use of good error-correcting codes can be our only hope for service continuity. It is as if algebra comes to the rescue of physics. Furthermore, it is increasingly recognized that FEC, which merely depends on electronic chip sets to implement mathematical algorithms, can be very cost effective. FEC is being implemented in network categories (such as long-reach PONs [Lin, 2006; Davey, 2006; Gilfedder, 2006]) for which it was not previously considered appropriate. To this end, its use is being incorporated into the initial design stage as a way to reduce the dependence on expensive optoelectronic hardware.

Conclusion

This chapter explained the main transmission techniques that are used in optical fiber systems. Its objective was to present engineering principles rather than to review recent results. The physical measure that is of greatest concern to network operators is the bit error rate (BER), which is defined as the ratio of errored bits at the receiver to the total number of bits transmitted. We showed that a wide range of effects, such as chromatic and polarization mode dispersion, nonlinear optical phenomena, amplifier noise and gain characteristics, and the nature of the transmitted pulses influence BER.

    System design normally requires compromises to satisfy many conflicting requirements. These become ever greater as the number of channels and the datarate per channel increase. Fortunately, there are numerous techniques to improve performance. They include optimization of amplifier placement; use of distributed Raman amplification; management and compensation of chromatic dispersion, using both optical and electronic methods; application of optimal modulation formats; reduction and compensation of polarization mode dispersion; appropriate choice of transmission fiber to minimize nonlinear effects; and use of advanced FEC codes. Clearly, the physical design of optical fiber systems and networks is a wide-ranging task that requires many skills.

Acknowledgments   The writing of this chapter was supported in part by the Spanish CICyT (Project TEC2004-05936-C02). I am grateful to Nuria Miguel Zamora and Oscar García López for Figures 6.8 and 6.9, respectively.

Resources

Afonso, A., et al. (2006) Dynamic Gain and Channel Equalizers or Wavelength Blocker Using Free Space: Common Elements and Differences, *IEEE J. Lightw. Technol.* 24(3):1534–1542.

Agrawal, G. P. (2002) *Fiber Optic Communication Systems*, Third Edition, Wiley.

Agrawal, G. P. (2001) *Nonlinear Fiber Optics*, Third Edition, Academic Press.

Ait Sab, O. (2001) FEC Techniques in Submarine Transmission Systems, *Proceedings Conference on Optical Fiber Communications (OFC)*, paper TuF.1.

Aozasa, S. (2007) Highly Efficient S-Band Thulium-Doped Fiber Amplifier Employing High Thulium-Concentration Doping Technique, *IEEE J. Lightw. Technol.* 25(8):2108–2114.

Barge, M., Battarel, D., and De Bougrenet de la Tocnaye, J. L. (2005) A Polymer-Dispersed Liquid Crystal Based Dynamic Gain Equalizer, *IEEE J. Lightw. Technol.* 23(8):2531–2541.

Bayvel, P., and Killey, R. (2002) Nonlinear Optical Effects in WDM Transmission, in Kaminow, I., and Li, T. (eds.), *Optical Fiber Telecommunications IVB: Systems and Impairments*, Academic Press.

Becker, P. C., Olsson, N. A., and Simpson, J. R. (1999) *Erbium Doped Fiber Amplifiers: Fundamentals and Technology*, Academic Press.

Beck Mason, T. G. (2002) Electroabsorption Modulators, in Dutta, A. K., Dutta, N. K., and Fujiwara, M. (eds.), *WDM Technologies: Active Optical Components*, Academic Press.

Belanger, M. P., and Cavallari, M. (2006) Network Cost Impact of Solutions for Mitigating Optical Impairments: Comparison of Methods, Techniques and Practical Deployment Constraints, *IEEE J. Lightw. Technol.* 24(1):150–157.

Bergano, N. S. (1997) Undersea Amplified Lightwave Systems Design, in Kaminow, I. P., and Koch, T. L. (eds.), *Optical Fiber Telecommunications IIIA*, Academic Press.

Bergano, N. S., Kerfoot, F. W., and Davidson, C. R. (1993) Margin Measurements in Optical Amplifier Systems, *IEEE Photon. Technol. Lett.* 5(3):304–306.

Berrou, C., Glavieux, A., and Thitimajshima, P. (1993) Near Shannon Limit Error Correcting Coding: Turbo Codes, *Proceedings International Conference on Communications*, 1064–1070.

Bickham, S. R., and Cain, M. B. (2002) Submarine Fiber, in Chesnoy J. (ed.), *Undersea Fiber Communication Systems*, Academic Press.

Bigo, S., et al. (2000) 5.12 Tbit/s (128 × 40 Gbit/s) WDM Transmission over 3 × 100 km of Teralight™ Fibre, Paper PD1.2, *Proceedings European Conference on Optical Communications (ECOC)*.

Bracewell, R. N. (1986) *The Fourier Transform and Its Applications*, Second Edition, McGraw-Hill.

Bristiel, B. (2006) *Bruit dans les Amplificateurs Raman*, Doctoral thesis, Ecole Nationale Supérieure des Télécommunications, Paris (see especially Section 2.3).

Brittain, J. E. (1995) Scanning the Past: Harald T. Friis, *Proceedings IEEE* 83(12):1674.

Bromage, J., Winzer, P. J., and Essiambre, R.-J. (2004) Multiple Path Interference and its Impact on System Design, in Islam M. N. (ed.), *Raman Amplifiers for Telecommunications 2: Sub-Systems and Systems*, Springer.

Bülow, H., and Lanne, S. (2005) PMD Compensation Techniques, in Galtarossa, A., and Menyuk C. R. (eds.), *Polarization Mode Dispersion*, Springer.

Butcher, P. N., and Cotter, D. (1990) *The Elements of Nonlinear Optics*, Cambridge University Press.

Chan, B., and Conradi, J. (1997) On the Non-Gaussian Noise in Erbium Doped Fiber Amplifiers, *IEEE J. Lightw. Technol.* 15(4):680–687.

Chandrasekhar, S., et al. (2006) Chirp-Managed Laser and MLSE-Rx Enables Transmission over 1200 km at 1550 nm in a DWDM Environment in NZ-DSF at 10 Gbit/s Without Any Optical Dispersion Compensation, *IEEE Photon. Technol. Lett.* 18(14):1560–1562.

Chesnoy, J. (ed.) (2002) *Undersea Fiber Communication Systems*, Academic Press.

Conradi, J. (2002) Bandwidth Efficient Modulation Formats for Digital Fiber Transmission Systems, in Kaminow, I. P., and Li, T. (eds.), *Optical Fiber Telecommunications IVB: Systems and Impairments*, Academic Press.

Damask, J. N. (2005) *Polarization Optics in Telecommunications*, Springer.

Davey, R. P., et al. (2006) Designing Long Reach Optical Access Networks, *BT Tech. J.* 24(2):13–19.

Desurvire, E., et al. (2002) *Erbium Doped Fiber Amplifiers: Device and System Developments*, Wiley.

Desurvire, E. (1999) Comments on "The Noise Figure of Optical Amplifiers," *IEEE Photon. Technol. Lett.* 11(5):620–621.

Desurvire, E. (1994) *Erbium Doped Fiber Amplifiers: Principles and Applications*, Wiley.

Downie, J. D., Sauer, M., and Hurley, J. (2006) Flexible 10.7 Gbit/s DWDM Transmission over up to 1200 km without Optical In-Line or Post Compensation of Dispersion Using MLSE-EDC, *Proceedings Conf. OFC*, paper JThB5.

Dutta, N. (2003) Semiconductor Optical Amplifiers (SOA), in Dutta, A. K., Dutta, N. K., and Fujiwara, M. (eds.), *WDM Technologies: Passive Optical Components, Volume III*, Academic Press.

Ellis, A. D., et al. (1997) Dispersion Compensation, Reconfigurable Optical Add-Drop Multiplexer Using Chirped Fibre Bragg Gratings, *Electron. Lett.* 33(17):1474–1475.

Ellis, A. D., Smith, K., and Patrick, D. M. (1993) All-Optical Clock Recovery at Bit Rates up to 40 Gbit/s, *Electron. Lett.* 29(15):1323–1324.

Ennser, K., Laming, R. I., and Zervas, M. N. (1998) Analysis of 40 Gbit/s TDM Transmission over Embedded Standard Fiber Employing Chirped Fiber Grating Dispersion Compensators, *IEEE J. Lightw. Technol.* 16(5):807–811.

Essiambre, R-J., Mikkelsen, B., and Raybon, G. (1999) Intra-Channel Cross Phase Modulation and Four Wave Mixing in High Speed TDM Systems, *Electron. Lett.* 35(18):1576–1578.

Fibre Channel Industry Association. *http://www.fibrechannel.org/*.

Gallion, P. (2002) Basics of Digital Optical Communications, in Chesnoy, J. (ed.), *Undersea Fiber Communication Systems*, Academic Press.

Galtarossa, A., et al. (2005) Low-PMD Spun Fibres, in Galtarossa, A., and Menyuk, C. R. (eds.), *Polarization Mode Dispersion*, Springer.

García López, O. (2008) *Estudio de la Utilización de EDFA Bombeados Remotamente en Redes de Sensores con Protección*, Report of final year engineering diploma project, Departamento de Ingeniería Eléctrica y Electrónica, Universidad Pública de Navarra, Spain.

Gautschi, W. (1972) Error Function and Fresnel Intervals, in Abramowitz, M., and Stegan, I. A., *Handbook of Mathematical Functions*, Ninth Edition, Dover Publications.

Gilfedder, T. (2006) Deploying G-PON Technology for Backhaul Applications, *BT Technol. J.* 24(2):20–25.

Gu, X., et al. (1994) 10 Gbit/s 138 km Uncompressed Duobinary Transmission over Installed Standard Fibre, *Electron. Lett.* 30(23):1953–1954.

Guruswami, V., and Sadan, M. (1999) Improved Decoding of Reed-Solomon and Algebraic-Geometry Codes, *IEEE Trans. Inf. Theory* 45(6):1757–1767.

Gutiérrez-Castrejón, R., and Filios, A. (2006) Pattern-Effect Reduction Cross-Gain Modulated Holding Beam in Semiconductor Optical In-Line Amplifier, *IEEE J. Lightw. Technol.* 24(12):4912–4917.

Halabi, S. (2003) *Metro Ethernet*, Cisco Press.

Haus, H. A. (2000) Noise Figure Definition Valid from RF to Optical Frequencies, *IEEE J. Sel. Topics Quantum Electron.* 6(2):240–247.

Haus, H. A. (1998) The Noise Figure of Optical Amplifiers, *IEEE Photon. Technol. Lett.* 10(11):1602–1604.

Haus, H. A., and Phua, P. B. (2005) Three Representations of Polarization Mode Dispersion, in Galtarossa, A., and Menyuk C. R. (eds.), *Polarization Mode Dispersion*, Springer.

Hecht, J. (1999) *City of Light: The Story of Fiber Optics*, Oxford University Press.

Heegard, C., and Wicker, S. (1998) *Turbo Coding*, Klewer Academic Publishers.

Ho, K-P. (2001) *Phase-Modulated Optical Communication Systems*, Springer.

Horne, J. (1989) Marine and Maintenance (From Inception to the Grave), in Chesnoy, J. (ed.), *Undersea Fiber Communication Systems*, Academic Press.

Islam, M. N., DeWilde, C., and Kuditcher, A. (2004) Multichannel Raman Amplifiers, in Islam, M. N. (ed.), *Raman Amplifiers for Telecommunications 2: Sub-Systems and Systems*, Springer.

Ito, T., Sekiya, K., and Ono, T. (2002, September) Study of 10 G/40 G Hybrid Ultra Long Haul Transmission Systems with Reconfigurable OADMs for Efficient Wavelength Usage, *Proceedings ECOC*, vol. 1, paper 1.1.4.

ITU-T *Recommendation G.985, 100 Mbit/s Point-to-Point Ethernet-Based Optical Access System.*

ITU-T *Recommendations G.984.1–G.984.6, Gigabit-capable Passive Optical Networks (G-PON) Standards.*

ITU-T *Recommendation G.959.1, Optical Transport Network Physical Layer Interfaces.*

ITU-T *Recommendation G.780, Terms and Definitions for Synchronous Digital Hierarchy (SDH) Networks.*

ITU-T *Recommendation G.707, Network Node Interfaces for the Synchronous Digital Hierarchy (SDH).*

ITU-T *Recommendation G.694.2, Spectral Grids for WDM Applications: CWDM Wavelength Grid.*

ITU-T *Recommendation G.694.1, Spectral Grids for WDM Applications: DWDM Frequency Grid.*

ITU-T *Recommendation G.691, Optical Interfaces for Single-Channel STM-64 and Other SDH Systems with Optical Amplifiers, Appendix I.*

ITU-T *Recommendation G.655, Characteristics of a Non-Zero Dispersion-Shifted Single-Mode Optical Fibre and Cable.*

ITU-T *Recommendation G.654, Characteristics of a Cut-Off Shifted Single-Mode Optical Fibre and Cable.*

ITU-T *Recommendation G.652, Characteristics of a Single-Mode Optical Fibre and Cable.*

ITU-T *Recommendations G.983.1 to G.983.5, Broadband Passive Optical Networks (BPON) Standards.*

Jiang, S., et al. (2007) Bit Error Rate Evaluation of the Distributed Raman Amplified Transmission Systems in the Presence of Double Rayleigh Backscattering Noise, *IEEE Photon. Technol. Lett.* 19(7):468–450.

Jones, G., Nijhof, J., Forysiak, W., and Killey, R. (2005, December) Comparison of Optical and Electronic Dispersion Compensation Strategies, *IEE Seminar on Optical Fibre Communications and Electronic Signal Processing*, ref. 2005–11310.

Kartalopoulos, S. V. (2004) *Optical Bit Error Rate: An Estimation Methodology*, IEEE Press/Wiley.

Kasamatsu, T., Yano, Y., and Ono, T. (2002) 1.49 μm-Band Gain-Shifted Thulium-Doped Amplifier for WDM Transmission Systems, *IEEE J. Lightw. Technol.* 20(16):1826–1838.

Kashyap, R. (1999) *Fiber Bragg Gratings*, Academic Press.

Keiser, G. (2006) *FTTX Concepts and Applications*, Wiley.

Keiser, G. (2002) *Local Area Networks, Second Edition*, McGraw-Hill.

Killey, R. I., et al. (2006) Electronic Dispersion Compensation by Signal Predistortion, Paper OWB3, *Proceedings Conf. OFC.*

Kobayashi, M., and Mikawa, T. (2002) Avalanche Photodiodes, in Dutta, A. K., Dutta, N. K., and Fujiwara, M. (eds.), *WDM Technologies: Active Optical Components*, Academic Press.

Kobayashi, Y., et al. (2000) A Comparison among Pure-RZ, CS-RZ, and SSB-RZ Format in 1 Tbit/s (50×20 Gbit/s), 0.4 nm Spacing WDM Transmission over 4000 km, Paper PD1.2, *Proceedings, European Conference on Optical Communications (ECOC).*

Kudou, T., et al. (2000) Theoretical Basis of Polarization Mode Dispersion Equalization up to the Second Order, *IEEE J. Lightw. Technol.* 18(4):614–617.

Kumar, P. V., et al. (2002) Error Control Coding Techniques and Applications, in Kaminow, I. P., and Li, T. (eds.), *Optical Fiber Telecommunications IVB*, Academic Press.

Lin, C (ed.) (2006) *Broadband Optical Access Networks and Fiber to the Home: Systems Technologies and Development Strategies*, Wiley.

Liu, J. M. (2005) *Photonic Devices*, Cambridge University Press.

Lo, Y. L., and Kuo, C.-P. (2003) Packaging a Fiber Bragg Grating with Metal Coating for an Athermal Design, *IEEE J. Lightw. Technol.* 21(5):1377–1383.

Lo, Y. L., and Kuo, C.-P. (2002) Packaging a Fiber Bragg Grating without Preloading in a Simple Athermal Bimaterial Device, *IEEE Trans. Adv. Packag.* 25(1):50–53.

Ludwig, R., et al. (2002) Enabling Transmission at 160 Gbit/s, Paper TuA1, *Proceedings Conf. OFC.*

Macleod, H. A. (2001) *Thin Film Optical Filters, Third Edition*, Institute of Physics Publishing.

Madabhushi, R. (2002) Lithium Niobate Optical Modulators, in Dutta, A. K., Dutta, N. K., and Fujiwara, M. (eds.), *WDM Technologies: Active Optical Components*, Academic Press.

Masuda, H., Kawai, S., and Suzuki, K-I (1999) Optical SNR Enhanced Amplification in Long Distance Recirculating Loop WDM Transmission Experiment Using 1580 nm Band Amplifier, *Electron. Lett.* 35(5):411–412.

Matera, F. K., et al. (1999) Field Demonstration of 40 Gbit/s Soliton Transmission with Alternate Polarizations, *IEEE J. Lightw. Technol.* 17(11):2225–2234.

McGhan, D. (2006) Electronic Dispersion Compensation, Tutorial Paper OWK1, *Proceedings Conf. OFC.*

McGhan, D., et al. (2006) 5120 km RZ-DPSK Transmission over G.652 Fiber at 10 Gbit/s with no Optical Dispersion Compensation, Paper PDP27, *Proceedings Conf. OFC.*

Méndez, A., and Morse, T. F. (eds.) (2007) *Speciality Optical Fibers Handbook*, Academic Press (See especially Chapters 3 and 5).

Menif, M., Rusch, L. A., and Karásec, M. (2001) Application of Pre-Emphasis to Achieve Flat Output OSNR in Time Varying Channels in Cascaded EDFAs Without Equalization, *IEEE J. Lightw. Technol.* 19(10):1440–1452.

Michie, C., et al. (2007) An Adjustable Gain-Clamped Semiconductor Optical Amplifier (GC-SOA), *IEEE J. Lightw. Technol.* 25(6):1466–1473.

Miguel Zamora, N. (2005) Diseño de Filtros de Capas Finas Dieléctricas para la Ecualización de Ganacia Raman con dos Bombeos, Report of final year engineering diploma project, Departamento de Ingeniería Eléctrica y Electrónica, Universidad Pública de Navarra, Spain.

Morita, I., et al. (2000) 40 Gbit/s Single Channel Transmission Over Standard Fibre Using Distributed Raman Amplification, *Electron. Lett.* 36(25):2084–2085.

Morito, K., et al. (2003) High Output Power Polarization-Insensitive Semiconductor Optical Amplifier, *IEEE J. Lightw. Technol.* 21(1):176–181.

Mynbaef, D. K., and Scheiner, L. L. (2001) *Fiber-Optic Communications Technology*, Prentice Hall.

Namiki, S., Emori, Y., and Oguri, A. (2005) Discrete Raman Amplifiers, in Headley, C., and Agrawal, G. P. (eds.), *Raman Amplification in Fiber Optical Communication Systems*, Academic Press.

Neslon, L. E., and Jopson, R. M. (2005) Introduction to Polarization Mode Dispersion in Optical Systems, in Galtarossa, A., and Menyuk C. R. (eds.), *Polarization Mode Dispersion*, Springer.

Neuhauser, R. E., et al. (2001) Impact of Nonlinear Pump Interactions on Broadband Distributed Raman Amplification, Paper MA4, *Proceedings Conf. OFC.*

Norimatsu, S., and Yamamoto, T. (2001) Waveform Distortion due to Stimulated Raman Scattering in Wideband WDM Transmission Systems, *IEEE J. Lightw. Technol.* 19(2):159–171.

Okuno, T., et al. (2002) Verification of Coarse WDM Transmission in the Zero Dispersion Wavelength Region, Paper 3.2.6, *Proceedings European Conference on Optical Communications (ECOC)*.

Olsson, N. A. (1989) Lightwave Systems with Optical Amplifiers, *IEEE J. Lightw. Technol.* 7(7):1071–1082.

Othonos, A., et al. (2006) Fibre Bragg Gratings, in Venghaus, H. (ed.), *Wavelength Filters for Fibre Optics*, Springer.

Philips, I. D., et al. (1998) Drop and Insert Multiplexing with Simultaneous Clock Recovery Using an Electroabsorption Modulator, *IEEE Photon. Technol. Lett.* 10(2):291–293.

Poole, C. D., and Nagel, J. (1997) Polarization Effects in Lightwave Systems, in Kaminow, I. P., and Koch T. L. (eds.), *Optical Fiber Telecommunications IIIA*, Academic Press.

Ramaswami, R., and Sivarajan, K. N. (2002) *Optical Networks: A Practical Perspective*, Second Edition, Morgan Kaufmann.

Riley, K. F., Hobson, M. P., and Bence, S. J. (2004) *Mathematical Methods for Physics and Engineering: A Comprehensive Guide*, Second Edition, Cambridge University Press (See Chapter 26).

Ritwitt, K. (2005) Distributed Raman Amplifiers, in Headley, C., and Agrawal, G. P. (eds.), *Raman Amplification in Fiber Optical Communication Systems*, Academic Press.

Sachiano, M. (2005) PMD Measurements on Installed Fibers and Polarization Sensitive Components, in Galtarossa, A., and Menyuk C. R. (eds.), *Polarization Mode Dispersion*, Springer.

Sexton, M., and Reid, A. (1997) *Broadband Networking: ATM, SDH and Sonet*, Artech House.

Shake, I., et al. (1998) Influence of Inter-bit Four Wave Mixing in Optical TDM Transmission, *Electron. Lett.* 34(16):1600–1601.

Shepard, S. (2001) *Sonet/SDH Demystified*, McGraw-Hill Education.

Sikora, E., et al. (2006) Mitigating the Risk of High Power Damage to the BT Network, *BT Tech. J.* 24(2):48–56.

Srivastava, A. K., and Sun, Y (2002) Advances in Erbium Doped Fibre Amplifiers, in Kaminow, I. P., and Li, T (eds.), *Optical Fiber Telecommunications IV-A: Components*, Academic Press.

Steele, R. C., Walker, G. R., and Walker, N. G. (1991) Sensitivity of Optically Preamplified Receivers with Optical Filtering, *IEEE Photon. Technol. Lett.* 3(6):545–547.

Taguchi, K. (2002) P-I-N Photodiodes, in Dutta, A. K. Dutta, N. K., and Fujiwara, M. (eds.), *WDM Technologies: Active Optical Components*, Academic Press.

Thiele, H-J., and Nebeling M. (eds.) (2007) *Coarse Wavelength Division Multiplexing: Technologies and Applications*, CRC Press.

Tilsch, M. K., Sargent, R. B., and Hulse, C. A. (2006) Dielectric Multilayer Filters, in Venghaus H. (ed.), *Wavelength Filters for Fibre Optics*, Springer.

Torres, P., and Guzmán, A. M. (2007) Complex Finite Element Method Applied to the Analysis of Optical Waveguide Amplifiers, *IEEE J. Lightw. Technol.* 15(3):546–550.

Urquhart, P. (2004) High Capacity Optical Transmission Systems, in Webb, C. E., and Jones J. D. C. (eds.), *Handbook of Laser Technology and Applications, Volume 3*, IoP Publishing/Taylor and Francis.

Walklin, M., and Conradi, J. (1999) Multi-Level Signalling for Increasing the Reach of 10 Gbit/s Lightwave Systems, *IEEE J. Lightw. Technol.*, 17(11):2235–2248.

Wandel, M., et al. (2002) Dispersion-Compensating Fibers for Non-Zero Dispersion Fibers, *Proceedings Conf. OFC*.

Wandel, M., et al. (2001) Dispersion Compensating Fiber with a High Figure of Merit, *Proceedings ECOC*, 52–53.

Wicker, S. B., and Bhargava, V. K. (1994) *Reed-Solomon Codes and their Applications*, IEEE Press.

Wilner, A. E., and Hoanca, B. (2002) Fixed and Tunable Management of Fiber Chromatic Dispersion Compensation, in Kaminow, I. T., and Li, T. (eds.), *Optical Fiber Telecommunications IVB: Systems and Impairments*, Academic Press.

Yablon, A. D. (2005) *Optical Fibre Fusion Splicing*, Springer (see especially Chapter 9).

Yu, Q., and Shanbhag, A. (2006) Electronic Data Processing for Error and Dispersion Compensation, *IEEE J. Lightw. Technol.* 24(12):4514–4525.

Zhu, Y., and Hadjifotiou, A. (2004) Nonlinear Tolerance Benefit of Modified CSRZ DPSK Modulation Format, *Electron. Lett.* 40(14):903–904.

Zhu, Y., et al. (2002) Polarisation-Channel-Interleaved CS-RZ Transmission at 40 Gbit/s with 0.8 bit-7 s-7 Hz Spectral Efficiency, *Electron. Lett.* 38(8):381–382.

Zimmerman, D., and Spiekman, L. H. (2004) Amplifiers for the Masses: EDFA, EDWA and SOA Amplets for Metro and Access Applications, *IEEE J. Lightw. Technol.* 22(1):63–70.

7 Electromagnetic Waves

J. H. Causebrook
Telecommunications Consultant

R. V. Goodman
Broadcasting and Telecommunications Engineering Consultant

Fundamental Laws of Electromagnetism

Many engineering aspects of propagation do not need a knowledge of historically derived principles, but these do provide a good background. The following are the fundamental laws of electromagnetism.

Gauss's electrostatic theorem: The normal component of electric displacement (flux), **D**, integrated over any closed surface equals the total charge, ρ, within the surface.
Gauss's magnetic theorem: The normal component of magnetic induction, **B**, integrated over any closed surface is 0.
Faraday's law of electromagnetic induction: The parallel component of an electric field, **E**, integrated around a closed path is proportional to the rate of change of the flux of magnetic induction enclosed by the path (Lenz's law: the constant of proportionality is −1).
Ampere's law for magnetomotive force: The parallel component of a magnetic field, **H**, integrated round any loop equals the total current, **J**, flowing through the loop.

Maxwell's Equations and Wave Equations

Maxwell combined these laws and added the concept of displacement current to give a set of equations. These are expressed below in differential form, using the rationalized M.K.S. units as in equations 7.1 to 7.4.

$$\nabla D = \rho \tag{7.1}$$

219

$$\nabla B = 0 \tag{7.2}$$

$$\nabla \Lambda E = -\frac{\partial B}{\partial t} \tag{7.3}$$

$$\nabla \Lambda H = J + \frac{\partial D}{\partial t} \tag{7.4}$$

For an infinite homogeneous medium, with no free charges, ρ, and zero conductivity, σ, these equations become as in equations 7.5 to 7.8, where ε is the permittivity (dielectric constant), which equals $\varepsilon_r\, \varepsilon_0$, the relative permittivity multiplied by the permittivity of free space, and μ is the magnetic permeability, which equals $\mu_r\, \mu_0$, the relative permeability multiplied by the permeability of free space.

$$\varepsilon \nabla E = 0 \tag{7.5}$$

$$\mu \nabla H = 0 \tag{7.6}$$

$$\nabla \Lambda E = -\mu \frac{\partial H}{\partial t} \tag{7.7}$$

$$\nabla \Lambda H = J + \varepsilon \frac{\partial E}{\partial t} \tag{7.8}$$

Elimination of H from this set of simultaneous equations produces a wave equation, as in equation 7.9.

$$\nabla^2 E = \mu \varepsilon \frac{\partial^2 E}{\partial t^2} \tag{7.9}$$

Equally, E could have been eliminated to give equation 7.10.

$$\nabla^2 H = \mu \varepsilon \frac{\partial^2 H}{\partial t^2} \tag{7.10}$$

If E or H is disturbed at a point within a homogeneous nonconductive medium, a "wave" is produced, which propagates away from the point with a velocity of v meters/sec. When this disturbance is a sinusoidal oscillation of frequency f hertz, the wave equations may be solved for a general wave front of any shape by considering it to be composed of a set of plane waves.

One of these plane waves can be expressed in a Cartesian frame, such that it progresses along the Z axis with the electric field along the Y axis. Then it has a solution, at time t, proportional to equation 7.11, where β is 2π divided by the wavelength in meters.

$$e^{-j(Z-vt)\beta} \tag{7.11}$$

Analysis shows that this plane wave has its magnetic component along the X axis; that is, the electric vector is at right angles to the magnetic vector, and both are perpendicular to the direction of propagation. The velocity is given by equation 7.12.

$$v = \frac{1}{\sqrt{\mu\varepsilon}} \tag{7.12}$$

If the relative permittivity and permeability are both unity, v is the velocity of the wave in free space, c, and λ is the free space wavelength, λ_0. The value of μ_0 is defined as $(4\pi)10^{-7}$ Henry per meter, and measurements have set ε at $(8.8547)10^{-12}$ Farad per meter. This leads to a velocity of the wave in free space of $(2.998)10^8$ meters per second. This is the same as the velocity determined by direct measurement, at least within the limits of experimental error. For most practical purposes it is adequate to take this velocity to be equal to 3.10^8 meters per second.

The refractive index of a medium is defined by equation 7.13.

$$n = \frac{c}{v} = \sqrt{\mu_r \varepsilon_r} \tag{7.13}$$

By analogy to Ohm's law, the impedance of the medium is given by equation 7.14.

$$\eta = \frac{E}{H} = \left(\frac{\mu}{\varepsilon}\right)^{1/2} \tag{7.14}$$

For free space this impedance is equal to 377 ohms.

The energy that is carried by the wave is given by the Poynting vector as $E\Lambda H$ watts per square meter. Historical understanding of the fields from electric charges and magnets led to the prediction of the electromagnetic wave traveling through space. Two major categories of receiving antenna (rods and loops) extract from the wave voltages related to the electric and magnetic fields. Furthermore, many other features of the wave can be accurately formulated from the theory (some of which are given here; others can be found in textbooks with mathematical derivation). Thus, there is a tendency to attach more significance to the theory than is justified. It does not prove that the wave is composed of distinct electric and magnetic vectors. It only shows that it is very convenient for us to make the distinction.

Boundary Conditions

If there are discontinuities in the electrical properties of the medium, determination of the wave behavior requires a knowledge of what are called "boundary conditions." The most relevant of these is that components of both E and H tangential to a surface must be continuous. Thus, it is possible to determine how a wave behaves if it is incident on a plane surface where there is a change of impedance from η_1 to η_2 and of refractive index from n_1 to n_2. If the incident wave-normal makes an angle, α_1, with the plane, as shown in Figure 7.1, Snell's law gives the angle α_2 for the wave passing into medium 2.

$$n_1 \cos\alpha_1 = n_2 \cos\alpha_2 \tag{7.15}$$

To determine the reflection, r, and transmission, t, coefficients it is necessary to consider two cases, as follows.

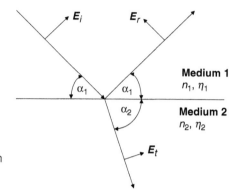

Fig. 7.1   Reflection and refraction at a boundary with the electric vector in the plane of incidence.

Case 1

The electric vector in the plane of incidence (equivalent to a vertically polarized transmitter), as shown in Figure 7.1, gives equations 7.16 and 7.17.

$$r_v = \frac{\eta_1 \sin\alpha_1 - \eta_2 \sin\alpha_2}{\eta_1 \sin\alpha_1 + \eta_2 \sin\alpha_2} \tag{7.16}$$

$$t_v = \frac{2\eta_2 \sin\alpha_2}{\eta_1 \sin\alpha_1 + \eta_2 \sin\alpha_2} \tag{7.17}$$

At high frequencies this can be simplified because it is possible to neglect magnetic permeability, giving equation 7.18.

$$r_v = \frac{\cot(\alpha_1 - \alpha_2)}{\cot(\alpha_1 - \alpha_2)} \tag{7.18}$$

For $n_1 < n_2$ Snell's law gives $\alpha_1 < \alpha_2$, so for $(\alpha_1 + \alpha_2)$:

- Greater than $\pi/2$: r_v is positive (i.e., no phase change).
- Less than $\pi/2$: r_v is negative (i.e., a π phase change).
- Equal to $\pi/2$: r_v is 0 and α_1 equals the Brewster angle.

Case 2

An electric vector perpendicular to the plane of incidence (equivalent to a horizontally polarized transmitter), as shown in Figure 7.2, gives equations 7.19 and 7.20.

$$r_h = \frac{\eta_2 \sin\alpha_1 - \eta_1 \sin\alpha_2}{\eta_2 \sin\alpha_1 + \eta_1 \sin\alpha_2} \tag{7.19}$$

$$t_h = \frac{2\eta_2 \sin\alpha_1}{\eta_2 \sin\alpha_1 + \eta_1 \sin\alpha_2} \tag{7.20}$$

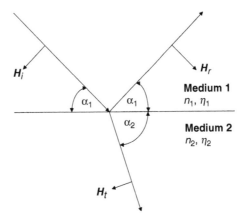

Fig. 7.2 Reflection and refraction at a boundary with the magnetic vector in the plane of incidence.

These results allowed for a relative magnetic permeability of nonunity, but conductivity was zero. For propagation over the Earth's surface, it is more appropriate to approximate magnetic permeability to unity but allow for the finite conductivity of the medium. It is also reasonable to assume unity for the refractive index of air in order to obtain formulas for reflections off the Earth's surface. Then, for a wave of angular frequency ω, equation 7.21 can be obtained.

$$n = n_2 = \left(\varepsilon_r - j\frac{\sigma}{\omega\varepsilon_0} \right)^{1/2} \tag{7.21}$$

Then the reflection coefficient modulus and phase are given by equations 7.22 and 7.23.

$$r_v = \frac{n^2 \sin\alpha_1 - (n^2 - \cos^2\alpha_1)^{1/2}}{n^2 \sin\alpha_1 + (n^2 - \cos^2\alpha_1)^{1/2}} \tag{7.22}$$

$$r_h = \frac{\sin\alpha_1 - (n^2 - \cos^2\alpha_1)^{1/2}}{\sin\alpha_1 + (n^2 - \cos^2\alpha_1)^{1/2}} \tag{7.23}$$

Examples of the use of these formulas are given by Figures 7.3 and 7.4 for the land and by Figures 7.5 and 7.6 for the sea.

Huygen's Principle/Fresnel Diffraction

An earlier solution to propagation phenomena came via Huygen's principle, which states that every point on a wave front acts as a source of a secondary wavelet. This provides an explanation for some propagation phenomena. Most important, it helps to quantify diffraction.

Diffraction effects can be formulated via the wave equation and Kirchhoff's integral (Longhurst, 1967), but the original derivation by Fresnel was via Huygens principle. A semi-infinite opaque screen is assumed to exist between transmitter and receiver, as shown in Figure 7.7. Then, Fresnel's integral states that the attenuation is related to free space, in

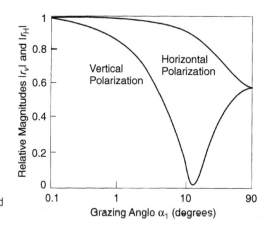

Fig. 7.3 Magnitude of the reflection coefficient over the ground (frequencies above 100 MHz).

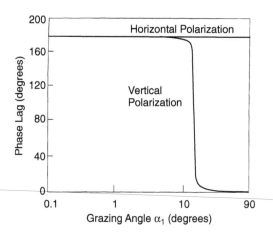

Fig. 7.4 Phase of the reflection coefficient over the ground (frequencies above 100 MHz).

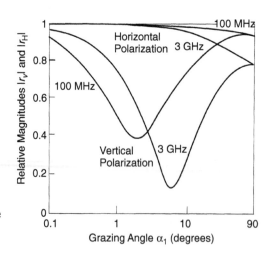

Fig. 7.5 Magnitude of the reflection coefficient over the sea.

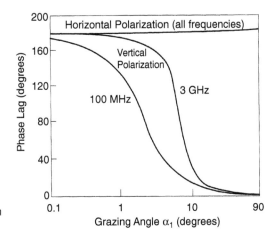

Fig. 7.6 Phase of the reflection coefficient over the sea.

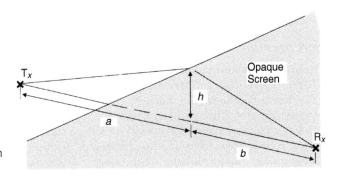

Fig. 7.7 Fresnel diffraction parameters.

amplitude and phase, by the complex function F, given by equation 7.24, where v is given by equation 7.25.

$$F(v) = \frac{1+j}{n} \int_v^\infty e^{-j\frac{\pi}{2}(t^2+v^2)} dt \tag{7.24}$$

$$v = h\left(\frac{2(a+b)}{\lambda ab}\right)^{1/2} \tag{7.25}$$

Propagation Mechanisms

The principles of propagation enable the communications engineer to produce convenient calculation techniques. This process is facilitated by recognizing three major propagation mechanisms, which tend to dominate in different wave bands:

- Space wave, below 10 meters (i.e., frequencies above 30 MHz)
- Sky wave, 10 to 200 meters (i.e., from 1.5 to 30 MHz)
- Surface wave, above 200 meters (i.e., below 1.5 MHz)

This division into three categories should not be regarded as rigidly defining the propagation mechanisms for the frequency ranges specified. First, it should be appreciated that the stated ranges are only approximate. Second, there can be a considerable overlap, where more than one mechanism is relevant. For example, MF broadcast stations normally provide coverage via the surface wave, but this coverage can be reduced, at night, by sky wave interference.

Generation of Electromagnetic Waves

Electromagnetic waves, as distinct from electrostatic fields or steady magnetic fields, involve a continuous net transfer of energy, or radiation, through a medium and are produced by current and charge distributions that vary with time.

The types of antenna employed to generate such waves are many, depending on the radio frequency to be transmitted, the radiation pattern required to fulfill the needs of the system, and the environment in which the antenna is required to operate.

The following sections deal with several fundamental antenna elements from which the performance of many practical antennas can be derived.

Radiation from an Infinitesimal Linear Element

The electromagnetic field equations for many complex antenna systems can be derived from a knowledge of the fields produced by an oscillating current in an infinitesimal linear element (Jordan & Balmain, 1968; Balanis, 1982). For such an element, the current can be assumed to be constant throughout its length.

Consider a current element of length δl carrying a current $I \cos \omega t$, located at the origin of a spherical coordinate system in a homogeneous, isotropic, nonconducting medium, as shown in Figure 7.8. The formulation of the electric and magnetic field magnitudes generated by such a current element is made more succinct by the introduction of two variables, as in equations 7.26 and 7.27, from which equations 7.28 to 7.30 are obtained.

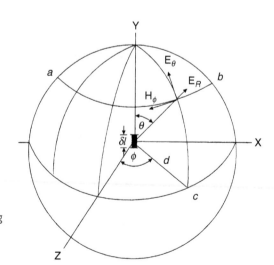

Fig. 7.8 Radiation from an oscillating current in an infinitesimal linear element at the center of a spherical coordinate system.

$$\Psi = \frac{I\delta I\beta \sin\theta}{4\pi d} \tag{7.26}$$

$$\alpha = \frac{j}{\beta d} \tag{7.27}$$

$$H_\varphi = \Psi(1-\alpha) \tag{7.28}$$

$$E_\theta = \eta\Psi(1-\alpha+\alpha^2) \tag{7.29}$$

$$E_r = 2\eta\Psi\alpha(1-\alpha)\cot\theta \tag{7.30}$$

Terms in α and α^2 are known as the induction and electrostatic fields, respectively. They are also known as the near-field terms, representing energy that ebbs and flows between the field and the current source. At $d = 16\lambda$, α is equal to 0.01 and can effectively be set to zero. This leaves us with the radiation field terms.

The total power radiated (P_{rad}) by the current element can be obtained by integration of the radial Poynting vector over a spherical surface with the element at the center, to give equation 7.31, where I is the peak current and I_o is the r.m.s. current.

$$P_{\text{rad}} = 10(\beta I\delta l)^2 = [20(\beta\delta l)^2]I_o^2 \tag{7.31}$$

From equation 7.31 it can be seen that the part in square brackets has the dimensions of resistance. This is known as the *radiation resistance* of the current element.

Radiation from an Oscillating Current in a Small Loop

The magnetic equivalent of the infinitesimal linear element is the small loop of radius a. Consider such a current loop at the center of a spherical coordinate system.

The electric and magnetic field components generated by such a loop can be derived, as for the infinitesimal linear current element, and are given by equations 7.32 to 7.34, where χ is given by equation 7.35.

$$E_\varphi = \eta\chi(1-\alpha) \tag{7.32}$$

$$H_\theta = \chi(1-\alpha+\alpha^2) \tag{7.33}$$

$$H_r = 2\chi\cot\theta\alpha(1-\alpha) \tag{7.34}$$

$$\chi = \frac{\beta^2 a^2 I \sin\theta}{4d} \tag{7.35}$$

Integration of the Poynting vector over the surface of a surrounding sphere gives the total radiated power as in equation 7.36.

$$P_{\text{rad}} = \eta\left(\frac{\pi}{12}\right)\left(\frac{c}{\lambda}\right)^4 I^2$$
$$= \left[20\pi^2\left(\frac{c}{\lambda}\right)^4\right]I_o^2 \tag{7.36}$$

Again the part in square brackets is the radiation resistance, which can be increased by constructing the loop from N turns, wound so that the magnetic field passes through all the turns. The radiation resistance is then given by equation 7.37.

$$R_{\text{rad}} = 20\pi^2\left(\frac{c}{\lambda}\right)^4 N^2 \tag{7.37}$$

Radiation from a Short Dipole

The derivation of the radiation characteristics of practical antennas, from the equations for the infinitesimal current element, requires a knowledge of the current distribution in the practical antenna. In the case of the center-fed short dipole (Figure 7.9), the current distribution can be regarded as linearly tapering from a maximum value at the center to zero at the tips, provided that the electrical length is small (i.e., less than 0.1 wavelength). In this figure the current distribution is given by equation 7.38.

$$\text{Current distribution } I = I_m\left(1 - \frac{2l}{L}\right) \tag{7.38}$$

The mean value of the current along the dipole is half the current at the center. Thus, the magnitudes of the far-field terms are half what would be given by equations 7.26 to 7.30 for a current element of the same length. Therefore, the total power radiated and the radiation resistance, expressed in terms of the r.m.s. current, I_o, at the center of the dipole, are a quarter of that given by equation 7.31.

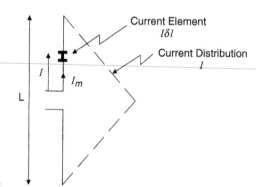

Fig. 7.9   Short dipole.

Radiation from a Half-Wavelength Dipole

The radiation characteristics of the half-wavelength dipole can be derived on the basis that, for a thin dipole, the current distribution along the antenna is approximately sinusoidal, as shown in Figure 7.10. In this figure the value of the current distribution is given by equation 7.39 and E_θ by equation 7.40.

$$I = I_m \cos \beta l \tag{7.39}$$

$$E_\theta = \frac{60\pi I_m \sin\theta}{\lambda d} e^{-j\beta d} \int_{-\frac{L}{2}}^{\frac{L}{2}} \cos \beta l \, e^{j\beta l \cos\theta} dl \tag{7.40}$$

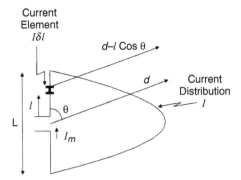

Fig. 7.10 Half-wavelength dipole.

Using the same system of spherical coordinates defined earlier, the far-field radiation terms in free space are given by equations 7.41 and 7.42.

$$H_\varphi = \frac{I_m}{2\pi d} \frac{\cos\left(\frac{\pi}{2}\cos\theta\right)}{\sin\theta} \tag{7.41}$$

$$E_\theta = \eta H_\varphi \tag{7.42}$$

The average power flux density through the surface of a surrounding sphere is given by the Poynting vector. The total power, P, radiated by the dipole is equal to the total power radiated through the surface of the sphere and is given by equation 7.43.

$$P = \frac{\eta I_m^2}{4\pi} \int_0^{\frac{\pi}{2}} \frac{\cos^2\left(\frac{\pi}{2}\cos\theta\right)}{\sin\theta} d\theta \tag{7.43}$$

Evaluation of this integral is dealt with by Jordan and Balmain (1968) and (for I_o given by equation 7.45) gives equation 7.44.

$$P = 73.1 I_o^2 \text{ watts} \tag{7.44}$$

$$I_o = \frac{I_m}{\sqrt{2}} \tag{7.45}$$

The radiation resistance of the thin half-wavelength dipole is given by the coefficient of I_o^2 in equation 7.46 and is therefore equal to 73.1 ohms.

It is now possible to determine the r.m.s. value of the electric field in the direction of maximum radiation by combining equations 7.41, 7.42, and 7.44, with θ equal to zero, giving equation 7.46, where E is in volts per meter, with P in watts and d in meters.

$$E \approx 7\frac{\sqrt{P}}{d} \tag{7.46}$$

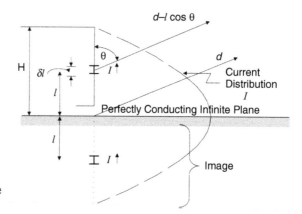

Fig. 7.11 Vertical monopole over a ground plane.

Radiation from a Short Monopole

Figure 7.11 shows a vertical monopole on a perfectly conducting, infinite, flat ground plane. The current distribution (given by equation 7.47), for a thin monopole is approximately sinusoidal and, as a result of the perfect image in the ground plane, the radiation pattern above the plane is identical to that of a dipole of length 2H.

$$\text{Current distribution } I = I_m \sin[\beta(H - l)] \tag{7.47}$$

Thus, for a short monopole ($H < 0.1\lambda$) the current can be assumed to taper linearly from a maximum at the base to zero at the top. The radiation pattern above the plane will be identical to that of the short dipole.

However, because power can only be radiated into the hemisphere above the reflecting plane, the total power radiated (for a given value of current at the base of the monopole) is half that for the short dipole of length 2H with the same current at its center.

Thus, in free space equation 7.48 can be obtained.

$$P = \left[40\pi^2 \frac{H^2}{\pi^2} \right] I_o^2 \tag{7.48}$$

Once again the radiation resistance is given by the term in square brackets.

The maximum field strength at a given distance from the monopole, in the far field, occurs at the surface of the conducting plane and is $\sqrt{2}$ times that for a short dipole radiating the same total power, because all the energy is confined to a hemisphere. By choosing units of power in kilowatts, distance in kilometers, and r.m.s. electric field in mV/m, the maximum electric field strength is given by equation 7.49.

$$E = \frac{300\sqrt{P}}{D} \tag{7.49}$$

The quantity $300\sqrt{P}$ is called the Cymomotive Force (CMF).

Radiation from a Quarter-Wavelength Monopole

The quarter-wavelength monopole above the conducting surface can be compared to the half-wavelength dipole with the same base current: It has the same radiation pattern above the surface and half the total radiated power, and the maximum field strength is given by equation 7.50, where E is in mV/m, P in kW, and D in km.

$$E = \frac{313\sqrt{P}}{D} \tag{7.50}$$

The Isotropic Radiator

The hypothetical isotropic radiator, which radiates power equally in all directions, is widely used as a reference for comparing the performance of practical antennas.

Such a radiator located at the center of a large sphere of radius, d meters, and radiating a total power, P watts, produces a uniform power flux density, $E\Lambda H$, over the surface of the sphere, as in equation 7.51.

$$E\Lambda H = \frac{P}{4\pi d^2} = \frac{E^2}{\eta} \tag{7.51}$$

Therefore, in free space the electric field strength in volts/meter, for P in watts and d in meters, is given by equation 7.52.

$$E = \frac{\sqrt{30P}}{d} \tag{7.52}$$

Antenna Gain and Effective Radiated Power

The term *antenna gain* defines the degree to which an antenna concentrates radiated power in a given direction, or absorbs incident power from that direction, compared to a reference antenna. The logical reference antenna is the isotropic radiator, in view of its lack of directional properties, though the half-wavelength dipole is also often used. Antenna gain is most easily derived by considering all antennas as transmitting antennas.

If the r.m.s. field strength in free space produced at a distance, d, is given by E, the total power radiated by the antenna is given by the surface integral, as in equation 7.53.

$$P = \frac{d^2}{120\pi} \int_0^{2\pi} \int_0^{\pi} E^2 \sin\theta \, d\theta \, d\varphi \tag{7.53}$$

If the field strength in a particular direction is E_n the gain relative to an isotropic antenna is given by equation 7.54.

$$G_i = \frac{4\pi E_n^2}{\int_0^{2\pi} \int_0^{\pi} E^2 \sin\theta \, d\theta \, d\varphi} \tag{7.54}$$

If a choice is made such that E_n is in the direction of maximum radiation, the maximum intrinsic gain of the antenna is obtained.

G_i is the gain solely attributable to the directional characteristics of the radiation pattern of the antenna and takes no account of dissipative losses in the antenna itself or in the transmission line between the antenna and the transmitter. These losses can be combined and expressed as a power ratio, K, so that the antenna gain referred to the transmitter output becomes KG_i for the gain in any direction. This gives the effective gain relative to an isotropic antenna. The product of transmitter power and effective antenna gain relative to an isotropic antenna is the Effective Isotropically Radiated Power (EIRP).

When the half-wavelength dipole is used as the reference antenna, instead of the isotropic antenna, the gain figures are reduced by the maximum intrinsic gain of the half-wavelength dipole. The product of transmitter power and effective antenna gain relative to a half-wavelength dipole is the Effective Radiated Power (ERP). The numerical power gain of a half-wave dipole relative to an isotropic source is 1.641 (2.15 dB).

The computation of the intrinsic or directive gain of an antenna requires a complete knowledge of the three-dimensional radiation pattern. Fortunately, for many practical cases a sufficiently accurate approximation to the three-dimensional pattern can be obtained from the radiation patterns measured in two mutually perpendicular planes. These patterns represent cross-sections of the three-dimensional pattern and are normally measured in the horizontal plane and the vertical plane. Under these conditions, the patterns are called the Horizontal Radiation Pattern (HRP) and the Vertical Radiation Pattern (VRP), respectively. They can also be referred to as the H-plane and E-plane radiation patterns, respectively, for a vertically polarized antenna and, vice versa, for a horizontally polarized antenna.

It is not uncommon for transmitting antennas to have the main lobe, or beam, of the VRP tilted down below the horizontal to reduce the power wasted in radiation above the horizon. In such cases, the HRP is measured for θ equal to the beam tilt angle, not 90 degrees.

Figure 7.12 shows a vertical radiation pattern for a typical high-power UHF television broadcasting antenna, and Figure 7.13 shows the associated HRP. The maximum intrinsic gain or directivity can be computed by performing an integration, provided the functional

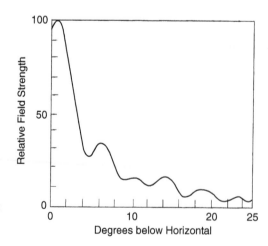

Fig. 7.12 Vertical radiation pattern of a typical UHF broadcast antenna.

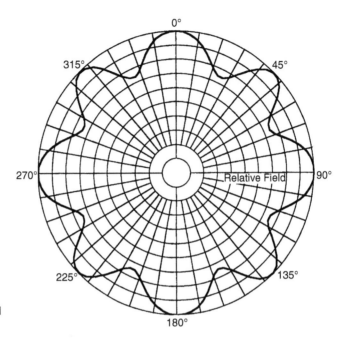

Fig. 7.13 Horizontal radiation pattern of a typical UHF broadcast antenna.

form of the radiation pattern is known. The maximum effective gain can then be derived by subtracting all the dissipative losses and the maximum effective radiated power then obtained for any particular value of transmitter power.

The ERP for any direction can be derived by measuring the relative field coefficients r and s from the radiation patterns and computing as in equation 7.55, where P_m is the maximum ERP.

$$P = r^2 s^2 P_m \qquad (7.55)$$

The three-dimensional radiation pattern can be more fully represented in two dimensions by a family of polar patterns, each contour line joining directions having equal ERP.

Polarization

The polarization of an electromagnetic wave defines the time-varying behavior of the electric field vector at a point.

Linear Polarization

When the electric vector lies wholly in a single plane, the wave is said to be linearly polarized. In terrestrial communications, when this plane is parallel to the surface of the Earth, the wave is defined as horizontally polarized and, when the plane is perpendicular to the surface of the Earth, the wave is defined as vertically polarized.

Circular Polarization

A wave is said to be circularly polarized where the electric field vector rotates in the plane normal to the direction of propagation such that the locus of the extremity of the vector

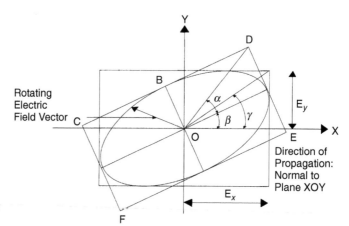

Fig. 7.14 Polarization ellipse and circumscribing rectangles.

describes a circle. The agreed convention for defining the sense of rotation of the electric vector is in terms of the direction of rotation for an observer looking; in the direction of propagation (i.e., toward the receiver).

Clockwise rotation is often called right-hand circular polarization; counterclockwise, left-hand circular polarization. The electric field vector rotates at the angular frequency, ω, of the wave and describes one revolution in the period $2\pi/\omega$.

Elliptical Polarization

A wave is defined as elliptically polarized where the electric field vector rotates in the plane normal to the direction of propagation while its amplitude varies such that the locus of the extremity of the vector describes an ellipse. The vector rotates by one revolution during the period of oscillation, $2\pi/\omega$, of the source, though the angular velocity is not constant throughout the revolution, as it would be for circular polarization.

Elliptical polarization should be regarded as the most general form of polarization, circular and linear being special cases of elliptical polarization with axial ratios of unity and zero, respectively. An elliptically polarized wave can be completely defined by three parameters: axial ratio, orientation angle, and sense of rotation. These are defined in the list that follows, using Figure 7.14 as a reference. In this figure γM is given by equation 7.56:

$$\tan \gamma = \frac{Ey}{Ex} \tag{7.56}$$

Axial ratio: The axial ratio is defined as the ratio of the minor axis to the major axis. This ratio is often expressed in decibels.

Orientation angle: The orientation angle, β, is defined as the angle between the major axis and the reference axis. In terrestrial communications this reference axis is normally taken as parallel to the surface of the Earth.

Sense of rotation: The convention for defining the sense of rotation of the electric vector is the same as for circular polarization; the commonly used terms are *clockwise* (or right-hand) and *counterclockwise* (or left-hand).

Ellipticity angle: This is an alternative way of defining the axial ratio and is equal to the angle, α, defined in Figure 7.14 for the circumscribing rectangle CDEF.

Reception of Electromagnetic Waves

Electromagnetic waves may be received using a simple rod antenna, a loop antenna, or an array of rod attennas (a "Yagi" aerial). The concept of a hypothetical istropic antenna is described because it is useful in understanding the performance of aerials.

The Simple Rod Antenna

It can be assumed that a receiving rod antenna consists of an infinite number of current generators of length dx driven by fields $E(x)$. Then, if the total impedance at the terminals is z and the current generators have a normalized value of $D(x)$ relative to the value at the terminals ($x = 0$), the elemental current generators will each contribute an amount, $E(x)D(x)dx/z$, to the current, I. The current distribution, $D(x)$, on a rod antenna is assumed to be analogous to the sinusoidal standing wave pattern obtained on an open-circuited transmission line of the same length as the antenna, along which a wave is propagating with the speed of light. This statement lacks some rigor, but is accurate enough for most practical purposes and has stood the test of time.

The current distribution, shown as a dotted line on Figure 7.15, is for a half-wavelength dipole antenna. In this case, if the field is considered constant along the antenna, the current is given by equation 7.57 where R is the load impedance and the intrinsic radiation resistance of the half-wavelength dipole is 73 ohms.

$$I = \frac{E\int_{-\frac{\lambda}{4}}^{\frac{\lambda}{4}} \cos \beta x dx}{73+R} = \frac{\lambda E}{\pi(73+R)} \tag{7.57}$$

From equation 7.57 it follows that the open-circuited voltage, e, across the terminals is given by equation 7.58.

$$e = \frac{\lambda E}{\pi} \quad \text{volts} \tag{7.58}$$

Figure 7.16 gives the simple equivalent circuit for a half-wavelength dipole receiving antenna. For maximum power transfer, R also equals 73 ohms; thus, the power absorbed by

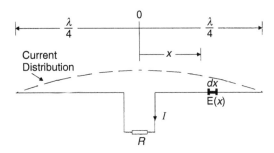

Fig. 7.15 Current distribution on a half-wavelength dipole.

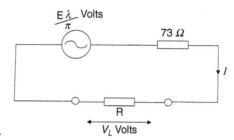

Fig. 7.16 Simple equivalent
circuit for a half-wavelength dipole.

the load is as in equation 7.59 and the voltage across the load is $\dfrac{E}{\beta}$, where β is given by equation 7.60.

$$P = \frac{1}{73}\frac{E^2}{\beta^2}$$ (7.59)

$$\beta = \frac{2\pi}{\lambda}$$ (7.60)

The effective capture area, or aperture, S, of an antenna is defined as the ratio of the received power to the incident power flux density. For the half-wavelength dipole this is given by equation 7.61.

$$S_d = \frac{5.15}{\beta^2}$$ (7.61)

The Hypothetical Isotropic Antenna

The concept of an isotropic receiving antenna creates conceptual problems because such an antenna has to be infinitesimally small. Nevertheless, an effective receiving aperture can be derived by taking that of a physically realizable antenna divided by its power gain relative to an isotropic antenna.

Equation 7.61 gives the effective aperture of a half-wavelength dipole that has a maximum power gain of 1.64, relative to an isotropic antenna. The effective aperture of an isotropic antenna is therefore given by equation 7.62.

$$S_i = \frac{\pi}{\beta^2}$$ (7.62)

The Loop Antenna

An application of Lenz's law gives the open-circuited voltage of a loop antenna as in equation 7.63, where N is the number of turns in the loop and A is its area.

$$e = \omega\mu HNA$$ (7.63)

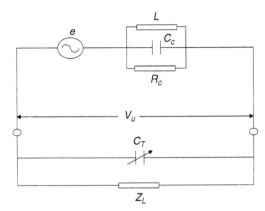

Fig. 7.17 Equivalent circuit of a loop antenna.

The equivalent circuit of the antenna is shown in Figure 7.17, with L being the self-inductance, C_c the self-capacitance, R_c the winding resistance, C_T a tuning capacitor, and Z_L the load impedance. This circuit will have a Q value such that the unloaded output voltage (V_u) at resonance is given by Q_e volts.

A sufficiently large voltage is only obtained from an air-filled loop by making its area large. Large voltages are often achieved by winding the loops on ferrite rods. In this way MF receivers can have antennas with cross-sectional areas some thousands of times smaller than would be necessary if they were air filled.

High-Gain Rod Antennas

Frequently a simple dipole is inadequate for practical reception purposes. More gain is required, together with the ability to reject unwanted signals coming from directions other than that of the wanted signal. This is often achieved with a Yagi antenna consisting of a driven element with a passive reflector and directors (Yagi, 1928; Balanis, 1982). The reflector is more efficient if it consists of a mesh of rods effectively forming a back screen. Gain can also be achieved by having several dipoles spaced and combined to give directivity. An additional requirement can be that the antenna have a wideband response as well as gain relative to a dipole; this is neatly achieved with a log-periodic antenna (Balanis, 1982).

The terminated voltage (V_μ) expressed in decibels relative to 1 μV, for a practical receiving system with antenna gain, G dB, relative to a half-wavelength dipole and attenuation of the feeder system, L dB, is given by equation 7.64, assuming perfect impedance matching to the load resistance, R.

$$V_\mu = 20\log E - L + G - 120 - 20\log \beta + 10\log \frac{R}{73} \tag{7.64}$$

The Space Wave

Earlier expressions were derived for field strength in free space. It is convenient to convert these to field strength expressed in decibels relative to 1 μV/m, as in equation 7.65, where D is the distance in kilometers and P is the ERP in watts.

$$E_\mu = 76.9 + 10\log_{10} P - 20\log_{10} D \qquad (7.65)$$

Some workers prefer to base an equation on the isotropic transmitting antenna. The difference is that $10\log_{10} P$ becomes 2.1 dB greater, because EIRP is used, which is compensated for by the constant 76.9 becoming 74.8.

Free-Space Path Loss

A transmission path can also be characterized by the free-space path loss, which is defined as the ratio of received power, P_r, to transmitted power, P. Expressed in decibels, in a non-dissipative medium, this is given by equation 7.66.

$$L = 17.7 + 20\log_{10} d - 20\log_{10} \lambda - G_t + L_t - G_r + L_r \qquad (7.66)$$

G_t and G_r are the antenna gains relative to a half-wavelength dipole, and L_t and L_r are the antenna feeder losses (subscripts t and r being for transmitter and receiver).

Again, some workers prefer to base an equation on the isotropic antenna, in which case the gains G_t and G_r are each increased by 2.1 dB to be relative to isotropic antennas instead of half-wavelength dipoles, which is compensated for by the constant 17.7 becoming 21.9. There can also be a preference to convert d in meters to D in kilometers, and λ (meters) wavelength to frequency in megahertz, which must be matched by an appropriate change of constant.

Variation from Free-Space Values

Free-space values rarely occur in practice because factors such as the Earth's surface, buildings, trees, and the atmosphere cause variations. It is, therefore, necessary to be able to predict an excess loss that is dependent on the particular propagation path under consideration. This loss prediction is complicated by the fact that the real world does not provide simple geometric shapes for the boundary conditions, and nearly all calculation systems have been based on such simple geometries. It is therefore convenient to stylize the real world into shapes that can be handled. The following sections discuss these shapes and the predictions based on them.

Knife Edge Diffraction

Fresnel's diffraction formula is still one of the most useful tools for prediction of the loss over a hilly terrain. A rigorous analysis for a path obstructed by many edges involves a multiple integral but, because this is difficult to solve, attempts have been made to get acceptable results from a sum of single integrals. It is now well established that a method originally devised by Deygout (1966) gives the best results in most cases. It consists of taking each hill as if it alone existed on the path to establish which one gives the largest predicted loss. This hill is then taken as a terminal to divide the path into two subpaths, and the process is repeated. Further subdivisions can then be made. The total diffraction loss is taken as the sum, in decibels, of the primary loss and the losses on each subpath.

This process is illustrated by the profile drawn in Figure 7.18, which is stylized into the construction shown in Figure 7.19. From this, three Fresnel limit parameters can be derived, of the type introduced in equation 7.24, as in equations 7.67 to 7.69.

$$v_1 = h_1 \left(\frac{2(a+b+c+e)}{\lambda(a+b)(c+e)} \right)^{1/2} \qquad (7.67)$$

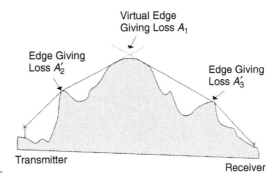

Fig. 7.18 Example of a path profile with multiple obstructions.

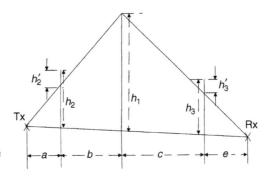

Fig. 7.19 Construction for the Deygout method.

$$v_2 = h_2\left(\frac{2(a+b+c+e)}{\lambda a(b+c+e)}\right)^{1/2} \tag{7.68}$$

$$v_3 = h_3\left(\frac{2(a+b+c+e)}{\lambda(a+b+c)e}\right)^{1/2} \tag{7.69}$$

Let us assume that v_1 is the largest of these so that we then get equations 7.70 and 7.71.

$$v_2' = h_2'\left(\frac{2(a+b)}{\lambda ab}\right)^{1/2} \tag{7.70}$$

$$v_3' = h_3'\left(\frac{2(c+e)}{\lambda ce}\right)^{1/2} \tag{7.71}$$

To avoid the complication of evaluating the Fresnel integral, it is possible to go directly from the above parameters to the required losses in dB via a graph of the type shown in Figure 7.20.

The diffraction attenuation assumed for the path is as in equation 7.72.

$$A_k = A_1 + A_2' + A_3' \tag{7.72}$$

As stated, this method is generally the most accurate of the approximations. However, it is liable to overestimate the loss by up to 6 dB if either A_2 or A_3 is nearly equal to A_1.

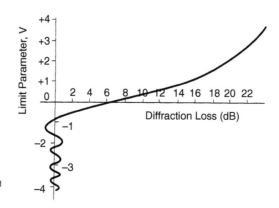

Fig. 7.20 Fresnel diffraction loss chart.

Surface Diffraction

The rounded nature of hilltops, and reflections from elsewhere on the Earth's surface, is liable to make losses greater than those derived from the simple screens assumed for the Deygout approximations. To overcome this, the profile can be assumed to consist of geometric shapes, for which calculations are possible, such as spheres, cylinders, or wedges.

Rounded Hilltops

One method of stylizing the profile is to assume a rounded top on each of the obstructions rather than a knife edge. This technique has a problem in its application because it requires the radius of a hilltop to be known and, with most maps in common use, it is not possible to determine this radius with sufficient accuracy. However, a useful analysis of this method is given in an IEE paper by Hacking (1970), where extra terms are derived to add to the knife edge loss.

Diffraction over a Sea Path

If it is only sea that interrupts the line-of-sight path, a spherical diffraction calculation can be made from a method that has its origins in work by Van der Pol and Bremmer (1937, 1949). These methods have a complicated formulation and will not be detailed here. However, for many applications a simplification, based on a paper by Vogler (1964), is of value for frequencies above 400 MHz.

Approximating a Land Path to a Cylindrical Surface

On a land path with many diffraction edges, a useful prediction is made with a cylindrical surface stylization of four radii joining each other with common tangents, as shown in Figure 7.21.

Point a is the transmitter horizon, and point b is the receiver horizon. Arc ac just clears the terrain between point a and the transmitter site, and it has a tangent to the horizon line, Ta, at point a. This condition defines the radius r_t. Radius r_r is defined similarly at the receiver end. The radii r_1 and r_2 are defined such that arcs ae and be form tangents with

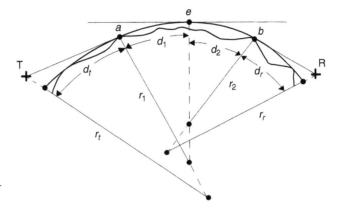

Fig. 7.21 Geometry for a curved-surface diffraction calculation.

horizon lines Ta and Rb at points a and b, respectively, and have a common tangent at e. The curved surface attenuation can then be approximated by equations 7.73 to 7.77.

$$A_c = G(Z_t + Z_r + Z_1 + Z_2) - R(Z_t) - R(Z_r) - 20 \tag{7.73}$$

$$Z_i = 448\lambda^{-1/3} r_i^{-2/3} d_i \tag{7.74}$$

$$G(Z) = 0.058Z - 10 \log Z \tag{7.75}$$

$$Z \le 2000: R(Z) = 0.058Z - 33 \tag{7.76}$$

$$Z > 2000: R(Z) = G(Z) \tag{7.77}$$

The calculation does not allow for the roughness of the terrain and as a result the loss of signal on the path is usually overestimated. A better result is obtained from an interpolation between this and the knife edge calculation. The form of interpolation can be determined by obtaining a best fit with a set of measured data. In the United Kingdom, it was found that an averaging of losses, expressed in decibels, gave good results.

Approximating a Single Hill to a Wedge Shape

On a land path with one diffraction edge, a prediction can be made by assuming a wedge-shaped profile. This wedge has its peak at the diffraction edge and surfaces that clear the terrain on both sides of the profile, as shown in Figure 7.22, with an angle between the horizon lines of θ.

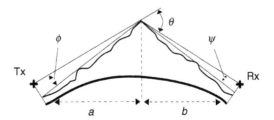

Fig. 7.22 Geometry for a wedge diffraction calculation.

The angles between the wedge surfaces and these horizon lines are φ and Ψ. The diffraction loss is given by equations 7.78 to 7.83, where F is given in equations 7.24.

$$A_D = -20\log[F(v_1) + \rho_r F(v_2) + \rho_t F(v_3) + \rho_r \rho_t F(v_4)] \qquad (7.78)$$

$$v_1 = \theta u \qquad (7.79)$$

$$v_2 = (\theta + 2\varphi)u \qquad (7.80)$$

$$v_3 = (\theta + 2\Psi)u \qquad (7.81)$$

$$v_4 = (\theta + 2\varphi + 2\Psi)u \qquad (7.82)$$

$$u = \left(\frac{2ab}{\lambda(a+b)}\right)^{1/2} \qquad (7.83)$$

The simulated reflection coefficients ρ_r and ρ_t depend on how the actual profile deviates from the wedge surface, and the values should range from about −0.9 for a very smooth wedge-like profile to about −0.1 for a rough, "peaky" edge.

Attenuation by Buildings and Trees

Significant attenuation is caused by buildings and trees. For television reception with rooftop antennas (about 10 m above ground level), the attenuation can be obtained by reasonably simple methods without specific details of buildings and trees. A clutter loss ACL is given approximately by Table 7.1. However, today the ability to predict for lower antennas has become increasingly important. This is the case for receivers in cars and for the mobile phone industry.

For predictions to lower receivers, it is necessary to have a good idea of the location and height of buildings and trees. The attenuation to the rooftop close to the receiver can then be obtained by adding a further term to the Deygout knife edge diffraction method given in the Knife Edge Diffraction section. This edge is chosen to represent the most intrusive building along the path. Then it is necessary to determine the attenuation due to the diffraction from the roof above the street. Again this can be treated as a knife edge diffraction. The loss might be less than would be expected from the angle of roof to receiver because the predominant signal often comes from the walls of the houses opposite. This would produce a smaller diffraction angle.

We turn now to loss caused by trees. The predominant signal in this case might be by passage through the trees. This can be determined by figures published for loss per distance

Table 7.1 Clutter Loss

	VHF (dB)	UHF (dB)
Urban	8	12
Suburban	6	8
Wooded	4	8
Rural	2	4
Rx foreground slope:	Change preceded by	
Facing Tx	−2	−4
Away from Tx	2	4

traveled through trees. However, care must be taken not to obtain a figure that is greater than that calculated for diffraction over trees.

There are now many cellular telephone sites where the base station is also below building height. In this case even more detail is required to make coverage and interference predictions. This detail is to enable calculations to be made of propagation via multiple reflections and diffraction off walls and around the buildings.

Ground Reflections

If the path between transmitter and receiver is well clear of intervening terrain and obstacles, the Earth's surface is still liable to modify the received field because of reflections. It is often difficult to accurately predict the effect of such reflections because the resultant field strength is a vector addition between the direct and reflected waves, so a small uncertainty about terrain can result in a significant uncertainty about relative phases, especially for higher frequencies.

The strength and phase of the reflected wave depends on

- The surface conductivity
- The surface dielectric constant
- The general shape of the ground profile (e.g., flat, convex, or concave)
- The roughness of the ground
- The presence of buildings and trees

If the surface of the ground is rough, the reflected energy is scattered. Smoothness can be judged by an application of the so-called Rayleigh criterion: For small angles of Ψ radians between the reflected ray and the ground plane, this criterion states that the undulations shall not be greater than $\dfrac{\lambda}{16\Psi}$ for the surface to be treated as smooth.

Of course, the irregular shape of the terrain can result in many reflected waves arriving at a reception point.

Despite all these difficulties, it is instructive to consider the formulation for a simple case, with the reflected path longer than the direct by δ and a reflection coefficient of magnitude ρ and a phase change of φ. From a summation of vectors (applying the cosine rule), the received field is reduced from the free-space value by equation 7.84.

$$A_r(\text{dB}) = -10\log[1 + \rho^2 - 2\rho\cos(\beta\delta - \varphi + \pi)] \qquad (7.84)$$

For horizontal polarization and angles of incidence below 10 degrees at frequencies above 400 MHz, it is reasonable to assume $\rho = 1$ and $\varphi = \pi$.

Tropospheric Refractive-Index Effects

A further influence on the received field strength is the refractive index of the troposphere. Irregularities in this region bend, scatter, and reflect the wave. This is the main cause of time variability, particularly on a long-distance path.

The refractive index, n, of the atmosphere differs by only a small amount from unity, and it is more convenient to use the so-called N factor. This factor is dependent on the pressure, P (millibars), temperature, $T(\text{K})$, and relative humidity, W, of the atmosphere, as given by equation 7.85.

$$N = (n-1)10^6 = \frac{77.6}{T}\left(P + \frac{4810W}{T}\right) \tag{7.85}$$

Under "normal" conditions, the refractive index of the atmosphere decreases with height by an amount that slightly bends the wave back toward the Earth. However, it is still more convenient to think of the wave traveling in straight lines, which can be done theoretically if the Earth's radius is suitably modified. Under normal atmospheric conditions, this modification involves multiplying the true Earth radius by 4/3.

Changes of meteorological conditions can greatly alter the refractive index from the normal and hence influence the mode of propagation of signals through the troposphere.

If the refractive index decreases from ground level at a sufficiently high rate to bend the wave with a curvature greater than that of the Earth, propagation of high field strengths to great distances can occur via the ground-based ducts created by such meteorological conditions. Under these conditions the wave is refracted back to the Earth, where it is reflected back toward the troposphere to be refracted back again toward the Earth, and so on, to produce a waveguide type of propagation. Where such a rapid decrease in refractive index occurs at a higher level in the atmosphere, it can produce an elevated propagation duct, without involving reflection off the ground.

If the change of refractive index is very rapid over a small height interval, the wave can be effectively reflected from it and again off the ground, followed by reflection back from the layer, to produce high fields at large distances.

In addition, the refractive index changes can be patchy and bloblike, in which case energy is returned to the Earth in the form of scattered waves. In this case, the resultant received signal level changes rapidly with time as the tropospheric discontinuities move. A satisfactory parameter for the prediction of such scatter fields is the angle between the receiver and transmitter horizons, as shown in Figure 7.23. An empirical relationship can then be determined to relate the field strength to this angle.

The signals resulting from abnormal tropospheric conditions usually represent unwanted interference to services sharing the same frequencies at some considerable distance away. However, they are made use of for some purposes, such as over-the-horizon radar and communications systems.

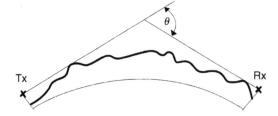

Fig. 7.23 Angle-of-scatter calculations.

Atmospheric Absorption

Electromagnetic waves propagating in the nonionized region of the atmosphere around the Earth are subject to attenuation due to absorption and scattering by hydrometeors (rain, snow, ice particles, fog, and clouds) and absorption by atmospheric gases. Curves showing the variation of specific attenuation with frequency are published by the CCIR (1990f). For most practical purposes, this source of attenuation can be ignored for frequencies below 1 GHz.

Waves passing through rain precipitation are subject to attenuation due to absorption and scattering, the attenuation being a function of frequency, polarization, the microstructure of the precipitation, and the distance the wave travels through the precipitation cell.

In practice, because of air resistance, falling raindrops, other than the very small droplets, assume an approximately oblate spheroidal shape and this causes the specific attenuation coefficient to be higher for horizontal polarization than for vertical polarization. The calculation of the total rain attenuation along a radio path requires a knowledge of the rainfall rate distribution along the path and an allowance for inclined paths. The total attenuation is obtained by integrating the specific attenuation coefficient over the path length. In most practical cases, predictions are required for the attenuation likely to be exceeded for small percentages of time, and this therefore requires a knowledge of the cumulative long-term statistics for rainfall in the geographic area concerned. In the absence of such measured rainfall data for the area concerned, the rainfall data presented by the CCIR, for various geographic areas throughout the world, can be used.

The attenuation, in dB/km, due to a cloud of small droplets of a diameter less than 0.1 mm, is given by the product of the specific attenuation factor and the condensed water content of the cloud in g/m^3.

Effect of Atmospheric Absorption on Sky Noise Temperature

The nonionized atmosphere surrounding the Earth, in addition to causing attenuation of electromagnetic waves, acts as a source of radio noise. For many terrestrial communications systems, this is of little importance because the receiving-antenna noise temperature is dominated by the noise temperature of the Earth. However, for satellite and space communications systems using low noise temperature antennas and receivers, the sky noise temperature plays an important role in determining the received carrier–to–noise power ratio.

If the temperature of the absorbing medium can be assigned a mean value, T_m (degrees kelvin), and the total atmospheric absorption at a particular frequency for a given radio path is A (dB), the effective sky noise temperature, T_S, for the ground station receiving antenna is given by equation 7.86, where T_E is the extraterrestrial noise temperature (usually taken as 2.7°K for the frequency range 2 to 50 GHz, unless the receiving antenna is pointing near the sun).

$$T_S = T_m(1 - 10^{-A/10}) + T_E 10^{-A/10} \qquad (7.86)$$

Figure 7.24 shows the variation of effective sky noise temperature for clear air with 7.5 g/m^3 water vapor density for various elevation angles (θ), for the frequency range 1 to 60 GHz. Noise contribution due to absorption by rain or clouds can also be calculated from equation 7.86, provided that the temperature and specific attenuation variations along the radio path are known. Figure 7.25 shows an example of the theoretical variation in sky noise temperature due to attenuation from clouds on a radio path.

Depolarization due to Hydrometeors

The use of orthogonal polarization in communications systems is an important technique by which economy of frequency spectrum usage can be improved by reducing mutual interference between systems sharing the same, or adjacent, frequency channels. This

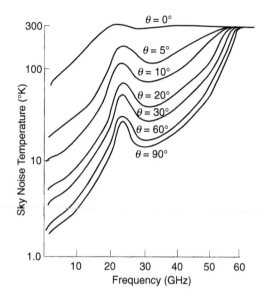

Fig. 7.24 Sky noise for clear air and infinitesimal antenna beamwidth. (Water vapor concentration = 7.5 g/m^3; surface temperature = 15°C; surface pressure = 1023 mb).

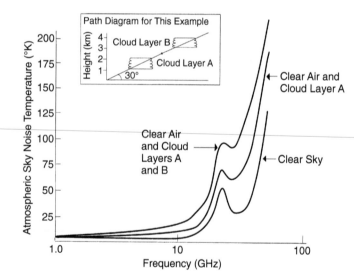

Fig. 7.25 Example of the effect of clouds on sky noise temperature. (Clear air: water vapor = 7.5 g/m^3 at surface; cloud layer: 0.5 g/m^3 liquid water).

technique relies on the Cross-Polarization Discrimination (XPD) achieved by the careful design and alignment of the transmitting and receiving antennas involved.

A significant factor affecting cross-polarization discrimination achieved in practice at SHF is the depolarization caused by hydrometeors, in particular rain and ice crystals.

Depolarization due to Rain

For rain-induced depolarization, there is a strong statistical correlation between the degradation in XPD and the attenuation of the wanted polarization, usually called Copolarized Path Attenuation (CPA).

The relationship between XPD (XdB) and CPA (CdB) is, however, complex and depends on frequency, polarization, angle of inclination of the path and raindrop size, shape, and canting angle distributions along the path.

Data is continually being amassed from propagation experiments, in many cases using communications satellites, and semiempirical relationships based on this data are in common use. One such semiempirical relationship is given in CCIR (1990b), with further parameters in CCIR (1990c).

Depolarization due to Ice Crystals

Depolarization due to clouds of ice crystals can be considerable and can occur in the absence of significant attenuation and depolarization due to rain. Because the major axes of the crystals are normally close to the horizontal, depolarization tends to a minimum for horizontal or vertical polarization. Abrupt changes in XPD have been observed to coincide with lightning strikes, and this appears to be due to rapid changes in the orientation of the axes of the crystals as the electrostatic charge distribution changes.

The occurrence of depolarization due to ice crystals and the magnitude of the effect on XPD vary with climate and geography, tending to be greatest in maritime climates. Further information on this subject can be obtained from CCIR (1990c) and Chu (1980).

The Sky Wave

Ultraviolet and X-radiation from the sun and, to a lesser extent, particle radiation from sunspots and galactic cosmic rays, cause the gases of the upper atmosphere to become ionized. The degree of ionization does not change uniformly with height, but there are relatively dense regions in layers above the Earth's surface. The properties of the four principal layers are

- D: About 50 to 90 km in altitude. This is greatly dependent on the sun's intensity and absorbs low-frequency radio waves on passage through the layer.
- E: About 90 to 130 km in altitude. It depends on the sun's intensity, with a small residual level at night, and exhibits a "sporadic" nature.
- $F1$: About 130 to 210 km in altitude. It merges with the $F2$ layer at sunset and depends on the sun's intensity.
- $F2$: About 210 to 400 km in altitude. It merges with the $F1$ layer at sunset and is influenced by Earth's magnetic field; and it depends on auroral zones, particularly in sunspot activity.

Reflection and Refraction of Waves in the Ionosphere

Radio waves cause free electrons and ions to oscillate and reradiate, increasing the phase velocity of the wave but reducing the group velocity. Thus, the ionosphere behaves as a medium with a refractive index of less than 1. This value is formulated later, when other influences have been discussed, but for now, it is sufficient to know that the refractive index of the ionosphere deviates further from 1 for the lower frequencies and for the higher electron densities.

The change of electron density with height results in a refractive index gradient, symbolized by dn/dh. This gradient will cause both reflection and refraction of the wave. Reflection dominates when the refractive index changes rapidly over a height range, which is small compared with the wavelength. Thus, reflection is the main mechanism for the low frequencies, where the refractive index changes more rapidly with real height and, more specifically, height measured in wavelengths.

On the other hand, for the higher frequencies refraction dominates over reflection, even though the amount of refraction decreases as the frequency increases. If the angle between the ionospheric layer and the direction of propagation is α, the wave will follow a path whose radius of curvature, r, is given by equation 7.87.

$$\frac{1}{r} = -\frac{\cos\alpha}{n}\frac{dn}{dh}$$ (7.87)

This radius of curvature might be sufficient to return the wave to Earth, depending on the angle, α, frequency, intensity of ionization, and other factors discussed later.

Ionospheric Attenuation

Collisions occur between free electrons and gas molecules, which destroy the oscillatory energy acquired by the electron, converting it to random thermal energy. Thus, energy is taken from the wave that is consequently attenuated. At low heights, there are more gas molecules and so more collisions (i.e., more attenuation). The number of collisions also depends on the amplitude of the electron vibrations, which increases with increasing wavelength and field strength.

In the LF and MF bands (i.e., 150 kHz to 1.6 MHz), the D layer almost totally attenuates the wave so that ionospheric propagation is mainly restricted to nighttime, when the D layer is absent. Then reflections take place off the lower edge of the E layer. This is rarely used as a means of communication, but it is a source of interference to MF broadcast services.

At still lower frequencies, reflection off the lower edge of the ionosphere occurs before much attenuation takes place.

Gyrofrequency—The Ordinary and Extraordinary Wave

In the absence of a magnetic field, electrons would oscillate in the direction of the electric vector of the wave. The presence of the Earth's magnetic field produces an additional force on the electron. The direction of this force is at right angles to the direction of motion of the electron and to the magnetic field. In general, a charged particle moves in a helix around the lines of a magnetic field. The frequency of rotation depends only on the charge of the particle, its mass, and the strength of the magnetic field. Thus, a resonance exists that is called the *gyrofrequency*.

If the wave frequency equals the gyrofrequency, the velocity of the electrons increases to large values and the wave attenuation is increased. The frequency at which this occurs varies between 0.8 MHz in equatorial regions to 1.6 MHz near the magnetic poles. In the United Kingdom it is about 1.3 MHz. Under these conditions, the electrons reradiate with two polarizations, giving rise to the ordinary and extraordinary waves. The ordinary wave propagates as though the Earth's magnetic field were absent, but the electron motion of the extraordinary wave is in the plane perpendicular to the direction of the Earth's magnetic field.

Polarization Coupling Loss

The energy in the extraordinary wave suffers more attenuation than the ordinary wave in its passage through the ionosphere, in fact, it is largely lost. Also, the ordinary wave might suffer a polarization change in its passage through the ionosphere and, where the receiving antenna is of the same polarization as that transmitted, there is a further loss. These factors give rise to so-called *polarization coupling loss*.

Polarization coupling loss is small in temperate latitudes because the Earth's magnetic field is nearer the vertical. Near the equator, the magnetic field is nearly horizontal. This means that on north–south paths, there might be as much as 6 dB of coupling loss. The effect is at its greatest on east–west paths near the equator, where the polarization coupling loss can be very large.

The Refractive Index of the Ionosphere

The refractive index of an ionized medium without a magnetic field and negligible loss of energy is given by equation 7.88, where N is the number of electrons per cubic centimeter and f is the frequency in kilohertz.

$$n^2 = 1 - \frac{81N}{f^2} \tag{7.88}$$

If the Earth's magnetic field and electron collisions are included, the complex refractive index, μ, is given by the Appleton-Hartree formula, as in equation 7.89, where Y is the gyrofrequency divided by the wave frequency, θ is the angle between magnetic field and ray, and $2\pi Z$ is the collision divided by wave frequencies.

$$\mu^2 = 1 - (1-n^2)\left[1 - jZ - \frac{Y^2 \sin^2\theta}{2(n^2 - jZ)}\right]\left[\pm\left(\frac{Y^4 \sin^4\theta}{4(n^2 - jZ)^2} + Y^2 \cos^2\theta\right)^{1/2}\right]^{-1} \tag{7.89}$$

The square root term takes the positive sign for the ordinary wave and the negative sign for the extraordinary wave.

The Critical Frequency

The maximum electron density of an ionospheric layer determines the critical frequency, f_c kHz. This is the highest frequency for which a vertically incident ray can be returned to Earth and corresponds to a refractive index of zero. The Earth's magnetic field and collisions can usually be neglected, resulting in equation 7.90.

$$f_c = 9\sqrt{N} \tag{7.90}$$

Ionospheric Cross Modulation

The oscillatory energy given to an electron by a wave is normally small compared with the ambient thermal energy, but for a very strong wave this might not be so. This increases the probability of electrons colliding with molecules so that all other waves entering the ionosphere will have increased attenuation. If the strong wave is amplitude modulated, the ionospheric attenuation will vary with the instantaneous amplitude of this wave and thus

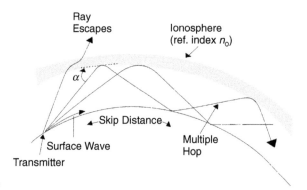

Fig. 7.26 Sky wave and surface wave transmission showing the skip distance.

the modulation. All other waves are therefore modulated in amplitude by the strong wave. This is called *ionospheric cross-modulation*.

Skip Distance and Maximum Usable Frequency

When a wave has a frequency greater than the critical frequency of the layer, it can still be returned to Earth if the angle of incidence, α, is such that inequality 7.91, where n_o is the refractive index associated with the maximum electron density, is satisfied.

$$\cos\alpha > n_o \tag{7.91}$$

This results in no sky wave energy being returned to Earth closer to the transmitter than a range referred to as the "skip distance," as shown in Figure 7.26.

For any given distance at a particular time, there is a maximum frequency that can be used, such that the signal will arrive at the receiver via the ionosphere. This frequency is the Maximum Usable Frequency (MUF). This frequency is also liable to be the one giving the strongest sky wave signal. The MUF is an important parameter in the propagation of short waves (i.e., frequencies from 1.6 to 30 MHz). It is given by equation 7.92.

$$\text{MUF} = f_c \cosec\alpha \tag{7.92}$$

The maximum skip distance is about 1700 km for the E layer and 3300 km for the F layer. Transmission can also take place by two or more hops, to be received at much greater distances.

Sporadic E Propagation

Sometimes waves are returned to Earth by the E layer at frequencies greater than the nominal MUF. This phenomenon is called *sporadic E*. It is assumed that this is caused by strongly ionized "clouds" in the lower part of the E region. By this means it is possible for propagation to occur over long distances at frequencies as high as about 100 MHz.

Sunspots

HF propagation is strongly influenced by sunspot activity. It can cause loss of communication or permit very extensive communication. The presence of sunspots is accompanied by a large increase in ultraviolet radiation.

Sunspots vary cyclically in number and size with a periodicity of about 11 years. This activity is approximately quantified by the relative Zurich sunspot number, also called the Wolf number, derived from observations of the number and size of sunspots.

Prediction of Field Strengths Received via the Ionosphere

So far, descriptions have been given of a variety of ionospheric phenomena. This is helpful in qualitative terms, but most engineers require an ability to make quantitative predictions of the field strength likely to be received in given circumstances. Such calculations should allow for each of the ionospheric features, which means they are complicated and lengthy. It is therefore not practicable to give a comprehensive description of the calculations here, but references listed below can be used. It must also be realized that these predictions are largely empirically derived; it does not seem possible to be deterministic about such a problem.

For HF predictions, CCIR (1970) is recommended. This, in its turn, has a comprehensive set of references and lists a lengthy computer program for completing the predictions described in the text.

For MF and LF predictions, CCIR (1982a) is recommended. Of course, a problem exists because of the large differences associated with different geographic locations. For this reason it is also recommended that CCIR (1982b) be consulted.

Ground Waves and Surface Waves

The terminologies of *ground wave* and *surface wave* are often considered synonymous, but it is useful to make a distinction. The influence of the ground in producing a reflected ray was discussed previously for elevated antennas and high frequencies. This ray or wave can usefully be described as the *ground wave*. With frequencies below about 1.5 MHz, and with antennas at ground level, another mechanism of propagation tends to dominate. This is a wave that is guided along the boundary between the surface of the Earth and the air above it. The best term to be applied to this mechanism is the *surface wave*.

Reference Field Strength

Equation 7.49 gave the field strength for a short monopole on a perfectly conducting plane. In practice, allowance must be made for

1. The power gain, g, of a particular antenna in a given direction, relative to the short monopole, assuming a perfect ground plane.
2. The imperfect conductivity of the ground plane, which causes a significant fraction of transmitter power to be lost as heat in the ground and not radiated as electromagnetic energy. This is taken into account by the transmitter antenna system efficiency, τ.
3. The reduction in field strength for the real path relative to that for a perfectly conducting flat plane. This is taken into account by the path attenuation factor, A.

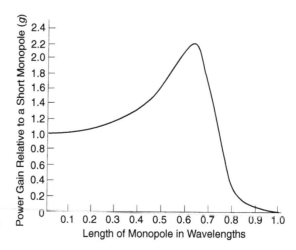

Fig. 7.27   Relative power gain of a monopole antenna.

The quantity P' given by equation 7.93 is the Effective Monopole Radiated Power (EMRP) in kilowatts, which is analogous to the ERP defined previously for the space wave.

$$P' = P_{kW}g\tau \tag{7.93}$$

Thus, an EMRP of 1 kW gives a field strength of 0.3 volts/meter at 1 km on a perfectly conducting ground.

Monopole antennas that are longer than 0.1 wavelength have gains given by the graph of Figure 7.27. A practical antenna can obtain additional gain, in a given direction, by many means. This can be allowed for, as with the space wave, by the use of a Horizontal Radiation Pattern (HRP).

The field strength for the practical case now becomes as in equation 7.94.

$$E = \frac{300(P_{kW}g\tau)^{1/2}}{d} A \quad \text{volts/meter} \tag{7.94}$$

The Sommerfeld–Norton Theory

The Sommerfeld–Norton theory (Jordan & Balmain, 1986; Terman, 1943) gives a method for prediction of attenuation over a homogeneous smooth earth. This method will not be fully described here, but for a vertical antenna it is convenient to define this calculation via the surface impedance concept. Thus, for ground conductivity σ, relative permittivity ε_r, and wave of length, λ, the relative surface impedance at grazing incidence can be given by equation 7.95.

$$\eta = \frac{(\varepsilon_r - j60\sigma\lambda - 1)^{1/2}}{\varepsilon_r - j60\sigma\lambda} \tag{7.95}$$

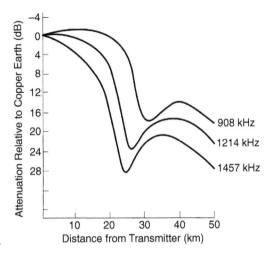

Fig. 7.28 Attenuations of MF signals through a built-up area.

For a given distance, d, a parameter, w, can be defined by equation 7.96.

$$w = -j\pi\eta^2\frac{d}{\lambda} \qquad (7.96)$$

From this, the complex surface wave attenuation factor, A, is given by equation 7.97, where *erfc* denotes the complementary error function (Abramowitz & Stegun, 1964).

$$A = 1 - j\sqrt{\pi w}e^{-w}erfc(j\sqrt{w}) \qquad (7.97)$$

The Realistic Terrain—A Solution via an Integral Equation

A realistic terrain is covered by clutter of human-made and natural objects that significantly modify the surface impedance, in ways described in Causebrook (1978a). This modification can result in attenuation varying as shown in Figure 7.28; that is, instead of a simple monotonic curve, there is an initial rise and a significant dip at $\frac{d}{\lambda}$ equal to about 100. In practice, it is desirable to have a calculation system that allows for

- The curvature of the Earth
- A terrain that varies in height
- A surface impedance modified by clutter

Hufford (1952) gives an integral equation that, at least in principle, provides a general solution for attenuation of radio waves. This can be written as in equation 7.98.

$$A(R) = 1 - \int_0^R (\eta + \Psi)e^{-j\beta\xi}A(r)\left(\frac{jR}{\lambda r(R-r)}\right)^{1/2}dr \qquad (7.98)$$

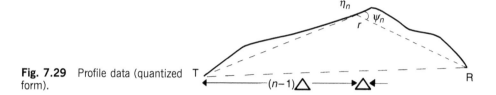

Fig. 7.29 Profile data (quantized form).

In equation 7.98, R is the distance of path over which attenuation is required; Ψ is the angle in radians between the line from the receiver to the surface at range r and the tangent to the surface at this point; ξ is the distance between terminals subtracted from the sum of distances from terminals to surface at r (see Figure 7.29); η is the relative surface impedance at r.

The integral contains quantities that do not have "mathematically closed" forms, so an analytical solution is unobtainable. To overcome this, it is first possible to divide the distance, R, into N quanta of length Δ to give equation 7.99, where I_n is given by equation 7.100.

$$A(r) = 1 - \sum_{n-1}^{N} Q_n I_n \tag{7.99}$$

$$I_n = \int_{(n-1)\Delta}^{n\Delta} A(r) \left(\frac{jN\Delta}{\lambda r(N\Delta - r)} \right)^{1/2} dr \tag{7.100}$$

Q_n is a part of the integrand that can be considered constant in the interval, Δ, to be given by equation 7.101, with quantities defined by Figure 7.29.

$$Q_n = (\eta_n + \Psi_n) e^{-j\beta \xi_n} \tag{7.101}$$

The next barrier to a simple solution is that $A(r)$ is initially an unknown function. Up to a distance of 2Δ this presents little problem because terrain variations will not greatly modify the result. Thus the Sommerfeld-Norton function can be used to determine attenuations A_1 and A_2 at distances Δ and 2Δ, respectively, by equation 7.102.

$$w = -j\pi \eta_1 \eta_2 \frac{n\Delta}{\lambda} \tag{7.102}$$

Also, I_1 can be determined analytically by putting $A(r)$ equal to a power series expansion of the Sommerfeld–Norton formula.

Further values of I_n and A_n can be obtained by assuming that $A(r)$ varies linearly within an interval, as in equation 7.103, where S_n and R_n can be derived from integrals that are amenable to analytic solution.

$$I_n = S_n A_{n-1} + R_n A_n \tag{7.103}$$

Of course, A_n is initially unknown for n greater than 2, but it can be solved by equation 7.104.

$$A_n = \frac{1 - \sum_{m=1}^{n-1} Q_m I_m - Q_n S_n A_{n-1}}{1 + Q_n R_n} \qquad (7.104)$$

This equation must be applied repeatedly from $n = 3$ to $n = N$, at which point the required attenuation at distance R is obtained

Ratio of Electric to Magnetic Field in Cluttered Environments

The attenuation, calculated by the method of the previous section, is only rigorously applicable just above the top of the clutter (buildings, trees, etc.). However, most reception of MF and LF signals takes place on antennas well in the clutter.

For antennas like the ferrite rod, designed to respond to the magnetic field, the attenuation experienced should, on average, be close to the calculation. This is not true for the electric field, like that received by a rod antenna on a car, because this can suffer more attenuation in the clutter.

In the open environment, the electric field is related to the magnetic field by free-space impedance. However, in the cluttered environment this is not the case. In fact, the impedance is less than the free space value. Further details on this subject are available in Causebrook (1978b).

Curves for Surface Wave Field Strength Prediction

For many practical purposes it can be considered that calculating the field strength with mathematical techniques is too difficult, cumbersome, or time consuming. In this case a prediction can be obtained from a set of published curves. These curves usually show the variation in surface wave field strength, with distance as a function of both frequency and ground conductivity (relative permittivity is a less relevant parameter). The trend is for attenuation with distance to be greater at higher frequencies and when ground conductivity is low. A comprehensive set of these curves is to be found in CCIR (1990d).

Ground Conductivity

It is necessary to add a few words about the way a conductivity figure is ascribed to the Earth for the present purposes. In the case where the ground is uniform and composed of similar material at all depths and composed of material having a conductivity that remains constant independent of frequency, the meaning of the term is fairly clear. However, the geologic structure of the real Earth has considerable variations in the constituent materials near the surface. To determine the depths to which the characteristics of the Earth's surface are significant, the "skin depth" of the current flow must be calculated. This skin depth varies between 70 m for low frequencies and poor ground conductivity to 20 m for high frequencies and more conductive ground. For the sea, with its greater conductivity, the skin depth at medium frequencies is less than 1 m. The effective conductivity for land is best taken as a mean value averaged over the values obtainable down to the skin depth.

Electricity undertakings have measured conductivity, for 50 Hz, to a depth of about 20 m. This gives a good guide for values that can be used in the MF band. Various methods are quoted in published literature for the determination of ground conductivity (CCIR, 1990e). Also, a method developed in Finland (Koskenniemi & Laiho, 1975) appears to be accurate and can be measured quickly from an aircraft.

In the United Kingdom, ground conductivity can be as high as 20 mS/m (milli-siemens per meter) but is more likely to be 10 mS/m in fertile regions and about 1 mS/m in mountainous regions. If the ground conductivity changes by a small amount over the path in question, a mean value will normally produce a reasonable result, but for large variations calculations must take account of the different characteristics in different segments of the path.

References

Balanis, C. A. (1982) *Antenna Theory Analysis and Design*, Harper & Row.

Bremmer, H. (1949) *Terrestrial Radio Waves: Theory of Propagation*, Elsevier.

Causebrook, J. H. (1978a) Medium-Wave Propagation in Built-up Areas, *Proceedings IEE* 125(9):804–808.

Causebrook, J. H. (1978b) Electric/Magnetic Field Ratios of Ground Waves in a Realistic Terrain, *IEE Electr. Lett.* 14(19):614.

CCIR (1970) *CCIR Interim Method for Estimating Sky-Wave Field Strength and Transmission Loss at Frequencies between the Approximate Limits of 2 and 30 mHz*, CCIR Report 252-2.

CCIR (1982a) *Prediction of Sky-Wave Field Strength between 150 kHz and 1600 kHz*, CCIR Recommendation 435-4.

CCIR (1982b) *Analysis of Sky-Wave Propagation Measurements for the Frequency Range 150 to 1600 kHz*.

CCIR (1990b) *Cross-Polarisation Due to the Atmosphere*, CCIR Report 722.

CCIR (1990c) *Propagation Data Required for Space Telecommunications Systems*, CCIR Report 564.

CCIR (1990d) *Ground-Wave Propagation Curves for Frequencies between 10 khz and 30 mhz*, CCIR Recommendation 368.

CCIR (1990e) *Electrical Characteristics of the Surface of the Earth*, CCIR Report 229.

CCIR (1990f) *Attenuation by Atmospheric Gases*, CCIR Report 719.

Chu, T. S. (1980) Analysis and Prediction of Cross-Polarisation on Earth-Space Links, *Proceedings URSI Int. Symp. (Commission F)*.

Deygout, J. (1966) Multiple Knife-Edge Diffraction of Microwaves, *IEEE Trans. Antennas Propag.* 14(4):480–489.

Gautschi, W. (1964) Error Function and Fresnel Intervals, in Abramowitz, M., and Stegun, I. A. (eds.), *Handbook of Mathematical Functions*, Dover, 227–232

Hacking, K. (1970) UHF Propagation over Rounded Hills, *Proceedings Inst. Elect. Engrs.* 117(3):499–511.

Hufford, H. A. (1952) An Integral Equation Approach to Wave Propagation over an Irregular Surface, *Q. J. Appl. Math.* 9 391–404.

Jordan, E. C., and Balmain, K. G. (1968) *Electromagnetic Waves and Radiating Systems, Second Edition*, Prentice-Hall, 317–321.

Koskenniemi, O., and Laiho, J. (1975) Measurement of Effective Ground Conductivity at Low and Medium Frequencies in Finland, *EBU Tech. Rev.* 153(October):237–240.

Longhurst, R. S. (1967) *Geometrical and Physical Optics, Second Edition*, Longmans.

Terman, F. E. (1943) *Radio Engineers' Handbook*, McGraw-Hill, 675–682.

Tsukiji, T. and Tou, S. (1980) On Polygonal Loop Antennas, *IEEE Trans. Antennas Propag.* 28(4):571–575.

Van Der Pol, B., and Bremmer, H. (1937) The Diffraction of Electromagnetic Waves from an Electrical Point Source Round a Finitely Conducting Sphere, with Applications to Radiotelegraphy and the Theory of the Rainbow, *Phil. Mag.* 24(1):141–176; 24(2):825–863.

8 Radio Wave Propagation

Mark Holker
Hiltek Ltd.

Introduction

Electromagnetic energy radiates outward from the source, usually an antenna, at approximately the speed of light and is attenuated and influenced by the medium through which it travels. Radio communications necessitate launching RF energy into the propagation medium, detecting its presence at some remote point, and recovering the information contained within it while eliminating noise and other adverse factors introduced over the transmission path. An understanding of radio wave propagation is therefore essential in the planning and operation of radio communications systems to ensure that communications can be established and that there is an optimum solution between costs (capital and running) and link availability.

This chapter examines radio wave propagation in the frequency bands from VLF (10 kHz) to the millimetric band (100 GHz) and the influence of the Earth, the atmosphere, and the ionosphere on such transmissions. It is assumed that propagation is in the far field (i.e., several wavelengths from the antenna), where the electric and magnetic components of the wavefront are at right angles and normal to the direction of propagation. The process by which energy is launched into the propagation medium in the near field is described in Chapter 9.

Radio waves can be propagated in one or more of five modes depending on the medium into which they are launched and through which they pass. These modes are

1. Free-space propagation, where radio waves are not influenced by the Earth or its atmosphere
2. Ground-wave propagation, where radio waves follow the surface of the Earth
3. Ionospheric propagation, where radio waves are refracted by ionized layers in the atmosphere
4. Tropospheric propagation, where transmission is "line of sight," with some atmospheric refraction occurring

5. Scatter propagation, where natural phenomena such as tropospheric turbulence or
   ionized meteor trails, are used to scatter radio waves

The Radio Frequency Spectrum

The radio frequency spectrum is divided into a number of bands that have been given designations such as LF, MF, HF, and the like, for ease of reference. These are shown in Figure 8.1 and are presented on a logarithmic scale. The microwave band, usually taken to be from 300 MHz to 30 GHz, has been subdivided into a number of subbands, which have been given letter designations such as X band (8 to 12 GHz), but differing definitions are used and can cause confusion. GHz in the range of 30 to 100 are commonly referred to as the *millimetric band*. Allocations have been made up to 275 GHz by the ITU, but there is little activity other than experimental work above 100 GHz.

Frequency and wavelength are shown in Figure 8.1. As radio waves propagate at 3×10^8 meters per second in free space, frequency and wavelength are related by equation 8.1, where λ is wavelength in meters and f is frequency in cycles per second.

$$\lambda \times f = 3 \times 10^8 \qquad \text{(8.1)}$$

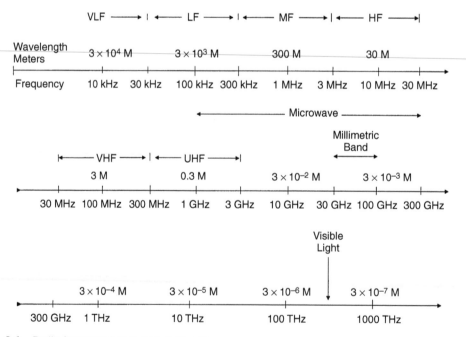

Fig. 8.1   Radio frequency bands and designations.

Free-Space Propagation

Free-space propagation is seldom a practical situation, but it occurs when both transmitting and receiving antennas are situated away from the influence of the Earth's surface or other reflecting and absorbing objects, including the transmitter and receiver themselves. If power is fed to an isotropic antenna (i.e., an antenna that radiates equally in all directions in azimuth and elevation), the wavefront will radiate outward from the antenna in an ever expanding sphere at 3×10^8 meters (186,282 miles) per second. The strength of the signal obviously decreases with distance as a given amount of power is spread over a greater area, and the incident power density at a remote point can be calculated as in equation 8.2, where P_r is received power density in watts per meter^2, P_t is transmitted power in watts, and r is distance in meters.

$$P_r = \frac{P_t}{4\pi r^2} \tag{8.2}$$

Because of the very large differences in power density over long propagation paths, particularly in the microwave bands, it is usual to measure power density in decibels (dB) relative to 1 watt (i.e., dBW), or 1 milliwatt (i.e., dBm).

At frequencies below the microwave range, field strength in volts per meter is a more usual measurement than power density, partly because volts per meter have historically been used for the measurement of signal strength and partly because the thermal heating effect of power absorption is easier to use for microwave measurements. The conversion can be made for power density to field strength using the simple derivation of Ohm's law (Braun, 1986), as in equation 8.3, where V is voltage, R is resistance in ohms, and P is power in watts.

$$V^2 = RP \tag{8.3}$$

In this case, R is substituted for Z_0, the characteristic impedance of free space, which is a constant of 87 ohms.

Field strength can therefore be converted to power density by equation 8.4 (Maslin, 1987), where E is field strength in volts per meter, Z_0 is the impedance of free space (87 ohms), and P_d is power density in watts per meter^2.

$$E^2 = Z_0 P_d \tag{8.4}$$

Equation 8.5 can be used to convert directly from dBW to dBμV, where E is in dB.

$$E = P_d \text{dB}(1 \text{W}/\text{m}^2) + 145.6 \tag{8.5}$$

On the left side of this equation, dB represents a voltage and not a power ratio.

The voltage at the center of an unterminated half-wave dipole in the path of a radiowave and aligned along the axis of the electric vector (i.e., no polarization loss) is given by equation 8.6, where V is the voltage at the center of the dipole, E is the field strength in volts per meter, and λ is the wavelength in meters.

$$V = \frac{E\lambda}{\pi} \qquad (8.6)$$

If the dipole is connected to a feeder of matching impedance, E is halved. If polarization and feeder losses are taken into account, the input to a receiver can be calculated.

A useful formula for calculating free-space attenuation, derived from the power density equation, 8.1 and from equation 8.6, and published in ITU-R recommendations, is given by equation 8.7, where L_{fr} is the free-space loss in decibels, f is the frequency in megahertz, and d is distance in kilometers.

$$L_{fr} = 32.44 + 20\log f + 20\log d \qquad (8.7)$$

It will be noted that equation 8.7 introduces a frequency component into the calculation.

As an example, the power density from a 500-watt transmitter working into a lossless isotropic antenna at a distance of 1 kilometer is 0.039 W^{-3}/m^2, and at 16 kilometers it is 0.155 W^{-6}/m^2. This is a power ratio of 24 dB, which is to be expected because there is a 6-dB power loss every time that the distance is doubled. The ITU-R formula (equation 8.7) gives a free-space loss of 72.4 dB at 1 km and 96.5 dB at 16 km. Again, that is a difference of 24 dB. Figure 8.2 shows the free-space loss for frequency and distance based on the equation's formula.

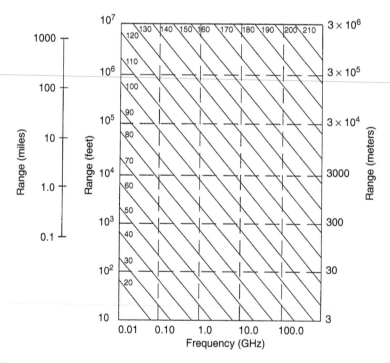

Fig. 8.2 Free-space loss versus distance for frequencies from 1 MHz to 100 GHz (units on the diagonal lines are in dB).

The Propagation Medium

Almost all propagation involves transmission through some of the Earth's atmosphere, and the structure of the atmosphere and features within it are shown in Figure 8.3. Most meteorological activity and cloud formation occur in the first 10 km, and jet aircraft cruise at between 10 and 15 km. Air pressure falls with height, and at 30 km radiation from the Sun is sufficient to generate some free electrons, but the first distinct ionized layer, the D layer, occurs at 70 km. Above the D layer, temperature and incident radiation increase and the E, $F1$, and $F2$ layers are formed between 120 km and 450 km. The ionosphere, which is the region in which the ionized layers are formed, spans the region from 50 to 600 km. Ionized trails from meteors occur at around 100 km, and the lowest satellite orbit is at about 150 km.

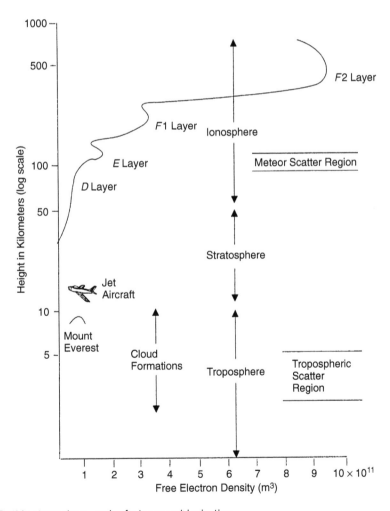

Fig. 8.3 Earth's atmosphere: major features and ionization.

The primary influence on the atmosphere is radiation from the Sun, which causes the ionized layers to form and creates global climatic and regional weather patterns. Solar radiation follows an 11-year sunspot cycle, which has a direct correlation with ionization and is recorded on a scale from 0 to about 200. This is a smoothed average of the number of sunspots, which are observed disturbances on the Sun's surface. A sunspot number is issued by the Sunspot Data Center in Brussels and by the Telecommunications Services Center in Boulder, Colorado, and is a smoothed 12-month average of solar activity indicating the degree of ionization that can be expected. High sunspot numbers are an indication of better conditions for long-range HF communication, and this might also lead to unwanted long-distance propagation in the VHF bands, creating interference in mobile communications and FM and TV broadcasting services.

Because the sunspot number is a running average, it might not be an accurate indication of daily or hourly conditions. Solar flux is a measure of solar activity taken at 2.8 GHz and is a better indication of real-time ionospheric conditions. It is measured continuously and quoted on a scale usually in the range from 60 to 260. The U.S. National Bureau of Standards radio station WWV transmits ionospheric data information hourly.

The signal-to-noise ratio (SNR) is often the determining factor in establishing communications, and it depends on the absolute level of signal, the external noise in the propagation medium, and the internal noise in the transmitting and receiving equipment. At HF, communications are normally externally noise limited so that it is the noise in the propagation medium that predominates, while at VHF and above the noise generated within the first stages of the receiver is usually the determining factor. Therefore, there might be no benefit in trying to increase the received strength of HF signals by using larger and higher-gain antennas because noise can increase proportionally with no improvement in SNR.

External noise is from three sources: galactic noise, atmospheric noise, and man-made noise. Noise power in a communication link is given by equation 8.8, where N is the noise power in watts, T is the temperature in degrees absolute, often taken as 290 K because this corresponds to the normal room temperature of 17°C, and k is Boltzmann's constant (1.38×10^{-23} J·K^{-1}).

$$N = kTB \qquad (8.8)$$

In communications links, noise is measured in terms of the noise power available from a lossless antenna and can be expressed as in equation 8.9, where P_n is the total power in dBW, F_a is the antenna noise figure in dB, $B = 10\log$ bandwidth (in Hertz), and 204 is $10\log kT°$, assuming $T° = 290$ K.

$$P_n = F_a + B - 204 \qquad (8.9)$$

Atmospheric and man-made noise information is given in ITU-R publications, and typical values are shown in Figure 8.4. Galactic noise originates from sources outside the Earth's atmosphere, such as the Sun and the stars, and extends from about 15 to 100 MHz. It is limited by ionospheric absorption below this frequency range and atmospheric absorption above it. Atmospheric noise is the major source of noise in the MF and HF bands and is primarily due to lightning discharges, so it is particularly severe in the rainy season in tropical regions, such as equatorial Africa, and at its lowest value in high latitudes at night. It is transmitted over long distances by sky wave paths. Human-made noise can be similarly

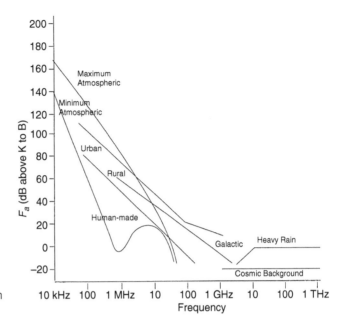

Fig. 8.4 Noise in the propagation medium: typical figures.

transmitted and emanates from power lines, industrial machinery, and fluorescent tubes. Four standard levels for business, residential, rural, and quite rural sites are defined in the ITU-R reports.

Analysis of a propagation path usually requires calculation of distance from transmitter to receiver, and it is useful to know the great circle bearing. This information can be calculated from equations 8.10 and 8.11 for two points, A and B, on the surface of the Earth, where A is point-A latitude in degrees; B is point-B latitude in degrees; L is the difference in longitude between A and B; C is the true bearing of the receiver from the transmitter, which can be 360°; and D is the distance along the path in degrees of arc, which can be converted to kilometers by multiplying by 111.111 (i.e., 1 degree of arc = 111.111 kilometers).

$$\cos D = \sin A \sin B + \cos A \cos B \cos L \qquad (8.10)$$

$$\cos C = \frac{\sin B - \sin A \cos D}{\cos A \sin D} \qquad (8.11)$$

Low- and Medium-Frequency Ground Waves

In the low- and medium-frequency bands up to 3 MHz, ground wave (rather than sky wave) propagation is used because sky wave is heavily absorbed in daytime by the D layer. LF and MF antennas are generally short in terms of wavelength, so vertical mast radiators are used. These have the advantages of maximum radiation at low angles to the horizon and also radiating vertically polarized transmissions.

The attenuation of vertically polarized ground wave transmissions is very much less than for horizontal polarization—for example, the ground wave attenuation of a 2-MHz

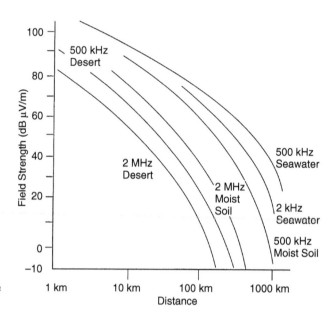

Fig. 8.5 Ground wave propagation: field strength for different frequencies and surface conditions.

vertically polarized transmission over medium soil is 45 dB at 30 km, whereas a horizontally polarized signal is attenuated by nearly 95 dB. Vertical polarization is therefore almost always employed.

Because currents are induced in the ground by vertically polarized ground wave transmissions, attenuation is also dependent on ground conductivity and dielectric constant. Salt water offers low attenuation whereas desert sand or polar ice offer high attenuation and consequently reduced communications coverage. Terrain irregularities, such as hills and mountains, and rough (as compared with calm) seas also reduce coverage.

Received, field strength at any distance can be calculated theoretically (Braun, 1986; Maslin, 1987); however, ITU-R published a series of graphs showing received field strength in microvolts per meter and in dBs relative to 1 microvolt for distances up to 1000 km. Frequencies from 500 khz to 10 MHz and homogeneous surface conditions from seawater to dry soil are shown. Some of the curves are shown in Figure 8.5 and are calculated on the basis of a 1-kW transmitter working into a short omnidirectional monopole antenna. The field strength values are proportional to the square root of the power and must be adjusted accordingly for different powers or for higher-gain or directional antennas.

The ITU-R curves shown in Figure 8.5 assume propagation over homogeneous terrain, but discontinuities occur in paths over land and sea or over different types of soil, and the field strength over such a path can be calculated by a method developed by Millington (1949). An example of the field strength over a land and sea path is shown in Figure 8.6, and it is interesting to note that the field strength rises after the wavefront passes from land to sea.

MF ground wave propagation offers the advantage of predictable but limited communications coverage, which is largely independent of ionospheric conditions and diurnal or seasonal variations. Distances of up to 1000 km are achievable over seawater, but in desert

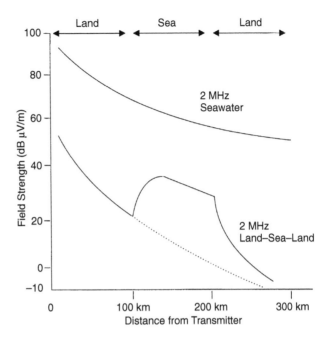

Fig. 8.6 Ground wave field strength curve for a hypothetical nonhomogeneous path.

conditions this might be limited to tens of kilometers unless very high powers and directional antennas are used. A limitation that often occurs in medium-frequency broadcasting is interference between ground wave and sky wave from the same transmitter during darkness because of the absence of the absorbing D layer, causing deep and rapid fading in received signals. The only practical solution is limiting power radiated at night, when the D layer is no longer present, and designing the antenna to minimize high angle radiation. At LF and VLF, very-long-distance and global communications can be achieved, but this necessitates the use of high transmitter powers and very large antenna systems.

High-Frequency Sky Wave Propagation

Propagation in the HF band from 3 to 30 MHz is probably the most variable and least predictable of all transmission modes because it depends on the height and intensity of the ionized layers in the ionosphere. A wealth of data has been collected and incorporated in comprehensive computer programs to enable predictions to be made with a reasonable degree of statistical accuracy, and modern ionospheric sounders can evaluate conditions on a real-time basis. However, unpredictable events, such as solar flares, mean that HF propagation always contains some degree of uncertainty (as does most radio wave propagation), and predictions can only be made on a statistical probability basis.

The ionosphere, which is the primary influence on HF propagation, extends from approximately 80 to 300 km above the Earth's surface, and it divides into a number of distinct ionized layers. Variations in the height and intensity of the layers occurs on a diurnal and seasonal basis—the result of the rotation and position of the Earth in relation to the Sun—and on the longer-term 11-year sunspot cycle. The diurnal variation is shown in

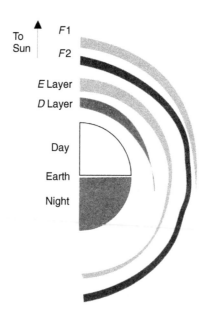

Fig. 8.7 Diurnal variation in the
Earth's ionized layers.

Figure 8.7, and it will be seen that at night, when incident radiation is at a minimum as the ionosphere is in the Earth's shadow, the ionosphere comprises two comparatively thin layers at 110 km and 210 km, called the *E* and *F* layers, respectively.

During daylight, these two layers increase in thickness and intensity, and the *F* layer divides into the separate *F*1 and *F*2 layers at 210 km and 300 km. In addition, the *D* layer forms at 80 km during daylight but disappears at night. Similar variations occur with the seasons, so there is a greater level of ionization in northern latitudes during July (summer) than in January (winter), but the reverse applies in the southern hemisphere.

The *D* layer actually spans an altitude from 50 to 90 km and absorbs frequencies in the MF and lower HF bands. Higher frequencies pass through the *D* layer suffering some attenuation and can be reflected back to Earth by the *E* layer at distances up to 2000 km. Still higher frequencies within the HF band will pass through the *E* layer and can be reflected by one of the *F* layers, providing very-long-distance communications. Frequencies in the VHF bands and above (i.e., above 30 MHz) generally pass through all the layers and into space, except in unusual conditions of strong ionospheric activity.

Although a radio wave can be visualized as being reflected by the ionosphere, the process is in fact one of refraction, and the angle of refraction is proportional to both the angle of incidence and the frequency. When the wavefront enters an ionized layer, it excites free electrons into oscillation, which reradiates electromagnetic energy, and this reradiation modifies the direction of the wavefront, tending to bend it back to Earth. For a given frequency, this tendency increases as the angle of incidence is reduced, so at the critical angle the wave will be refracted back to Earth, but at greater angles (i.e., nearer the vertical) only partial or no refraction occurs. This is illustrated in Figure 8.8.

Similarly, for a given angle of incidence, refraction decreases with an increase in frequency, so at a critical frequency the wavefront will pass through the layer. As the wavefront passes through an ionized layer, it imparts energy to the electrons, and a small amount of this energy is lost as heat. In the *D* layer, this transfer of energy is sufficient to completely

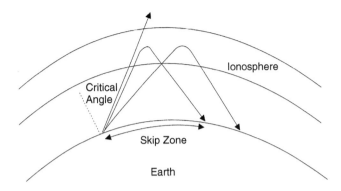

Fig. 8.8 Critical angle of ionospheric refraction.

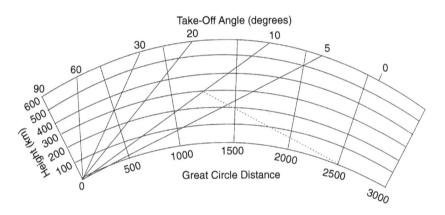

Fig. 8.9 Take-off angle for single-hop sky wave paths.

absorb medium frequencies during daylight, but at higher frequencies it causes some attenuation in the refraction process. Refracted waves also undergo a change of polarization that is due to the complex movement of the free electrons; this phenomenon is called *Faraday rotation*.

In planning an HF sky wave link, it is necessary to know the distance between the transmitter and receiver, or the area to be covered, and this can be calculated by equations 8.10 and 8.11. It is desirable to minimize the number of "hops" (i.e., refractions from the ionosphere and reflections from the Earth's surface) to minimize path attenuation and the inherent variability associated with ionospheric refraction. Maximum radiation from the antenna should therefore be at an angle (the take-off angle) that results in refraction from the ionosphere onto the target area. Figure 8.9 is a diagrammatical representation of the surface of the Earth from which the optimum take-off angle of radiation from the antenna can be derived if the distance and refracting layer height are known. A lower take-off angle is required for longer distances, and for distances of greater than 2000 km two or more hops are normally required.

In practice, a number of different propagation modes can be established on multihop links, which makes field strength predictions for the receive site difficult and uncertain.

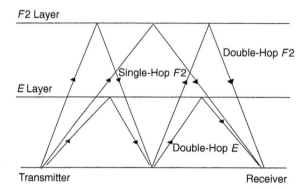

Fig. 8.10 Possible multihop sky wave paths.

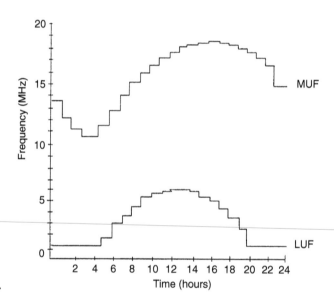

Fig. 8.11 Frequencies for an HF link from London to Lisbon.

Three possible modes are shown in Figure 8.10. A first approximation can be obtained by calculating the free-space loss using equation 8.7, taking into account increased distance due to ionospheric refraction and applying this to the field strength at the transmitter site. Additional losses of up to 20 dB will result from ionospheric and ground reflections, but these are very approximate values.

Actual sky wave refraction losses can be calculated (Braun, 1986), as can received field strength (Damboldt, 1975). Computer propagation prediction programs are now generally employed for analyzing HF paths and predicting received signal strength because of the large number of cancelations required rather than their inherent complexity. Such programs run on desktop PC computers (Hitney, 1990), although many research establishments have larger and more sophisticated programs (Dick, 1990).

Figure 8.11 shows a typical printout of a link from London to Lisbon, a great circle distance of 1656 kilometers and a single-hop, *F2*, distance of 1757 kilometers. The program shows the Maximum Usable Frequency (MUF) and the Lowest Usable Frequency (LUF), and at times when the LUF exceeds the MUF, HF sky wave communication is not possible. The MUF is determined by the degree of layer ionization, while the LUF is generally determined by the multihop path attenuation and the noise level at the receive site. The Frequency of Optimum Transmission (FOT) is the frequency with maximum availability and minimum path loss, and it is generally taken as 90 percent of the MUF.

HF ground wave and sky wave communications are still widely used for long-distance low-capacity services such as aeronautical ground-to-air transmissions, despite the increasing availability of satellite services. Defense forces are a major user, but the effects of the Electromagnetic Pulse (EMP) released by a nuclear explosion on the ionosphere and on equipment require special consideration in military systems. Much of the information on this topic is classified, but some papers are available.

Terrestrial Line-of-Site Propagation

In the VHF, UHF, and microwave bands, the ionosphere has a minimal effect on propagation, although anomalous conditions, such as sporadic E propagation, do affect the lower frequencies in this range. The frequencies are generally well above the critical frequency so that transmissions pass through the ionized layers and out into space. This offers the considerable benefit that the same frequencies can be used and reused many times without causing mutual interference, provided sensible frequency planning is carried out and adequate physical separation is provided.

Communications link calculations can be carried out using the Friis power transmission formula as in equation 8.12, where P_r is received power in watts, P_t is transmitted power in watts, G_t is the gain of the transmitting antenna in the direction of the receiving antenna, G_r is the gain of the receiving antenna in the direction of the transmitting antenna, and r is the distance in meters between antennas.

$$P_r = \frac{P_t G_t G_r \lambda}{(4\pi r)^2} \tag{8.12}$$

For systems calculations, the formula can be expanded to include transmit and receive antenna $VSWR$ and polarization mismatch, as in equation 8.13, where ρ_r is the magnitude of the voltage reflection coefficient at the receive antenna, ρ_t is the magnitude of the voltage reflection coefficient at the transmit antenna, and p is the polarization mismatch.

$$\rho_r = \frac{P_t G_t G_r \lambda^2 (1-\rho_r^2)(1-\rho_t^2)p}{(4\pi r)^2} \tag{8.13}$$

The voltage reflection coefficient can be calculated from the $VSWR$ as in equation 8.14.

$$p = VSWR - \frac{1}{VSWR} + 1 \tag{8.14}$$

The mismatch between two elliptically polarized waves is given by equation 8.15, where R_t is given by equation 8.16 and R_r by equation 8.17. R_t is the transmit antenna axial ratio, and R_r is the receive antenna axial ratio.

$$p = \frac{1 + R_t^2 R_r^2 + R_t R_r \cos 2\theta}{(1 + R_t^2)(1 + R_r^2)} \qquad (8.15)$$

$$R_t = r_t + \frac{1}{r_t} - 1 \qquad (8.16)$$

$$R_r = r_r + \frac{1}{r_r} - 1 \qquad (8.17)$$

Equation 8.13 assumes free-space conditions; however, absorption occurs because of rain and fog and water vapor and oxygen in the air, as shown in Figure 8.12. It should be noted that the first water absorption band occurs at 2.45 GHz, a phenomenon that is put to good use in microwave ovens. The attenuation figure taken from this graph must be added to the figure calculated from the Friis formula.

Although VHF, UHF, and microwave point-to-point communications are frequently referred to as "line of sight," the change in the refractive index of the atmosphere with height does in fact cause radio waves to be bent in the same direction as the Earth's curvature. This in effect extends line of sight, and the relationship between the surface refractivity and the

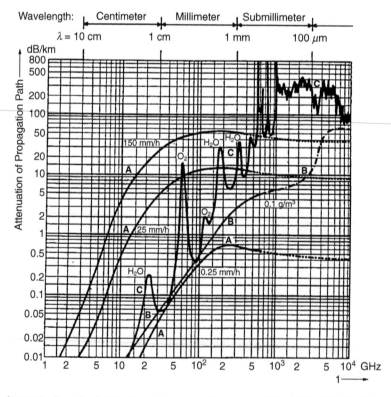

Fig. 8.12 Atmospheric and rainfall attenuation versus frequency: (A) rainfall at 0.25, 25, and 150 mm per hour; (B) fog at 0.1 gram per m; (C) molecular absorption by water vapor and air.

effective Earth radius is shown in Figure 8.13. The refractive index of air does, in fact, depend on atmospheric pressure, water vapor pressure, and temperature, and these factors all vary. An average effective Earth radius factor of 1.33 or "four thirds Earth" is therefore assumed for most link assessments, and this is shown diagrammatically in Figure 8.14 for a hypothetical link. This presentation has the advantage that transmission paths can be plotted as straight lines.

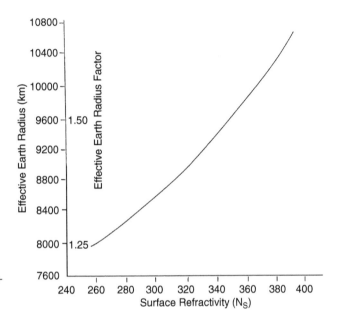

Fig. 8.13 Atmospheric surface-refractive index versus effective Earth radius.

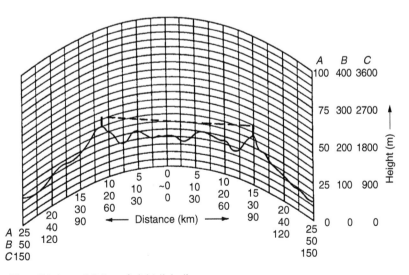

Fig. 8.14 "Four thirds earth" line-of-sight link diagram.

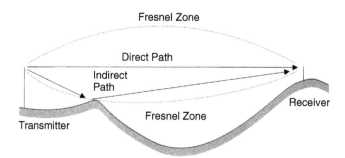

Fig. 8.15   Fresnel zone.

The profile of the topology between transmitting and receiving antennas can also be plotted. It is desirable to avoid any obstructions within the first Fresnel zone, which is defined as the surface of an ellipsoid of revolution with the transmitting and receiving antennas at the focal points in which the reflected wave has an indirect path half a wavelength longer than the direct paths. Figure 8.15 illustrates this, and the radius of the first Fresnel zone at any point (P) is given by equation 8.18, where R is the radius and d_1 and d_2 are the distances from point P to the ends of the path.

$$R^2 = \frac{\lambda d_1 d_2}{d_1} + d_2 \tag{8.18}$$

The height of the transmitting and receiving antennas above the intervening terrain and the roughness of the terrain have a marked effect on path attenuation and received field strength in the VHF and UHF bands. The Friis formula generally yields results that are too optimistic; that is, attenuation is too low and field strength is too high. Figure 8.16, reproduced from ITU-R Recommendations, shows received field strength values at 10 meters above ground level for a radiated power of 1 kW for different transmit antenna heights.

The field strengths are modified by correction factors for the terrain irregularity that is between the transmit and receive antennas, which can increase attenuation by up to 18 dB or reduce it by 7 dB depending on the topology. Average terrain irregularity is defined as the differences in height above and below 50 meters in only 10 percent of the path length. Figure 8.17 displays the attenuation correction factors for field strength for various terrain height differences.

Frequencies at 900 MHz and 1800 MHz are used for comparatively short-distance "cellular" services, and communications are often limited by multipath interference caused by single or multiple reflections, particularly in urban environments. The only practical solution is to try different positions for the transmit antenna to minimize this problem. Multipath interference can also be a problem on microwave links for TV outside broadcasts, where transmissions might be from moving vehicles or aircraft. This can be minimized by using circular polarization because the reflected signals undergo a polarization reversal, which enables the receive antenna to discriminate against them.

Because the propagation medium is continuously changing, the figures derived from ITU-R curves are based on statistical probabilities, generally 50 percent signal levels for 50 percent of the time, and therefore diversity transmission and a larger link margin might be required to increase channel availability.

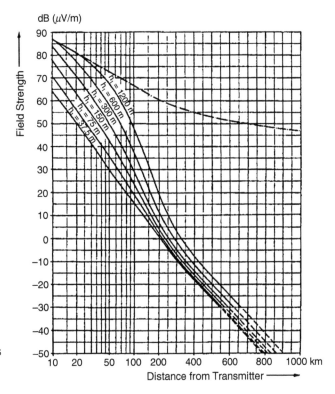

Fig. 8.16 Received field strengths for different transmit antenna heights, VHF and UHF bands.

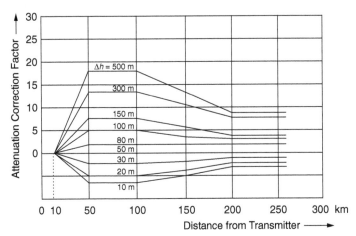

Fig. 8.17 Attenuation correction factors for different terrain roughness.

Over-the-Horizon Transmissions

The four thirds Earth radius does extend transmission beyond the horizon in normal conditions, but a natural phenomenon known as ducting can extend this considerably. Ducting happens in stable weather systems when large changes in refractive index with height occur and cause propagation with very low attenuation over hundreds of kilometers. The

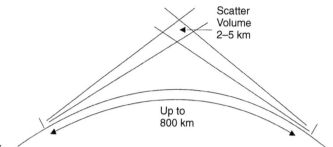

Fig. 8.18 Tropospheric
scatter communications link.

propagation mode is similar to that in a waveguide. Ducting is not a very predictable form of propagation and therefore cannot be used commercially.

Tropospheric scatter and meteor scatter communications both provide over-the-horizon communications on a regular and predictable basis. Tropospheric scatter relies on small cells of turbulence in the troposphere between 2 km and 5 km above the Earth's surface, which scatter incident radiation, as illustrated in Figure 8.18.

Frequencies in the lower part of the microwave range from 1.0 to 5.0 GHz are generally used because they give the maximum amount of forward scatter; nevertheless, the amount of energy that is scattered forward is extremely small, and the losses in the scatter volume are on the order of 50 to 70 dB in addition to the free-space loss. Large short-term variations following the Rayleigh law also occur, and therefore quadruple or multiple diversity is almost always used. This is achieved by employing two antennas, horizontal and vertical polarization, different scatter volumes (angle diversity), and different frequencies. Space and polarization diversity generally yield the best results. Tropospheric scatter can provide communications up to 500 km, but paths as long as 800 km have been operated.

Meteor scatter communications make use of ionized trails of meteors burning up on entering the Earth's atmosphere. This is not such a rare event as might be thought, and there is a steady shower of material entering the Earth's atmosphere, although with random and predictable variations. Meteors that leave usable trails have diameters between 0.2 mm and 2 mm, and the trails, which last for around half a second to several seconds, occur at a height of about 120 km. They can extend up to 50 km in length, although 15 km is a typical value. The number of particles entering the Earth's atmosphere is inversely proportional to size, so larger meteors occur too infrequently to be of use, while the numerous very small particles do not generate sufficient ionization. The variation in the number of usable ionized trails occurs on a daily basis, with more occurring during daylight as the Earth's rotations tend to sweep up more meteors than at night. There are also predictable showers occurring throughout the year.

Meteor scatter communications can be established for paths from 200 to 2000 km, but information has to be transmitted in bursts when a link is established, so the medium cannot be used for services such as speech. Radio waves are reflected from the trails by different mechanisms depending on the density of ionization; however, frequencies in the 30- to 100-MHz range are most effective, with 50 MHz being the optimum frequency. Horizontal polarization is generally preferred. When a link is established, path attenuation is high because of the scattering process and is typically 175 dB for a 40-MHz, 1000-km path (Cannon, 1987). This is approximately 50 dB greater than the free-space attenuation of the same distance.

Propagation for Satellite Communications

Most communications satellites are placed in geostationary orbits 36,000 km above the equator; therefore, transmitting and receiving Earth stations can fix their antenna positions with only minor required adjustments for small shifts in satellite position or changes in atmospheric propagation conditions. Such orbits also have the advantage of providing potential coverage of almost one-third of the Earth's surface, but the disadvantage of high free-space loss compared with lower nonstationary orbits. The systems planner will need to calculate the link budget, taking into account such factors as the satellite EIRP (Equivalent Isotropically Radiated Power) and receiver noise performance.

The free-space loss can be calculated using equation 8.7, and distance will have to take into account the difference in both latitude and longitude of the position of the satellite on the Earth's surface from that of the transmitting or receiving station. If the distance and great circle bearing are calculated using equations 8.10 and 8.11, the elevation and distance of the satellite can also be calculated. In addition to the free-space loss, the loss due to atmospheric attenuation must be taken into account, and this will depend on precipitation conditions in the earth station area. Typical values at 11 GHz would be 1.0 dB for an "average year," increasing to about 1.5 dB for the worst month. The actual figures to be used should be calculated from local meteorological data and the attenuation curves given in Figure 8.12.

Calculation of satellite paths is not often required because operators usually publish "footprint" maps showing the received power contours in dBW, taking into account the path loss and the radiation pattern of the satellite transmitting antenna.

Conclusion

Radio wave propagation uses frequencies of up to 100 GHz and higher. The following sections subdivide the frequency range into broadbands and summarize their areas of application.

VLF and LF: 10 to 300 kHz

This range is for long-distance (greater than 1000-km) ground wave transmission and is used for communications and radio navigation. Propagation is unaffected by ionospheric conditions. High-power and large antennas are required, and there is very limited spectrum availability.

MF: 300 kHz to 3 MHz

This range is for medium- to short-distance (up to 1000-km) ground wave transmission and is used for sound broadcasting and mobile communications. Propagation is little affected by ionospheric conditions, but nighttime interference can be a problem.

HF: 3 to 30 MHz

Worldwide communications use comparatively low power (1 kw), but they are heavily dependent on ionospheric conditions. There is limited channel availability, and it is used for sound broadcasting, point-to-point, and mobile maritime and aeronautical communications.

VHF and UHF: 30 MHz to 1 GHz

Officially, UHF extends to 3 GHz. Lower frequencies are affected by anomalous propagation conditions. UHF is used for line-of-sight communications, typically 80 km (meteor scatter on 50 MHz can provide services up to 2000 km), and multichannel point-to-point communications, FM sound, and TV broadcasting and mobile communications.

Microwave: 1 to 30 GHz

Microwaves are unaffected by ionospheric conditions, but there is some attenuation at higher frequencies because of rain. They are used for line-of-sight communications, typically of 50 km (tropospheric scatter can provide services up to 500 km). Satellite services provide worldwide coverage. Wideband multichannel communication is available. Microwaves are extensively used for terrestrial point-to-point and satellite communications and radar. Wi-Fi and WiMAX use some frequency bands within this range.

Millimetric: 30 to 100 GHz

This range is unaffected by ionospheric conditions but has moderate to severe attenuation due to atmospheric conditions. It is used for limited line-of-sight communications, typically 10 km. There is limited usage for this range—mainly short-distance speech and data links, but increasing usage is likely, especially for indoor applications.

References

Braun, G. (1986) *Planning and Engineering of Short-Wave Links, Second Edition*, John Wiley & Sons.
Cannon, P. S. (1987) The Evolution of Meteor Burst Communications Systems, *J. IERE.* 57(3).
Damboldt, T. (1975) A Comparison between Deutsche Bundespost Ionospheric HF Propagation Predictions and Measured Field Strengths, *Proceedings NATO AGARD, Radio Systems and The Ionosphere Conf.* 173(May).
Dick, M. I. (1990) *Propagation Model of CCIR Report 894*, Appleton Laboratories.
Hitney, H. V. (1990) *IONOPROP, Ionospheric Propagation Assessment Software and Documentation*, Artech House.
Maslin, N. M. (1987) *H.F. Communications: A Systems Approach*, Pitman.
Millington, G. (1949) Ground Wave Propagation over Inhomogeneous Smooth Earth, *Proceedings IRE* 96:53–64.

Resources

Boithias, L. (1987) *Radio Wave Propagation*, North Oxford Academic Publishers.
Budden, K. G. (1985) *The Propagation of Radio Waves*, Cambridge University Press.
Friis, H. T. (1946) A Note on a Simple Transmission Formula, *IRE Proc.* (May):254–256.
Johnson, R. C., and Jasick, H. (eds.) (1984) *Antenna Engineering Handbook*, McGraw-Hill.
Picquenard, A. (1984) *Radio Wave Propagation*, Philips Technical Library, McMillan Technical Press.
Stanniforth, J. A. (1972) *Microwave Transmission*, English University Press.
Stark, A. (1986) Propagation of Electromagnetic Waves, Rohde & Schwarz.

9 Antennas

Martin Smith
Nortel Ltd.

Types of Antenna

Antennas form the link between the guided parts and the free-space parts of a communications system. The purpose of a transmitting antenna is to efficiently transform the currents in a circuit or waveguide into radiated radio or microwave energy. The purpose of a receiving antenna is to efficiently accept the received radiated energy and convert it into guided form for detection and processing by a receiver. The design and construction of an antenna usually involves compromises between the desired electromagnetic performance and the mechanical size, mass, and environmental characteristics.

Antennas for communications systems fall into two broad categories depending on the degree to which the radiation is confined. Microwave radio relay and satellite communications use pencil beam antennas, where the radiation is confined to one narrow beam of energy (see Figure 9.1). Mobile communications are more likely to require antennas with omnidirectional patterns in the horizontal plane and toroidal patterns in the vertical plane, as shown in Figure 9.2.

Pencil beam antennas usually consist of one or more large to medium reflectors that collimate the signals from a feed horn at the focus of the reflector. Both reflector and feed horn fall within the generic class of aperture antennas because they consist of an aperture that radiates into space. The design problem is first to determine the aperture fields that will yield the specified radiation characteristics, and second to design the reflectors and horns to produce the aperture fields. Aperture antennas can be designed to meet very stringent specifications. Omnidirectional antennas consist of elements that are small in wavelength, such as dipoles and monopoles. The radiation characteristics are influenced by the presence of surrounding objects. Nonelectromagnetic factors, such as size, are often as important in the design as radiation performance. For this reason the design of omnidirectional antennas is partly an empirical process in which expertise and previous experience play an equal part with theoretical knowledge.

277

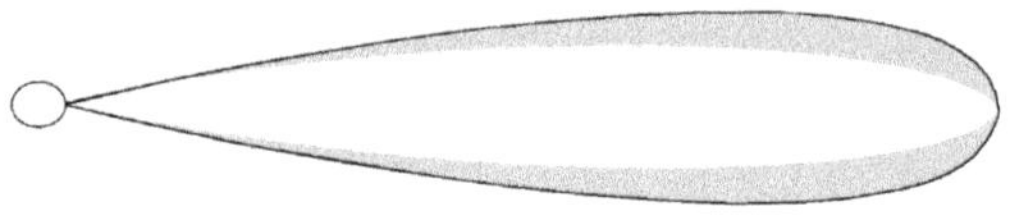

Fig. 9.1   Pencil beam radiation pattern.

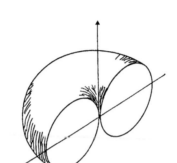

Fig. 9.2   Toroidal radiation pattern.

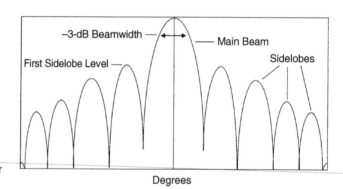

Fig. 9.3   Typical rectangular radiation plot.

Between the large aperture antennas and the small element antennas lie array antennas that consist of two or more elements. The radiation from an array antenna is determined principally by the physical spacing and electrical signals driving the elements rather than the radiation characteristics of the elements themselves, such as that shown in Figure 9.3.

The detailed theory of antennas can be found in Stutzman and Thiele (1981), Elliot (1981), Balanis (1982), Silver (1984), and Collin and Zucher (1969). Design information and descriptions of particular types can be found in a number of handbooks, such as Rudge (1986), Johnson and Jasik (1984), Lo (1988), and Milligan (1985).

Antennas Used in Communications

Table 9.1 lists the principal types of antenna that are used in communications. Under each category of communication system, the specific type of antenna used is given. The last column gives the generic type of aperture antenna, array antenna, or small element antenna. The generic type describes the general radiation characteristics that can be obtained from the antenna and is useful because it makes the explanation of performance easier.

Table 9.1 Antennas Used in Communications Systems

Use	Specific Type	Generic Type
Microwave line-of-sight radio	Prime-focus reflector with small feed	Aperture
Earth Stations (large)	Dual reflector with corrugated horn feed	Aperture
Earth Stations (medium)	Offset reflector with corrugated or dual-mode horn feed	Aperture
Direct Broadcast Satellite Receiving Antennas	Prime-focus symmetric or offset reflector	Aperture
	Flat plate antennas	Array
Satellite Antennas (spot beams)	Offset reflector with single feed	Aperture
Satellite Antennas (multiple beams)	Offset reflector with array feed	Aperture and Array
Satellite Antennas (shaped beams)	Shaped reflectors	Aperture
	Offset reflector with array feed	Aperture and Array
VHF/UHF Communications	Yagis, dipole arrays, slots	Element and Array
Mobile Communications (base stations)	Dipole arrays	Element and Array
Mobile Communications (mobile)	Monopoles, microstrip	Element
HF Communications	Dipoles, monopoles	Element

The following sections describe first generic antenna characteristics, then the specific antenna types, and then, briefly the practical implementation of antennas in communications systems.

Basic Properties

The principle of reciprocity is one of the most important properties of an antenna. It means that the properties of an antenna when acting as a transmitter are identical to the properties of the same antenna when acting as a receiver. For this to apply, the medium between the two antennas must be linear, passive, and isotropic, which is always the case for communications systems.

The directional selectivity of an antenna is represented by the radiation pattern. It is a plot of the relative strength of the radiated field as a function of the angle. A pattern taken along the principal direction of the electric field is called an *E-plane cut*; the orthogonal plane is called an *H-plane cut*. The most common plot is the rectangular decibel plot (Figure 9.3), which can have scales of relative power and angle chosen to suit the antenna being characterized. Other types of plots, such as polar (used for small and two-dimensional antennas), contour (or three-dimensional), and isometric are also used. A radiation pattern is characterized by the main beam and sidelobes. The quality is specified by the beamwidth between the −3-dB points on the main beam and the sidelobe level.

Communication antennas radiate in either linear polarization or circular polarization. In modern communications, cross-polarization is important. This is the difference between the two principal plane patterns and is specified relative to a reference polarization, called the *copolar pattern*. There are three definitions of cross-polarization. The one in normal use with reflector antennas and feed systems is Ludwig's third definition (Ludwig, 1973), which assumes that the reference polarization is that due to a Huygen's source. It most closely corresponds to what is measured with a conventional antenna test range.

The power gain in a specified direction is defined by the ratio of the power radiated per unit solid angle in direction θ, φ to the total power accepted from the source, as in equation 9.1.

$$G(\theta, \varphi) = 4\pi \frac{\text{Power radiated per unit solid angle in direction } \theta, \varphi}{\text{Total power accepted source}} \qquad (9.1)$$

This is an inherent antenna property and includes dissipation losses in the antenna. The dissipative losses cannot easily be predicted, so a related parameter, directivity, is used in calculations. The definition of directivity is similar to that of gain except that the denominator is replaced by the total power radiated. The terms *gain* and *directivity* are often used interchangeably in the literature. Normally only the peak gain along the boresight direction is specified. If the direction of the gain is not specified, peak value is assumed. The value is normally quoted in decibels. The definitions previously given are, in effect, specifying the gain relative to a lossless isotropic source. This is sometimes stated explicitly using the symbol *dBi*.

The efficiency of an aperture antenna is given by the ratio of the effective area of an aperture divided by the physical area. Normal aperture antennas have efficiencies in the range 50 to 80 percent.

As far as circuit designers are concerned, the antenna is an impedance. Maximum power transfer will occur when the antenna is matched to the transmission line. The impedance consists of the self-impedance and the mutual impedance. The mutual impedance accounts for the influence of nearby objects and of mutual coupling to other antennas. The self-impedance consists of the radiation resistance, the loss resistance, and the self-reactance. Loss resistance consists of the ohmic losses in the antenna structure. Radiation resistance measures the power absorbed by the antenna from the incoming plane waves. It is one of the most significant parameters for small antennas, where the problem is often matching very dissimilar impedances.

A receiving antenna is both a spatially selective filter (measure by the radiation pattern) and a frequency selective filter. The bandwidth measures the frequency range over which the antenna operates. The upper and lower frequencies can be specified in terms of a number of possible parameters: gain, polarization, beamwidth, and impedance.

A communications link consists of a transmitting antenna and a receiving antenna. If the transmitter radiates P_t watts, then the received power, P_r, at a distance, r, is given by equation 9.2, where G_t and G_r are the transmitter and receiver antenna gains, respectively.

$$P_r = P_t \frac{G_t G_r \lambda^2}{(4\pi r)^2} \qquad (9.2)$$

This formula is known as the Friis transmission equation. It assumes that the antennas are impedance and polarization matched. If this is not the case, then extra factors must be multiplied to the equation to account for mismatches.

Generic Antenna Types

Antenna design is based on several generic antenna types. These include apertures, small antennas, and array antennas. The properties of each are considered in this section.

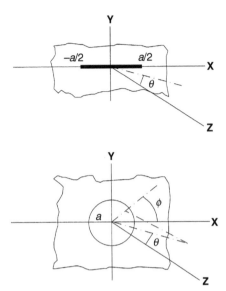

Fig. 9.4   Radiation aperture in a ground plane.

Radiation from Apertures

The radiation from apertures illustrates most of the significant properties of pencil beam antennas. The radiation characteristics can be determined by simple mathematical relationships. If the electric fields across an aperture (see Figure 9.4) are $E_a(x, y)$, then the radiated fields, $E_p(\theta, \varphi)$, are given by equation 9.3, where $f(\theta, \varphi)$ is given by equation 9.4 (Olver, 1986; Milligan, 1985).

$$E_p(\theta, \varphi) = \cos^2 \frac{\theta}{2}\left(1 - \tan^2 \frac{\theta}{2}\cos 2\varphi\right) f(\theta, \varphi) \qquad (9.3)$$

$$f(\theta, \varphi) = \int_{-\infty}^{\infty} \int_{-\infty}^{\infty} E_a(x, y)e^{jk(x\sin\theta\cos\varphi + y\sin\theta\sin\varphi)}dx\,dy \qquad (9.4)$$

For high- or medium-gain antennas, the pencil beam radiation is largely focused to a small range of angles around $\theta = 0$. In this case it can be seen from equation 9.3 that the distant radiated fields and the aperture fields are the Fourier transformation of each other. Fourier transforms have been widely studied, and their properties can be used to understand the radiation characteristics of aperture antennas. Simple aperture distributions have analytic Fourier transforms, while more complex distributions can be solved numerically on a computer.

The simplest aperture is a one-dimensional line source distribution of length $\pm\frac{a}{2}$. This serves to illustrate many of the features of aperture antennas. If the field in the aperture is constant, the radiated field is given by equation 9.3 as in equations 9.5 and 9.6.

$$E_p = \frac{\sin(\pi u)}{\pi u} \qquad (9.5)$$

$$u = \frac{a}{\lambda}\sin\theta \qquad (9.6)$$

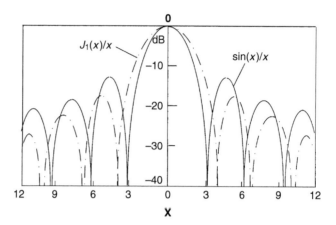

Fig. 9.5 Dipole and radiation pattern.

This distribution occurs widely in antenna theory. It is plotted in Figure 9.5.

The beamwidth is inversely proportional to the aperture width and is 0.88 λ/a. The first sidelobe level is at −13.2 dB, which is one disadvantage of a uniform aperture distribution. The level can be reduced considerably by a tapered aperture distribution, where the field is greatest at the center of the aperture and tapers to a lower level at the edge. For example, if equation 9.7 holds, then the first sidelobe level is at −23 dB.

$$E_a(x) = \cos\left(\frac{\pi x}{2a}\right) \tag{9.7}$$

The energy that was in the sidelobes moves to the main beam with the result that the beamwidth broadens to 1.2 λ/a. In practice almost all antennas have natural tapers across the aperture that result from boundary conditions and waveguide modes. Rectangular apertures are formed from two line source distributions in orthogonal planes.

Circular apertures form the largest single class of aperture antennas. The parabolic reflector is widely used in communications and is often fed by a conical horn. Both the reflector and the horn are circular apertures. For an aperture distribution, which is independent of azimuthal angle, the simplest case is uniform illumination, which gives a radiated field as in equation 9.8, where $J_1(x)$ is a Bessel function of zero order.

$$E_p = \frac{2J_1(\pi u)}{\pi u} \tag{9.8}$$

This can be compared to $\sin(x)/x$ and is also plotted in Figure 9.5. The first sidelobe level is at −17.6 dB.

Table 9.2 lists a number of circular aperture distributions and corresponding radiation pattern properties. The pedestal distribution is representative of many reflector antennas, which have an edge taper of about −10 dB corresponding to $E_{a(a)} = 0.316$. The Gaussian distribution is also important because high-performance feed horns ideally have Gaussian aperture distributions. The Fourier transform of a Gaussian taper that decreases to zero at the edge of the aperture gives a Gaussian radiation pattern, which has no sidelobes.

Table 9.2 Radiation Characteristics of Circular Apertures

Electric Field Aperture Distribution	3-dB Beamwidth	Level of First Sidelobe
Uniform	$1.02\dfrac{\lambda}{D}$	−17.6 dB
Taper to zero at edge $1-\left(\dfrac{2r}{D}\right)^2$	$1.27\dfrac{\lambda}{D}$	−24.6 dB
Taper on pedestal $0.5+\left[1-\left(2\dfrac{r}{D}\right)^2\right]^2$	$1.16\dfrac{\lambda}{D}$	−26.5 dB
Gaussian $\exp\left[-p\left(\dfrac{2r}{D}\right)^2\right]$	$1.33\dfrac{\lambda}{D}$	−40 dB (p = 3)

Radiation from Small Antennas

Small antennas are needed for mobile communications operating at frequencies from HF to the low microwave region. Most of these are derivatives of the simple dipole (see Figure 9.6), which is an electric current element that radiates from the currents flowing along a small metal rod. The radiation pattern is always very broad, with energy radiating in all directions. An important design parameter is the impedance of the dipole, which can vary considerably depending on the exact size and shape of the rod. This means that the impedance matching between the antenna and the transmitting or receiving circuit becomes a major design constraint.

The radiation fields from a dipole are obtained by integrating the radiation from an infinitesimally small current element over the length of the dipole. This depends on knowing the current distribution, which is a function not only of the length but also of the shape and thickness of the rod. Many studies have addressed obtaining accurate results (King, 1956; King, Mack, & Sandler, 1968). For most cases this has to be done by numerical integration. A simple case is a short dipole with length $a \ll 1/10\lambda$ when the current distribution can be assumed to be triangular. This results in radiated fields of the form given in equation 9.9.

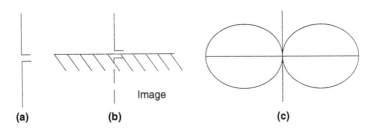

Fig. 9.6 Dipole and radiation pattern: (a) dipole; (b) monopole; (c) polar pattern of dipole.

$$E = j30\pi I \frac{a}{\lambda} \sin\theta \frac{e^{-jkr}}{r} \qquad (9.9)$$

The electric field is plotted in polar form in Figure 9.6. The radiation resistance is calculated by evaluating the radiated power and using $P = I^2 R$ to give equation 9.10.

$$R = 20\pi^2 \left(\frac{a}{\lambda}\right)^2 \qquad (9.10)$$

A dipole of length $a = \lambda/10$ has a radiation resistance of 2.0 ohms. This is low by comparison with standard transmission lines and indicates the problem of matching to the transmission line.

The half-wave dipole is widely used. Assuming a sinusoidal current distribution, the far fields are given by equation 9.11.

$$E = j60I \frac{\cos[\pi/2\cos\theta]}{\sin\theta} \frac{e^{-jkr}}{r} \qquad (9.11)$$

This gives a slightly narrower pattern than that of the short dipole and has a half-beamwidth of 78 degrees. The radiation resistance must be evaluated numerically. For an infinitely thin dipole it has a value of $73 + j42.5$ ohms. For finite thickness the imaginary part can become zero, in which case the dipole is easily matched to a coaxial cable of impedance 75 ohms. The half-wave dipole has a gain of 2.15 dB.

A monopole is a dipole divided in half at its center feed point and fed against a ground plane, as shown in Figure 9.6. The ground plane acts as a mirror, and consequently the image of the monopole appears below the ground. Because the fields extend over a hemisphere, the power radiated and the radiation resistance are half those of the equivalent dipole with the same current. The gain of a monopole is twice that of a dipole. The radiation pattern above the ground plane is the same as that of the dipole.

Radiation from Arrays

Array antennas consist of a number of discrete elements that are usually small in size. Typical elements are horns, dipoles, and microstrip patches. The discrete sources radiate individually, but the pattern of the array is largely determined by the relative amplitude and phase of the excitation currents on each element and the geometric spacing apart from the elements. The total radiation pattern is the multiplication of the pattern of an individual element and the pattern of the array, assuming point sources; it is called the *array factor*. Array theory is largely concerned with synthesizing an array factor to form a specified pattern. In communications most arrays are planar arrays, with the elements being spaced over a plane, but the principles can be understood by considering an array of two elements with equal amplitudes (see Figure 9.7(a)). This has an array factor given by equation 9.12, where ψ is given by equation 9.13.

$$E = E_1 + E_2\, e^{j\psi} \qquad (9.12)$$

$$\psi = \delta + kd\cos\theta \qquad (9.13)$$

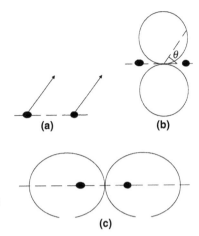

Fig. 9.7 Two-element array and radiation pattern: (a) two elements with equal amplitudes; (b) half-wavelength spacing; (c) 90° phase difference.

The pattern for small spacings will be almost omnidirectional, and as the spacing is increased the pattern develops a maximum perpendicular to the axis of the array. At a spacing of half a wavelength, a null appears along the array axis (see Figure 9.7(b)).

This is called a broadside array. If a phase difference of 90 degrees exists between the two elements, then the pattern shown in Figure 9.7(c) results. Now the main beam is along the direction of the array, and the array is called the *end-fire array*. This illustrates one of the prime advantages of the array: By changing the electrical phase it is possible to make the peak beam direction occur in any angular direction. Increasing the spacing above half a wavelength results in the appearance of additional radiation lobes, which are generally undesirable.

Consequently, the ideal array spacing is half a wavelength. However, if waveguides or horns are used this is not usually possible because the basic element is greater than half a wavelength in size. Changing the relative amplitudes, phases and spacings can produce a wide variety of patterns so that it is possible to synthesize almost any specified radiation pattern. The array factor for an N-element linear array of equal amplitude is given by equation 9.14.

$$E = N \frac{\sin(N\psi/2)}{\sin(\psi/2)}$$
(9.14)

This is similar to the pattern of a line source aperture (equation 9.5), and it is possible to synthesize an aperture with a planar array. There is a significant benefit to this approach. The aperture fields are determined by the waveguide horn fields, which are constrained by boundary conditions and are usually monotonic functions. This constraint does not exist with the array, so a much larger range of radiation patterns can be produced. Optimum patterns with most of the energy radiated into the main beam and very low sidelobes can be designed.

Adaptive Arrays

In a previous section it was shown that the radiation pattern of an aperture antenna is the spatial Fourier transform of the aperture distribution. Antenna radiation patterns can therefore, in principle, be synthesized by control of the aperture excitation. If the aperture excitation can also be varied electronically, an adaptable radiation pattern is available. Electronic beam steering or null steering (to minimize interference) can therefore be provided. There are limitations on beamwidth imposed by the aperture size, as the Fourier transform relationship implies. In practice it is very difficult to implement control of the excitation of a continuous aperture. An array antenna allows far more control by using a number of individual antenna elements grouped together to form a sampled aperture. The array elements themselves might not be much more expensive than a continuous aperture antenna, such as a reflector, but in general the control devices are a significant cost penalty. This has to be traded off against the various advantages of an array antenna system.

Because an array is a sampled aperture, there are limitations on the spacing of the array elements. A typical spacing is on the order of $\lambda/2$ (this will be explained in a later section), and so fairly small antennas are required as elements. Dipoles, monopoles, slots, patches, and open-ended waveguides can all be used.

The theory given here will be for a linear array of identical, equispaced elements. This is the simplest case and is also commonly used in practice. The effects of mutual coupling between elements will be ignored, and it is assumed that individual phase and amplitude, control of each element is available. There are various ways of implementing this control; this "beam-forming" process can be carried out at radio frequency or intermediate frequency or baseband.

The Radiation Pattern of a Linear Array

Figure 9.8 shows a linear array of equispaced antenna elements. First we will consider the radiation pattern of an array of (hypothetical) isotropic radiators, $F(\theta)$. This pattern is called the *array factor*. Note that the angle, θ, is defined relative to the array axis, rather than the normal to the aperture. This is because the array line is an axis of symmetry for the array factor, which is then a function of θ only.

The array in Figure 9.8 contains N elements with spacing d. In direction θ the path from the nth element to a distant point is $nd \cos \theta$ shorter than that from the 0th (reference) element to the same distant point. If the nth element is excited with amplitude $|a_n|$ and phase φ_n, the contribution from the element to the field at a great distance at the angle θ is proportional to $|a_n| \exp(j\varphi_n)$.

The total field is then given by equation 9.15, where a_n is a complex number equal to $|a_n| \exp(j\varphi_n)$.

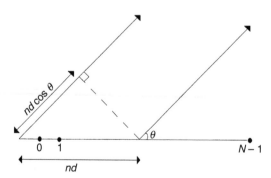

Fig. 9.8 Linear array of equispaced antenna elements.

$$F = \sum_{n=0}^{N-1} a_n \cdot \exp(jn\,kd\cos\theta) \tag{9.15}$$

A useful method of analyzing linear arrays, from Schelkunoff, consists of associating a polynomial with an array. Let $z = \exp(jkd\cos\theta)$. Then equation 9.15 is as shown in equation 9.16.

$$F(z) = a_0 + a_1 z + a_2 z^2 + \ldots + a_{N-1} z^{N-1} \tag{9.16}$$

The radiation pattern (array factor) is $F(z) = F(\exp(jkd\cos\theta))$. The complex coefficients of the polynomial $F(z)$ are the excitations of the individual elements in the array. In the general case, equation 9.15 is often used directly to evaluate $F(\theta)$.

As an example, consider an array where all the elements are excited with unit amplitude and zero phase. This is the sampled equivalent of a uniformly excited linear aperture. The radiation pattern for N elements is given by equation 9.17, which results in equation 9.18.

$$\begin{aligned}
F(z) &= 1 + z + z^2 + \ldots + z^{n-1} \\
&= \frac{(1-z^N)}{(1-z)} \\
&= \frac{1 - \exp(jkd/N\cos\theta)}{1 - \exp(jkd\cos\theta)}
\end{aligned} \tag{9.17}$$

$$\begin{aligned}
|F(z)|^2 &= F(z) \cdot F^*(z) \\
&= \frac{2 - 2\cos(Nkd\cos\theta)}{2 - 2\cos(kd\cos\theta)} \\
&= \frac{\sin^2(1/2 Nkd\cos\theta)}{\sin^2(1/2 kd\cos\theta)}
\end{aligned} \tag{9.18}$$

$$|F(z)| = \frac{\sin(1/2 Nkd\cos\theta)}{\sin(1/2 kd\cos\theta)} \tag{9.19}$$

The result of equation 9.19 can be compared with the radiation pattern of a uniform aperture of length a, as in equation 9.20.

$$P(\sin\alpha) = \frac{\sin(\frac{1}{2}kd\sin\alpha)}{(\frac{1}{2}ka\sin\alpha)} \tag{9.20}$$

The numerator in equation 9.20 is exactly the same if $a = Nd$, noting that $\sin\alpha = \cos\theta$. The denominators differ but are similar if $\sin(\frac{1}{2}kd\cos\theta) \approx \frac{1}{2}kd\cos\theta$—that is, if $\frac{1}{2}kd\cos\theta \ll \pi/2$ and if $d\cos\theta \ll \lambda/2$. (They also differ by a factor of N, but that is simply a constant of proportionality.)

If $d = \lambda/2$, this condition requires $\cos\theta \ll 1$, so that the patterns are similar near broadside to the aperture or array. The zero-to-zero beamwidth is the same for the two cases. In general, the pattern is similar to that of a diffraction grating in optics, and the denominator can cause "grating lobes." The condition for these is given by equations 9.21 and 9.22.

$$\frac{1}{2}kd\cos\theta = \pm\pi \tag{9.21}$$

$$\cos\theta = \pm\frac{\lambda}{d} \tag{9.22}$$

At such angles, the denominator in equation 9.19 becomes 0 simultaneously with the numerator, and a repeated main lobe occurs in the array factor radiation pattern. The first 0 of the pattern occurs where $x = \pi/N$, that is $\cos \theta \, (= \sin \alpha) = \pm \lambda/(Nd)$, while the first grating lobes occur where $x = \pi$—that is, $\cos \theta = \pm \lambda/d$. The main beamwidth is thus governed by the complete array length, while the grating lobes are governed by the spacing of the array elements. The relation of equation 9.22 implies that, if grating lobes are to be avoided, $d < \lambda$ is required. When beam steering is considered (see the Beam Steering section), further restrictions on d occur.

Now consider nonisotropic elements. The radiation pattern becomes as shown in equation 9.23.

$$F(\theta, \varphi) = \sum_{n=0}^{N-1} a_n \cdot f_n(\theta, \varphi) \cdot \exp(jnkd \cos\theta) \qquad (9.23)$$

Here $f_n(\theta, \varphi)$ is the radiation pattern of the nth array element. If the array elements have identical patterns and similar alignment, the element pattern can be factored out so that equation 9.24 is obtained.

$$F(\theta, \varphi) = f(\theta, \varphi) \cdot \sum_{n=0}^{N-1} a_n \cdot \exp(jnkd \cos\theta) \qquad (9.24)$$

This is simply the product of the element pattern and the array factor. The array factor is generally more directive than the element pattern so that it dominates properties such as main beamwidth and sidelobe levels. The element pattern will generally reduce far-out sidelobe levels, including any grating lobes that might be present.

Nulls

A null of the radiation pattern is here taken to be synonymous with a 0—that is, no transmission or reception at a particular angle. A null can also mean a minimum of the pattern, but in the present analysis exact 0s can be achieved. The analysis and synthesis of nulls can be achieved using Schelkunoff's representation.

Consider the complex variable given by equation 9.25.

$$z = \exp(jkd \cos\theta) \qquad (9.25)$$

A θ varies from 0 to π; z moves along a locus in the Argand diagram, as is shown in Figure 9.9.

The modulus of z is 1, while its phase (argument) is as in equation 9.26.

$$|z| = \frac{2\pi d}{\lambda} \cos\theta \qquad (9.26)$$

This therefore varies between $-2\pi d/\lambda$ and $+2\pi d/\lambda$. The locus of z is thus an arc of the unit circle to the Argand diagram. For a large value of d, z might move several times around the unit circle as θ varies from 0 to π. If $d = \lambda/2$, the locus of z as θ varies from 0 to π just closes to form a complete circle. If $d < \pi/2$, the locus of z is a small arc of the unit circle.

Now consider the polynomial representation. Any polynomial can be expressed as a product of linear factors. The polynomial of equation 9.16 will have $N = 1$ roots $z_1, z_2, \ldots z_{n-2}$, and can be written as in equation 9.27.

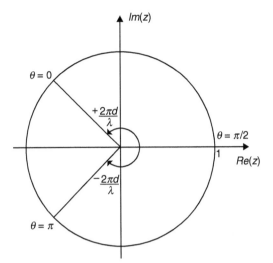

Fig. 9.9 Argand diagram for
$z = \exp(jkd\cos\theta)$.

$$F(z) = a_{N-1}(z - z_1)(z - z_2)\ldots(z - z_{N-1}) \tag{9.27}$$

Since only the relative radiation pattern is important, set a_{N-1} giving equation 9.28.

$$F(z)(z - z_1)(z - z_2)\ldots(z - z_{N-1}) \tag{9.28}$$

The roots of z_1, z_2, and so on (of $F(z) = 0$), are complex numbers, but do not necessarily lie on the unit circle. If they do lie on the unit circle, and d is such that that part of the unit circle is traversed by z as θ varies from 0 to π, a null will be present in the radiation pattern. Because these are $N - 1$ roots for an N-element array, there can be up to $N - 1$ independent nulls.

Beam Steering

Consider a general array excitation a_n ($n = 0, 1, \ldots N - 1$). The corresponding radiation pattern is as in equation 9.29, where $z = \exp(jkd\cos\theta)$.

$$F_a(z)\sum_{n=0}^{N-1} a_n z^n \tag{9.29}$$

Now apply a linear phase gradient to the array; that is, let the array excitations become a_v, $a_1\exp(j\varphi)$, $a_2\exp(j2\varphi)$, and so on.

Let b_n be given by equation 9.30; then the new radiation pattern is given by equation 9.31, where z' is given by equation 9.32.

$$b_n = a_n\exp(jn\varphi) \tag{9.30}$$

$$
\begin{aligned}
F_b(z) &= \sum_{N=0}^{N-1} a_n\exp(jn\varphi)z^n \\
&= \sum_{n=0}^{N-1} a_n(z')^n
\end{aligned}
\tag{9.31}
$$

$$z' = \exp(j(kd\cos\theta + \varphi)) \tag{9.32}$$

Then, if equations 9.33 and 9.34 are true, the radiation pattern keeps the same form but with an angular shift related to the incremental phase, φ.

$$kd\cos\theta_b + \varphi = kd\cos\theta_a \tag{9.33}$$

$$F_b(\cos\theta_b) = F_a(\cos\theta_a) \tag{9.34}$$

From these equations, equation 9.35 can be obtained. The right-hand side is proportional to the phase gradient $(d\varphi/dx)$ along the array.

$$\Gamma_b(\cos\theta_b) - \cos\theta_a = \frac{\varphi}{kd} \tag{9.35}$$

If the main beam peak is initially at $\theta = 90°$ broadside to the array, then $\cos\theta_a = 0$, and $\cos\theta_b = \sin\alpha$, where α is the beam deflection angle. Equation 9.36 is thus obtained.

$$\sin\alpha = \frac{-\varphi}{kd} \tag{9.36}$$

If $d = \lambda/2$ and α are small, so that $\sin\alpha = \alpha$, equations 9.37 and 9.38 are obtained.

$$\alpha(\text{radians}) = \frac{-\varphi(\text{radians})}{\pi} \tag{9.37}$$

$$\alpha^0 = \frac{-\varphi^0}{x} \tag{9.38}$$

Thus for an incremental phase value of 30°, the deflection angle is 10°. Using equation 9.36, an incremental phase of 90° for $\lambda/2$ spaced elements gives a deflection angle of 30°.

The use of electronically controlled phase shifters on each element of a "phased" array allows the beam-pointing direction of an array antenna to be varied rapidly and without any mechanical antenna movement. This is in contrast to a fixed-beam antenna, such as a simple horn-feed parabolic reflector, where the antenna is rotated mechanically to "scan" the beam. However, the latter is in general much cheaper!

The question of grating lobes arose in the Nulls section, where $d < \lambda$ was required to avoid them for a beam with a broadside peak. The radiation pattern of a uniformly excited N-element array is given by equation 9.39.

$$|F_a(z)| = \frac{\sin(1/2Nkd\cos\theta_a)}{\sin(1/2kd\cos\theta_a)} \tag{9.39}$$

Now consider this array with a phase gradient superimposed. Using equation 9.33 gives equation 9.40.

$$|F_b(z)| = \frac{\sin(1/2N(kd\cos\theta_b + \varphi))}{\sin(1/2(kd\cos\theta_b + \varphi))} \tag{9.40}$$

The peak of this radiation pattern is at $\theta = \theta_m$, where $\cos\theta_m = -\varphi/kd$. Equation 9.40 can be written as in equation 9.41.

$$|F(z)| = \frac{\sin(1/2Nkd(\cos\theta - \cos\theta_m))}{\sin(1/2kd(\cos\theta - \cos\theta_m))} \tag{9.41}$$

Grating lobes will appear if the denominator of equation 9.41 becomes 0 for some $\theta = \theta_m$. This will occur if equation 9.42 is true by comparison with equation 9.19.

$$\cos\theta - \cos\theta_m = \frac{\pm\lambda}{d} \tag{9.42}$$

Let the deflection angle from broadside ($\theta = 90°$) be α, so that $\theta = 90° + \alpha$. Equation 9.42 then becomes as in equation 9.43.

$$\sin\alpha_m - \sin\alpha = \frac{\pm\lambda}{d} \tag{9.43}$$

The extreme values of $\sin\alpha$ are ±1. If we consider $\alpha_m > 0$, the magnitude of the left-hand side of equation 9.43 has a maximum value of $\sin\alpha_m + 1$. Thus, with the inequalities in equations 9.44 and 9.45, grating lobes will not appear.

$$\frac{\lambda}{d} > 1 + \sin\alpha_m \tag{9.44}$$

$$\frac{d}{\lambda} < \frac{1}{(1+\sin\alpha_m)} \tag{9.45}$$

For example, if the maximum beam-steering angle is 30° from broadside, these inequalities require $d < 2/3\lambda$. If the beam can be steered by 90°, to "end fire," $d < 1/2\lambda$ is needed to avoid grating lobes.

Adaptive Null Steering

Communications antenna systems are susceptible to degradation in performance due to interference received in their main beam or sidelobes. The interference might be deliberate jamming or signals using the same frequency in a cellular system, originating from a reuse cell. In these cases the angle of arrival of the interfering signal needs to be determined and null formed in the antenna pattern in that direction.

An antenna array allows these functions to be performed, with up to $N - 1$ nulls for an N-element array. To independently control the angular location of the nulls, it is necessary to control the phase and amplitude of signals being fed to each array element. Both the angle-of-arrival determination and the control of the element excitations are difficult to perform accurately, which limits the performance of an "open-loop" system. For this reason adaptive "closed-loop" systems are often used, which can allow for statistical errors due to receiver noise and array tolerances.

In such an adaptive system, information on the desired null direction is obtained from the measured covariance matrix. This contains the mean values of the cross-products of the signals received by the array elements. For a two-element system, the covariance matrix M is given by equation 9.46, where $S*S$ is a complex conjugate and $<>$ denotes a time-averaged value.

$$M = \begin{bmatrix} <S_1^*S_1> & <S_1^*S_2> \\ <S_2^*S_1> & <S_2^*S_2> \end{bmatrix} \qquad (9.46)$$

For two isotropic elements and a single interferer at angle θ, equation 9.47 is obtained, where J is the jammer or interferer power, R is the receiver noise at each element (assumed uncorrected), and $\varphi = (2\pi d \sin \theta)/\lambda$.

$$M = \begin{bmatrix} J+R & J\exp(j\varphi) \\ J\exp(-j\varphi) & j+R \end{bmatrix} \qquad (9.47)$$

The direction of the interferer can be derived using the relation of equation 9.48.

$$\exp(2j\varphi) = \frac{M_{12}}{M_{21}} \qquad (9.48)$$

In principle, the $N \times N$ covariance matrix of an element array can be used to find the directions of up to $N-1$ interferers. For the commonly used maximum signal-to-noise adaptive algorithm, this step can be avoided because the optimum element excitations (weights) are found by inversion of the covariance matrix (Rudge et al., 1986).

The preceding theory has assumed that the wanted signal is absent when the adaptive weights are calculated. Wanted signal discriminations are an important topic in adaptive systems and take a number of forms. Possible discriminates include timing, spatial filtering, and coding.

Reflective Antenna Types

Reflective antennas are commonly used in terrestrial microwave and satellite applications. In this section, a simple parabolic reflector is first described, followed by "dual-symmetric" and "offset" antenna designs that are able to give improved performace. The properties of various forms of horn feeder, which are used with reflective antennas, are also described.

Prime-Focus Symmetric Reflector Antennas

The axisymmetric parabolic reflector with a feed at the focus of the paraboloid is the simplest type of reflector antenna. The geometry is shown in Figure 9.10. The paraboloid has the property that energy from the feed at F goes to the point P on the surface, where it is reflected parallel to the axis to arrive at a point A on the imaginary aperture plane. Equation 9.49, where F is the focal length, describes the surface.

$$r^2 = 4F(F-z) \qquad (9.49)$$

At the edge of the reflector, of diameter D, equation 9.50 applies.

$$\frac{F}{D} = \frac{1}{4}\cot\left(\frac{\theta_0}{2}\right) \qquad (9.50)$$

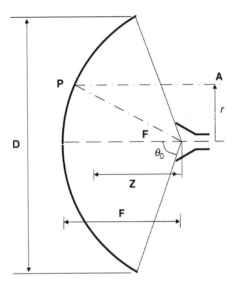

Fig. 9.10   Geometry of a reflector antenna.

The depth of the paraboloid is usually specified by its F/D ratio. Common sizes are between $F/D = 0.25$ ($\theta_0 = 90°$) to $F/D = 0.4$.

Aperture Fields and Radiation Patterns

Ray optics indicate that the path length from F to P to A is equal to twice the focal length. Hence the phase across the aperture is constant. The amplitude across the aperture plane will peak at the center and taper toward the edge for two reasons: First, the feed will have a tapered radiation pattern and, second, the action of a parabola in transforming a spherical wave from the feed into a plane wave across the aperture introduces a path loss, which is a function of angle θ. The aperture electric field is then given by equation 9.51, where $F(\theta, \varphi)$ is the pattern of the feed.

$$E_a(\theta, \varphi) = F(\theta, \varphi)\cos^2\left(\frac{\theta}{2}\right)$$
(9.51)

Feeds suitable for reflector antennas are discussed in a later section, but it is often convenient in initial design to take the feed pattern as being given by equation 9.52.

$$F(\theta, \varphi) = \cos^q(\theta)$$
(9.52)

Experience has shown that good-quality feeds approximate well to this function.

The radiation patterns can be predicted from the aperture fields using the Fourier transform relations described in the Radiation from Apertures section. This works well for large reflectors, but for detailed design of small to medium reflectors, it is necessary to take account of the precise form of the currents on the reflector surface and the diffraction that occurs at its edges. The former can be accomplished with physical optics theory (Rusch & Potter, 1970; Rusch, Ludwig, & Wong, 1986) and is good for predicting the main beam and near-in sidelobes. The diffracted fields influence the far-out sidelobes and can be predicted using the Geometrical Theory of Diffraction (GTD) (James, 1986).

Reflector Antenna Gain

The gain of a reflector antenna can be calculated from equation 9.53, where η is the efficiency of the reflector.

$$G = \eta \left(\frac{\pi D}{\lambda}\right)^2$$

(9.53)

The total efficiency is the product of six factors:

Illumination efficiency is the gain loss due to nonuniform aperture illumination.

Spillover efficiency is the gain loss caused by energy from the feed that radiates outside the solid angle subtended by θ_0. Called the spillover, this is the fraction of the power that is intercepted by the reflector. As the aperture edge taper increases, the spillover decreases and the spillover efficiency increases, while the illumination efficiency decreases. There is an optimum combination that corresponds to an edge illumination of about −10 dB.

Phase error efficiency is a measure of the deviation of the feed face front away from spherical and is usually nearly 100 percent.

Cross-polarization efficiency is a measure of the loss of energy in the orthogonal component of the polarization vector. For a symmetric reflector no cross-polarization is introduced by the reflector, so efficiency is determined by the feed characteristics. For good feeds this factor is also nearly 100 percent.

Blockage efficiency is a measure of the portion of the aperture that is blocked by the feed and the feed supports. The fields blocked by the feed do not contribute to the radiation, so it is desirable to keep the proportion of the area blocked to less than 10 percent of the total area of the aperture; otherwise, the sidelobe structure becomes distorted. The feed support blocking is more complicated because it depends on the shape and orientation of the supports (Lamb & Olver, 1986). It is electrically desirable to keep the cross-section of the supports small, which means that a compromise with mechanical constraints is needed.

Surface error efficiency is a measure of the deviations of the aperture wavefront from a plane wave due to surface distortions on the parabolic surface. Assuming that the errors are small and randomly distributed with a root mean square (r.m.s.) surface error, the efficiency is given by equation 9.54. This is a function of frequency and falls off rapidly above a certain value, which means that the upper frequency for which a reflector can be used is always given by the surface errors. The effect on the radiation pattern of random surface errors is to fill in the nulls and to scatter energy in all directions so that the far-out sidelobes are uniformly raised.

$$\eta_S = \exp\left[-\left(\frac{4\pi\varepsilon}{\lambda}\right)^2\right]$$

(9.54)

Dual Symmetric Reflector Antennas

The performance of a large reflector antenna can be improved and the design made more flexible by inserting a subreflector into the system (see Figure 9.11). There are two versions: the Cassegrain, where the subreflector is a convex hyperboloid of revolution placed on the inside of the parabola focus, and the Gregorian, where a concave elliptical subreflector is

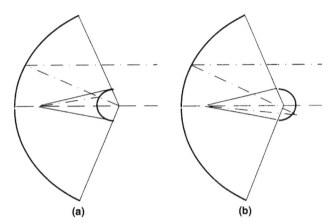

Fig. 9.11 Dual-reflector antennas:
(a) Cassegrain; (b) Gregorian.   **(a)**   **(b)**

placed on the outside of the parabola focus. In symmetric reflectors the Cassegrain is more common because it is more compact, but the electrical performance is similar for both systems.

The advantages of dual reflectors are as follows:

1. The feed is in a more convenient location.
2. Higher-performance feeds can be used because the subtend angle is such that wide aperture diameter feeds are needed.
3. Spillover past the subreflector is directed at the sky, which reduces the noise temperature.
4. The depth of focus and field of view are larger.

The study of radiation characteristics and efficiency dual reflectors is similar to that of the prime-focus reflector. Analysis of the radiation patterns depends partly on the size of the subreflector. If it is small then physical optics or GTD must be used. The main reflector is usually large, so geometric optics are adequate.

The limiting factor to obtaining high efficiency in a standard parabola is the amplitude taper across the aperture due to the parabola's feed pattern and the space loss (i.e., the illumination efficiency). By shaping the surfaces of a dual reflector antenna, it is possible to increase efficiency and produce a more uniform illumination across the aperture. A well-known method to produce a high-efficiency Cassegrain symmetric-reflector antenna is from Galindo (1964) and Williams (1965). It is a geometric optics technique in which the shape of the subreflector is altered to redistribute the energy more uniformly over the aperture. Then the shape of the main reflector is modified to refocus the energy and create a uniform phase across the aperture. After this process the reflector surfaces are no longer parabolic and hyperbolic. The method works well for large reflectors. For small- or medium-size reflectors, geometric optics are not adequate, and physical optics, including diffraction, must be used, at least on the subreflector.

Offset Reflectors

In recent years the growth in communications systems has led to a tightening of radiation pattern specifications and the consequent need to produce reflectors with low far-out

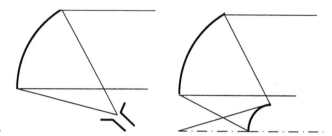

Fig. 9.12 Offset reflector antennas.

sidelobes. Symmetric reflectors cannot be made to have low sidelobes because of the inherent limitations caused by scattering from the feed and feed supports. This blockage loss can be entirely eliminated with the offset reflector (see Figure 9.12), which consists of a portion of a parabola chosen so that the feed is outside the area subtended by the aperture of the reflector. The projected aperture is circular, though the edge of the reflector is elliptical. The removal of the blockage loss also means that smaller reflector antennas can be made efficient, which has led to their widespread use as DBS receiving antennas.

In addition to the unblocked aperture, the offset reflector has other advantages (Rudge, 1986; Rahmat-Samii, 1986). The reaction of the reflector on the primary feed can be reduced to a very low order so that the feed VSWR is essentially independent of the reflector. Compared to a symmetric paraboloid, the offset configuration makes a larger F/D ratio possible, which in turn enables a higher-performance feed to be used. The removal of the feed from the aperture gives greater flexibility to use an array of feeds to produce multiple beams or shaped beams.

The offset reflector antenna also has some disadvantages. It is much more difficult to analyze and design because of the offset geometry, and it is only with the advent of powerful computers that this has become feasible. The lack of symmetry in the reflector means that when a linearly polarized feed is used, a cross-polarized component is generated by the reflector surface. When circular polarization is used, a cross-polarized component does not occur but the offset surface causes the beam to be "squinted" from the electrical boresight. Last, the construction of the offset reflector is more difficult. However, if the reflectors are made by fiberglass molding, this is not really significant. Also, the structural shape can be put to good use because it is convenient for deployable configurations on satellites or transportable Earth stations.

Horn Feeds for Reflector Antennas

A reflector antenna consists of the reflector plus the horn feed at the geometric focus of the reflector. Thus, the correct choice and design of the feed are important parts of the design of the entire reflector antenna. High-performance feeds are necessary to achieve high-performance antennas. The diameter of the feed in wavelengths will be determined by the angle subtended by the reflector at the feed. A prime focus reflector with an F/D between 0.25 and 0.5 will have a subtended half angle of between 90° and 53°. Application of the general rule that beamwidth is approximately equal to the inverse of the normalized aperture diameter shows that this means a feed with an aperture diameter of between about one and three wavelengths. Dual reflectors (Cassegrain or Gregorian) and offset reflectors have subtended angles between 30° and 7°, leading to feed diameters of between 3 and 10 wavelengths.

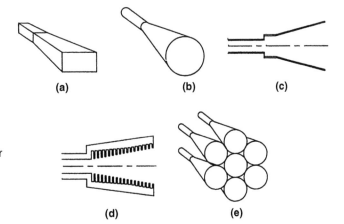

Fig. 9.13 Types of feed horns for reflector antennas: (a) rectangular or square; (b) small conical; (c) small conical with chokes; (d) conical corrugated; (e) array feed.

Of particular interest in horn feed design is the polarization performance, and the quality of a feed is usually expressed by the level of peak cross-polarization. The radiation characteristics of horns are predicted by a two-part process. First, the fields in the aperture are computed from a knowledge of the guided-wave behavior inside the horn. Second, the aperture fields are used to compute the radiated fields. The Fourier transform method has been found to work very well for the case of horns. The main types will now be briefly described. For more details, see studies done by Love (1976) and Love, Rudge, and Olver (1986).

Rectangular or Square Horns
Rectangular or square horns are the simplest horn type (see Figure 9.13(a)), but they are rarely used as feeds for reflectors because they have very high cross-polarization if the aperture size is not large.

Small Conical Horns
Small conical horns can have reasonably good cross-polarization performance (see Figure 9.13(b)). They are widely used as prime-focus feeds in small symmetric and offset reflectors. The basic design has an aperture diameter of about one wavelength and is essentially an open-ended circular waveguide propagating a TE_{11} mode. The radiation pattern can be improved by adding one or more rings or chokes around the aperture (see Figure 9.13(c)). These have the effect of changing the distribution of current on the flange and creating a more symmetric radiation pattern. The theoretical design of the open-ended waveguide is straightforward, but the analysis of the choked version is much more complicated. As a consequence most small feeds are designed empirically with measured data.

Multimode Conical Horns
Multimode conical horns improve the performance of conical horns by generating a second mode inside the horn in such a manner that the aperture fields are linearized. This second, TM_{11}, mode, is generated by a step change in the conical horn diameter, and the length of the horn is determined by the need to have the modes in the correct phase relationship at the aperture. The dual-mode horn gives low cross-polarization over a narrow band of frequencies. They are simple to make and of low weight.

The concept of adding higher-order modes in a horn can be extended for other purposes. In tracking feeds a higher-order mode is used to provide tracking information. The inherent cross-polarization that occurs in offset reflectors can be canceled by the appropriate addition of higher-order modes (Love, Rudge, & Olver, 1986). Finally, the main beam can be shaped to provide higher efficiency in prime-focus reflectors, although only over a narrow frequency band.

Conical Corrugated Horns

Conical corrugated horns are the leading choice for a feed for dual-reflector and medium-size offset reflectors (see Figure 9.13(b)). They have excellent radiation pattern symmetry and radiate very low cross-polarization over a broad range of frequencies.

A corrugated horn propagates a mixture of TE_{11} and TM_{11} modes, called a hybrid HE_{11} mode. The corrugations are approximately a quarter of a wavelength deep so that the electric short circuit at the base of the slot is transformed into a magnetic short circuit at the top of the slot. The result is that the azimuthal magnetic field is forced to zero at the corrugations and the azimuthal electric field is zero because of the ridges. Consequently, the boundary conditions of the TE and TM modes are identical, and the mutual propagating modes are linear combinations of the two parts. The design procedure for corrugated horns is well understood (Clarricoats & Olver, 1984), and it is possible to accurately predict the radiation characteristics.

Array Feeds

Array feeds are used to form multiple-beam and shaped-beam reflector antennas used on satellites (see Figure 9.13(e)). The individual elements of the array can be any type of horn, although for compactness small-diameter open-ended waveguides are preferred. The radiation patterns of the array are mainly determined by the element spacing and the amplitudes and phases of the signals sent to the individual elements. In addition to being able to form a wide range of multiple or shaped beams, the array has the advantage that the cross-polarization of the total array is lower than that of an individual element. However, the closeness of the array elements gives rise to mutual coupling between the aperture fields, which can distort the radiation patterns (Clarricoats, Tun, & Brown, 1984; Clarricoats, Tun, & Parini, 1984). A significant disadvantage of an array is that a beam-forming network of waveguide components must be used behind the array elements to produce the correct amplitudes and phases for the array. For large arrays this can be heavy, expensive, and a significant part of the design of the complete antenna system.

Microwave Line-of-Sight Radio

A typical microwave radio relay system consists of two axisymmetric parabolic reflector antennas on towers (see Figure 9.14), with a spacing of the order of 50 km in a line-of-sight path. The relationship between the transmitted and received powers and the antenna and path parameters is given by the Friis transmission formula (equation 9.2). The typical antenna gain is about 43 dBi, which means a diameter of about 3 meters at 6 GHz.

In addition to the pattern envelope specifications (which must be low because two or more antennas are normally mounted next to each other on a tower), there are a number of other important criteria for microwave radio antennas. The Front-to-back ratio must be high, and the cross-polar discrimination needs to be high for dual polarization operation—typically better than –25 dB within the main beam region over a bandwidth of up to 500 MHz. The VSWR needs to be low (typically 1.06 maximum) in a microwave radio relay system to

Fig. 9.14 Microwave line-of-sight reflector antenna.

reduce the magnitude of the round-trip echo. The supporting structure must be stable to ensure that the reflector does not move significantly in high winds. The reflectors must operate under all weather conditions, which means that a radome is often required.

This poses extra design problems because inevitably it degrades electrical performance. A long waveguide or coaxial cable feeder must be provided from the transmitter to the antennas. Not only must this be low loss, but it also must be well made so that there is no possibility of loose joins introducing nonlinear effects. Finally, the cost must be relatively low because a large number of reflectors are required in a microwave communication system.

The majority of antennas in use are prime-focus symmetric reflectors, often with shields and radomes. The design of the prime-focus reflectors follows the procedure discussed in earlier sections. The need to have a high front-to-back ratio means that either a low edge illumination or baffles and shields must be used. The latter methods are preferable, but they increase the weight and cost. The most common feed is a modified TE_{11} circular waveguide, which is designed to have a low VSWR and good pattern symmetry. Sometimes operation in two frequency bands is needed, in which case the feed must combine two waveguides and operate at the two frequencies. The VSWR can be reduced by replacing the center portion of the paraboloid with a flat plate, called a vertex plate. This minimizes the VSWR contribution from the dish, although it also degrades the near-in sidelobes and reduces the gain.

Earth Station Antennas

Earth station antennas are at the Earth end of satellite links. High gain is needed to receive the weak signals from the satellite or to transmit strong signals to it. The antennas can be divided into the following three types.

- Large antennas required for transmit and receive on the INTELSAT-type global networks with gains of 60 to 65 dBi (15 to 30 meters in diameter)

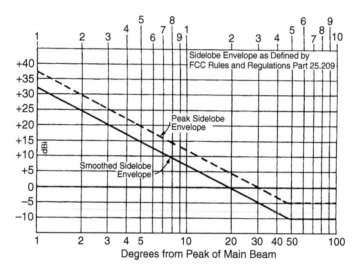

Fig. 9.15 FCC Earth station antenna pattern specification.

- Medium-sized antennas for cable head (TVRO) or data receive-only terminals (3 to 7 meters in diameter)
- Small antennas for direct broadcast reception (0.5 to 2 meters in diameter)

Types 1 and 2 have to satisfy stringent specifications imposed by various regulatory bodies. Previously, when the recommended spacing of geostationary satellites was 3 degrees, the pattern envelope was specified as 32 to 35 log θ. This requirement could be met with a symmetric-reflector antenna. With the new spacing of 2 degrees, the pattern specification has been improved to 29 to 25 log θ (see Figure 9.15). This can best be met with low-sidelobe, offset-reflector designs.

The minimum receivable signal level is set by inherent noise in the system. Earth stations are required to detect small signals, so control of noise parameters is important. The noise appearing at the output terminals of an Earth station used as a receiver has three components: noise received by the main beam of the reflector; spillover noise due to spillover from the feed; and receiver noise.

The first component can be due to natural sources or due to human-made interference. The natural noise emitters are Earth and sea absorption, galactic noise, isotropic background radiation, quantum noise, and absorption due to the oxygen and water vapor in the Earth's atmosphere. A minimum isotropic background radiation of about 3 K is always seen by any antenna. The value of the other factors depends on frequency. Spillover noise is the only component under the control of the antenna designer. Its value can be reduced by designing an antenna with very low sidelobes. Receiver noise is normally the dominant noise factor. It depends on the method of amplification and detection.

Early Earth stations all used cooled receivers, which have low noise temperatures. Modern Earth stations use uncooled receivers, which are dependent on the noise performance of the front-end transistor. This was improved dramatically in recent years, especially for small DBS terminals where the economies of scale support considerable research to reduce the noise temperature.

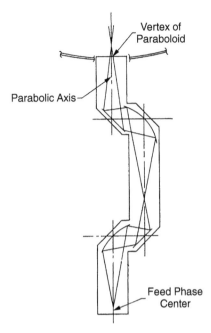

Vertex of
Paraboloid

Parabolic Axis

Feed Phase
Center

Fig. 9.16 Beam waveguide feed
system for large Earth stations.

The ratio of gain to noise temperature, the G/T ratio, is a useful measure of the influence of the noise components. Typical values are 40.7 dBK^{-1} for an INTELSAT A, 30-meter–diameter antenna operating at 4/6 GHz (Pratt & Bostian, 1986).

Large Earth station antennas are expensive to construct and maintain, so there is a premium in obtaining maximum efficiency from the system. The axisymmetric Cassegrain antenna (see the Nulls section) is the favorite choice for a number of reasons:

1. The gain can be increased over the standard parabola–hyperbola combination by shaping the reflectors. Up to an extra 1 dB is possible.
2. Low antenna noise temperatures can be achieved by controlling spillover using a high-performance corrugated horn and by using a beam waveguide feed system.
3. Beam waveguide feed systems place the low-noise receivers and high-power transmitters in a convenient, stationary location on the ground.

The beam waveguide feed system (Rudge, 1986), as shown in Figure 9.16, consists of at least four reflectors, whose shape and orientation are chosen so that the transmitter and receiver can be stationary while the antenna is free to move in two planes. The free-space beam suffers very little loss. The dual-polarized transmit and receive signals need to be separated by a beam-forming network placed behind the main feed horn. For 4/6-GHz operation, this also incorporates circular polarizers.

The narrow beam from the large antenna necessitates the incorporation of some form of tracking into the antenna because even a geostationary satellite drifts periodically. There are a number of schemes available, including monopulse, conical scan, and hill climbing. The favorite is a monopulse scheme using additional modes in the feed horn to electromagnetically abstract the tracking data.

The first generation of medium Earth station antennas were axisymmetric Cassegrain reflector antennas, sometimes shaped. However, the advent of tighter pattern specifications has led to the widespread use of single- or dual-offset reflector antennas (see the Offset Reflectors section). These can meet low-sidelobe specifications by removing blockage effects from the aperture. Very-high-efficiency designs have been produced by shaping the reflectors to optimize the use of the aperture (Bergman et al., 1988; Cha, 1983; Bjontagaard & Pettersen, 1983). For these high-efficiency designs, the r.m.s. surface error on the main reflector needs to be less than 0.5 mm for operation in the 11- to 14-GHz band. The feed is a high-performance corrugated horn. The offset reflector configuration lends itself to deployment, and portable designs have been produced where the offset reflector folds for transportation.

Cost is the main driver for small Earth station antennas for mass-market applications. Receive-only terminals in the 4/6-GHz band for data or TV reception are usually symmetric prime-focus paraboloid; they are made by spinning an aluminium sheet. In the 11-GHz communication band or the 12-GHz DBS band, prime-focus offset reflectors made from fiberglass molds are popular. A simple open-ended waveguide feed is incorporated on a sturdy feed support with the first-stage low-noise, converter incorporated directly into the feed. There is considerable interest in making flat-plate array antennas that can be mounted flush against buildings and that incorporate electronic scanning to receive the satellite signals. The technology for electronic scanning is available from military radars but not, so far, at a price that is acceptable to the domestic market.

Satellite Antennas for Telemetry, Tracking, and Command

The ideal Telemetry, Tracking, and Command (TT&C) antenna would give omnidirectional coverage so that the orientation of the satellite is irrelevant. Wire antennas are used for VHF and UHF coverage, but the spacecraft is a few wavelengths across at these frequencies and therefore considerable interaction between the antenna and the satellite distorts the radiation pattern. An alternative approach is to use a low-gain antenna to provide full Earth coverage. This is particularly useful for spin-stabilized spacecraft. The earth subtends 17 degrees from a geostationary satellite, which can be met with a small conical horn.

Spot Beams

Spot beam antennas are required to produce a beam covering a small region of the Earth's surface. The angular width of the beam is inversely proportional to the diameter of the antenna. Size considerations virtually dictate that some form of deployable mechanism is needed on the satellite, and this leads to the use of offset reflectors with a dual-mode or corrugated feed horn. The constraints of the launcher mean that the maximum size for a solid reflector is about 3.5 meters. Larger reflectors can only be launched by using some form of unfurlable mesh or panel reflector.

The trend toward smaller footprints on the Earth can be met either by using a larger reflector or by using a higher frequency, both of which involve higher costs. To date most spot beam communication satellites have used two prime-focus offset reflectors, one for transmit and one for receive, producing footprints on the Earth's surface that are elliptical because of the curvature of the Earth.

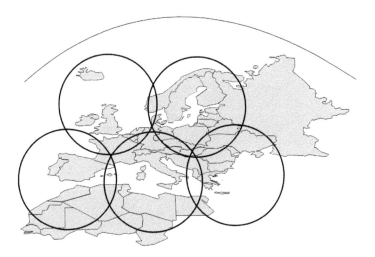

Fig. 9.17   Multiple spot beams generated by an array feed on a spacecraft.

Multiple Beams

It was early recognized that by using a single reflector and an array of feeds it was possible to produce multiple beams on the Earth (see Figure 9.17).

This has the advantage that most of the antenna subsystem is reused, with the penalty of having to design and make the array of feed horns and the beam-forming network behind the array. The array feed elements must be compact so that they occupy the minimum space in the focal plane of the offset reflector. At the same time the cross-polarization must be low. This tends to mean that corrugated horns cannot be used, and small-diameter dual-mode rectangular or circular horns are preferred. The maximum number of beams depends on the tolerable aberrations because array elements that are off-axis will have degraded performance.

Shaped Beams

It is desirable to optimize the shape of the satellite beam on the Earth's surface so as to conserve power and to not waste energy by illuminating portions of the oceans. An example is shown in Figure 9.18. Shaped beams can be produced in two ways. Multiple, overlapping beams produced by an offset parabolic reflector and an array of feeds can be used. This approach is an extension of multiple beams and has the advantage that it is possible to design for reconfiguration by incorporating switching systems into the beam-forming network. The alternative approach is to use a single, high-performance feed and to physically shape the surface of the reflector so that power is distributed uniformly over a shaped beam region. Both approaches have received considerable attention in recent years.

The multiple-beam approach is well illustrated by the INTELSAT VI communication satellite (see Figure 9.19), which produces multiple shaped beams to cover the main population regions of the earth (see Figure 9.20). To be able to use the same satellites over the Atlantic, Indian, or Pacific oceans, the array feed consists of 146 elements that can be switched to produce the appropriate shaped beams (Bennett & Braverman, 1984).

Fig. 9.18 European contoured beam.

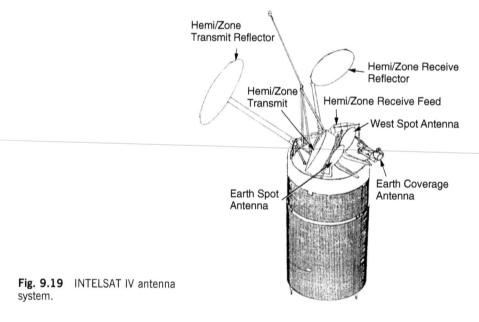

Fig. 9.19 INTELSAT IV antenna system.

The shaped-reflector approach has the advantage of mechanical simplicity and lower weight, at the penalty of fixed beams. The theoretical design process is quite extensive and involves a synthesis with the input of the required beam shape and the output of the contours of the reflector surface. A single offset reflector constrains the possible shapes because it is not possible to arbitrarily specify the amplitude and the phase of the synthesized pattern. This constraint is removed with a dual-reflector design.

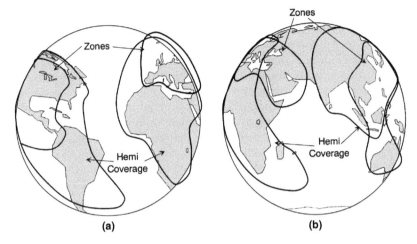

Fig. 9.20 Shaped beam generated by INTELSAT VI antenna: (a) Atlantic/Pacific Ocean; (b) Indian Ocean.

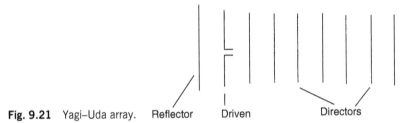

Fig. 9.21 Yagi–Uda array.   Reflector   Driven   Directors

VHF and UHF Communications

Antennas for VHF and UHF communications systems take on a wide variety of specific forms, but the vast majority are derivatives of the generic dipole-type antenna. The physical, mechanical, and environmental aspects are generally more significant than for microwave antennas because the smaller size of the antenna means that the radiation and impedance characteristics are partly determined by these aspects.

A comprehensive survey of VHF and UHF antennas can be found in both Rudge (1986) and Johnson and Jasik (1984). Antennas that give near uniform coverage in one plane can be obtained from half-wave dipoles or monopoles. Complementary antennas, such as loops and slots, will work equally well, and the actual shape will be determined more by application than by basic electromagnetic performance. The bandwidth of these simple elements is limited by the impedance characteristics, although most communication applications require only relatively narrow bandwidths. With small elements, some form of impedance matching network is required. One problem with balanced dipole-type antennas is that they are required to be fed by an unbalanced coaxial cable. A balun is needed to match the balanced to the unbalanced system, and this is inevitably frequency sensitive.

Antennas for point-to-point links need to be directional and have as high a gain as possible. This is achieved with a Yagi-Uda array (see Figure 9.21), which consists of one driven element, one reflector element, and a number of director elements.

Only the driven element is connected to the feed line; the other elements are passive, and currents are induced in them by mutual coupling; the spacing ensures that this is in the correct amplitude and phase to give a directional radiation pattern. Gains of up to about 17 dBi are possible from one Yagi-Uda array. Higher gains can be obtained by multiple arrays. The Yagi-Uda array is inherently linearly polarized. Circular polarized arrays can be made either from crossed dipoles or from helixes.

Antennas for mobile communications can be divided into those for base stations and those for mobiles. Base station antennas are mounted on towers and usually require nearly uniform patterns in the horizontal plane, with shaping in the vertical plane to conserve power. This can be achieved with a vertical array of vertical dipoles or other paneled dipoles. The influence of the tower on the antenna must be taken into account in the design.

Mobile antennas on vehicles, ships, and aircraft, or near humans present challenging problems to the antenna designer. In most cases the physical, mechanical, and environmental aspects take precedence over the electromagnetic design. In consequence the ingenuity of the antenna designer is required to produce an antenna that works well in adverse conditions. For example, antennas on aircraft must not disturb the aerodynamic profile, so they cannot protrude from the aircraft body. The effects of corrosion, temperature, pressure, vibration, and weather are other factors to be taken into account. Antennas for personal radios are constrained by the role of the operator and by the need for very compact designs commensurate with satisfying radiation safety levels. The human body acts partly as a director and partly as a reflector depending on the frequency of use and the relative position of the antenna to the body. Portable radio equipment has to be considered a part of the antenna system, including the radio circuits, batteries, and case. In general, improved performance results when the antenna is held as far from the body as possible and as high as possible.

HF Communications

HF antennas are used in the range of frequencies from 2 to 30 MHz for mobile communications and some fixed communications. Space precludes more than a brief mention of HF antenna types. Surveys can be found in Rudge (1986) and Johnson and Jasik (1984). HF antenna design is constrained by ionospheric propagation characteristics, which change daily, seasonally, and with the sunspot cycle. Antennas can receive either the sky wave reflected from the ionosphere or the ground wave if the transmitter and receiver are close together. The wavelength in the HF band is such that antennas are usually only a fraction of a wavelength in size; this in turn means that the local environment around the antenna will have a major impact on performance, which is particularly true for antennas mounted on vehicles, ships, and aircraft. The analysis of the antenna must take account of the environment by techniques such as wire grid modeling (Mittra, 1975). This is computer intensive and inevitably approximate, which means that much HF antenna design is empirical.

Most HF antennas are based on dipoles, monopoles, or wires. Complementary elements, such as loops or slot antennas, are also used. Directionality or gain is achieved by arrays of elements. A prime requirement of most HF antennas is that they are broadband so that the optimum propagation frequency can be used. The radiation patterns of the basic elements are wideband, but the input impedance or VSWR is narrowband. To overcome this limitation, a tuning unit has to be incorporated into the system. Wideband operation is achieved with automatic tuning units.

References

Balanis, C. A. (1982) *Antenna Theory: Analysis and Design*, Harper & Row.

Bennett, S. B., and Braverman, D. J. (1984) INTELSAT IV—A Continuing Evolution, *Proceedings IEEE* 72:1457.

Bergman. J., Brown, R. C., Clarricoats, P. J. B., and Zhou, H. (1988) Synthesis of Shaped Beam Reflector Antenna Patterns, *Proceeding, IEE* 135(H)(1):48.

Bjontagaard, G., and Pettersen, T. (1983) An Offset Dual Reflector Antenna Shaped from Near-Field Measurements of the Feed Horn; Theoretical Calculations and Measurements, *IEEE Trans.* 31:973.

Cha, A. G. (1983) An Offset Dual Shaped Reflector with 84.5 Percent Efficiency, *IEEE Trans.* AP-31: 896.

Clarricoats, P. J. B., and Olver, A. D. (1984) *Corrugated Horns for Microwave Antennas*, Peter Peregrinus (IEE).

Clarricoats, P. J. B., Tun, S. M., and Brown, R. C. (1984) Performance of Offset Reflector Antennas with Array Feeds, *Proceedings IEE* 131(H)(1):172.

Clarricoats, P. J. B., Tun, S. M., and Parini, C. G. (1984) Effects of Mutual Coupling in Conical Horn Arrays, *Proceedings IEE* 131(H)(1):165.

Collin, R. E., and Zucher, F. J. (1969) *Antenna Theory, Parts I & II*, McGraw-Hill.

Elliot, R. S. (1981) *Antenna Theory and Design*, Prentice-Hall.

Galindo, V. (1964) Design of Dual Reflector Antennas with Arbitrary Phase and Amplitude Distribution, *IEEE Trans.* 12, 403.

James, G. L. (1986) *Geometrical Theory of Diffraction for Electromagnetic Waves*, Peter Peregrinus (IEE).

Johnson, R. C., and Jasik, H. (eds.) (1984) *Antenna Engineering Handbook*, McGraw-Hill.

King, R. W. P. (1956) *The Theory of Linear Antennas*, Harvard University Press.

King, R. W. P., Mack, R. B., and Sandler, S. S. (1968) *Arrays of Cylindrical Dipoles*, Cambridge University Press.

Lamb, J. W., and Olver, A. D. (1986) Blockage Due to Subreflector Supports in Large Radiotelescope Antennas, *Proceedings IEE* 133(H)(1):43.

Lo, Y. T., and Lee, S. W. (1988) *Antenna Handbook*, Van Nostrand.

Love, A. W. (1976) *Electromagnetic Horn Antennas*, Selected Reprint Series, IEEE Press.

Love, A. W., Rudge, A. W., and Olver, A. D. (1986) Primary Feed Antennas, in Rudge, A. W., et al. (eds.), *The Handbook of Antenna Design*, Peter Peregrinus (IEE).

Ludwig, A. G. (1973) The Definition of Cross-Polarisation, *IEEE Trans.* AP-211(1):116.

Milligan, T. (1985) *Modern Antenna Design*, McGraw-Hill.

Mittra, R. (1975) *Numerical and Asymptotic Methods in Electromagnetics*, Springer-Verlag.

Olver, A. D. (1986) Basic Properties of Antennas, in Rudge, A. W., et al. (eds.), *The Handbook of Antenna Design*, Peter Peregrinus (IEE).

Pratt, T., and Bostian, C. W. (1986) *Satellite Communications*, Wiley.

Rahmat-Samii, Y. (1986) Reflector Antennas, in Lo, Y. T., and Lee, S. W. (eds.), *Antenna Handbook*, Van Nostrand.

Rudge, A. W., et al. (eds.) (1986) *The Handbook of Antenna Design*, Peter Peregrinus (IEE).

Rusch, W. V. T., and Potter, P. D. (1970) *Analysis of Reflector Antennas*, Academic Press.

Rusch, W. V. T., Ludwig, A. C., and Wong, W. G. (1986) Analytical Techniques for Quasi-Optical Antennas, in Rudge, A. W., et al. (eds.), *The Handbook of Antenna Design*, Peter Peregrinus (IEE).

Silver, S. (ed.) (1984) *Microwave Antenna Theory and Design*, Peter Peregrinus (IEE).

Stutzman, W. L., and Thiele, G. A. (1981) *Antenna Theory and Design*, Wiley.

Williams, W. F. (1965) High Efficiency Antenna Reflector, *Microwave J.* 8:79.

Part 3
Network Technologies

This part deals with the primary core network technologies that operate over the physical media. It explains both the transmission network and data overlay network technologies. The transmission network is closely related to the particular media and is used as a common network able to carry a wide variety of signal types. Above the transmission network are a number of overlay networks that are designed to provide a particular range of services to network users. A number of different overlays have been developed; for example, PSTN is an overlay network delivering voice services, while Frame Relay is an overlay network offering low speed data services. These networks are called overlays because they overlay the transmission network.

At the transmission level, both SONET and SDH technologies were largely developed by 1990 as a means of providing high-capacity circuit-switched transport when voice and private circuit traffic dominated. Both have a strictly hierarchical structure that well suited the needs of voice transmission. However, as demand for data grew and customer networks widely adopted Ethernet standards, it became clear that SONET and SDH in their original form did not have the flexibility to provide efficient transport for these data signals. For example, SDH bitrates increased by factors of 4, while Ethernet increased by factors of 10—a mismatch that resulted in inefficient fill of SDH channels when data was being carried. These restrictions were overcome with the introduction of Next Generation SDH, which provided much greater flexibility in the way SDH capacity could be used. These two technologies are described in Chapters 10 and 11.

The main "legacy" overlay data network technologies are described in Chapter 12. This starts with private circuits, which have been widely used for providing Wide-Area Network interconnections using either SONET/SDH or PSTN-like technology. For access, ISDN is also discussed.

While these technologies have been widely used, for secure and reliable interconnections, physical overlays are expensive because they provide a dedicated set of resources to each customer. Packet network technologies allow resources to be shared and so offer a more economical solution. Frame Relay (which itself was developed from the earlier 64 kbit/s X.25 network standard) typically operates over 2-Mbit/s lines, but it can be made to operate at higher speeds if needed. ATM (which uses short packets called cells) operates at line speeds of up to 155 or 622 Mbit/s and is structured to support realtime traffic.

The operation of IP networks was described in Volume 1. Here the transport of IP data across a core network using MPLS switching is described in Chapter 13. IP/MPLS has proved to be successful and is now widely adopted in the center of core networks. However, it is less easy to scale to large numbers of meshed nodes.

An alternative solution for metropolitan areas based on Ethernet principles is being developed through the IEEE and is known as "Carrier-Grade Ethernet." Ethernet itself does not scale to carrier network applications nor does it have suitable management facilities. However, the modified version currently under development through the IEEE will have a scalable architecture and full operations, administration, and management facilities; it will have a speed target of up to 100 Gbit/s. This is described in Chapter 14. It remains to be seen to what extent Carrier-Grade Ethernet can displace a firmly embedded IP/MPLS infrastructure.

10 Synchronous Optical Networks and Synchronous Digital Hierarchy

Mark Matthews
Nortel Ltd.

Goff Hill
GTel Consultancy Ltd.

Introduction

The Synchronous Optical Network (SONET) and Synchronous Digital Hierarchy (SDH) are standards that were designed for multiplexing together many low-rate digital traffic channels (in particular, 64-kbit/s voice channels) into higher-rate channels so that the low-rate channels can be more efficiently transported around a telecommunications network. SONET is both the forerunner and the North American equivalent of SDH.

SONET and SDH have many similarities, but they use different terminologies. In the following sections, the main differences between the two standards are first described and then, to avoid confusion, *SDH* is used to explain the principles of synchronous networking.

Table 10.1   SONET and SDH Bitrates

SONET Optical Carrier level	SONET Synchronous Transport Signal level	SDH Synchronous Transport Module	Payload Rate (Mbit/s)	Line Rate (Mbit/s)
OC-1	STS-1	(STM-0)	48,960	51,840
OC-3	STS-3	STM-1	150,336	155,520
OC-12	STS-12	STM-4	601,344	622,080
OC-48	STS-48	STM-16	2,405,376	2,488,320
OC-192	STS-192	STM-64	9,621,504	9,953,280
OC-768	STS-768	STM-256	38,486,016	39,813,120

SONET and SDH: Similarities and Differences

SONET was developed through the American National Standards Institute (ANSI) for use in North America, while SDH was developed through the International Telecommunications Union T (ITU-T) for use elsewhere in the world. Both standards are based on byte-interleaved multiplexing and share common bitrates as shown in Table 10.1.

However, they use different terminology. SONET describes its signals as *Synchronous Transport Signals*, while SDH describes its signals as *Synchronous Transport Modules*. A number is associated with each of these signal types to designate the level in the multiplexing hierarchy that is related to the bitrate. SDH does not have an STS-1 equivalent, although this rate is sometimes referred to as STM-0. Instead the SDH hierarchy starts at STM-1, which is the equivalent of SONET's STS-3. The highest rate in use today is STS-768 (STM-256), although higher rates have been defined in readiness for further developments.

Other terminology within SONET also differs from SDH. For example, SONET refers to *Virtual Tributaries* (VTs) whereas SDH refers to *Virtual Containers* (VCs), and the equivalent of the VC-3, referred to above, is known as a *Synchronous Payload Envelope* (SPE). Another distinction also makes a distinction between the STS logical signal and its optical manifestation at a Network Node Interface (NNI), in which case it is referred to as an *Optical Carrier* (OC). North American transmission systems are usually referred to by their transmission rates in OCs (e.g., OC-3 or OC-48), and in practice they are always constructed to use the same transmission rates as SDH systems, as illustrated in Table 10.1. In fact, the higher the transmission rates involved, the more closely the SONET and corresponding SDH systems come to resemble each other.

There are also some more detailed differences. For example, overhead bytes are defined slightly differently for the two systems. The similarity between an OC-3 and an STM-1 is very close but not exact. The difference is to be found in two bits within the pointer bytes of the SDH, known as the "SS" bits (for an explanation of the pointer mechanism, see The Concept of Pointers section). In SDH the default value is 11, and in SONET the default value is 00.

Although the STS-M bitrates are exact multiples of the STS-1 bitrates, and STM-N bitrates are exact multiples of the STM-1 bitrate, when multiplexing takes place each synchronous transport module is broken down to recover the VCs, and the outgoing STS-Ms or STM-Ns are reassembled with new overheads. Figure 10.1 shows the frame and header structure.

Both systems enable the performance of traffic paths to be monitored across the network and network management to be automated. A key driver for their introduction was that a

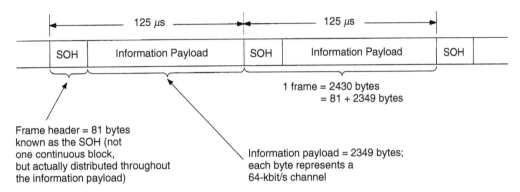

Fig. 10.1 Basis of the SDH frame structure.

PTO could have a consistent mechanism for partitioning, monitoring, and controlling the raw transport capacity of the whole network.

The Basis of SONET and SDH

Both SONET and SDH standards are designed to transport isochronous traffic channels and are focused very much on layer 1 of the well-known ISO seven-layer OSI protocol hierarchy (see Volume 1, *The Cable and Telecommnuications Professionals' Reference*). They are based on a hierarchy of continuously repeating, fixed-length frames originally designed to support predominantly voice traffic and to interwork with the earlier Plesiochronous Digital Hierarchy (PDH) standards. As the standards were developed, it was apparent that the increasingly widespread deployment of optical fibers would rapidly reduce the cost of raw transmission bandwidth. Hence, during the development of SDH standards, there was relatively little pressure to restrict the proportion of bandwidth devoted to multiplexing overheads, as opposed to the traffic payload.

It was necessary to adopt not only a synchronous frame structure but one that also preserved the byte boundaries in the various traffic bitstreams. Consequently, the basic frame structure is one that repeats at intervals of 125 *μs*; that is, SDH and SONET are tailor-made for the transport of 64-kbit/s channels or any higher-rate channels that are an integer multiple of 64 kbit/s.

Mapping PDH into Synchronous Containers

The SONET standards were designed around the need for efficient transport of the existing North American PDH rates, the most important of which are the 1.5- and 45-Mbit/s rates. This leads to a standard where a 45-Mbit/s signal is loaded into an Synchronous Transport Signal-1 (STS-1), which runs at 51.84 Mbit/s—that is, precisely one-third of the STM-1 rate. As with SDH, there is also a concatenation mechanism for dealing with customer signals that do not easily fit inside any of the already defined VC payload areas. This mechanism is particularly useful for carrying rates such as 140 Mbit/s, where the concatenation results in a SONET NNI signal known as an STS-3c, which has a line rate and payload that are identical to those of an STM-1. In contrast, SDH originally focused on the transport of

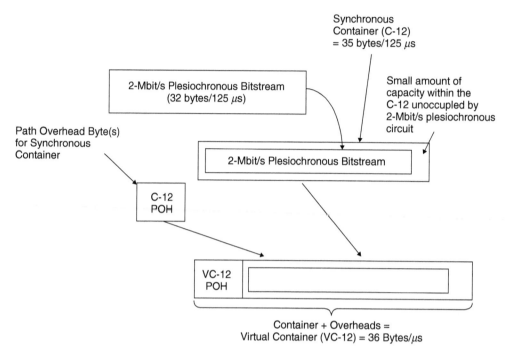

Fig. 10.2   Example of the creation of a VC (2-Mbit/s channel being loaded into a VC-12).

2.048-Mbit/s, 34-Mbit/s, and 140-Mbit/s circuits, plus their North American counterparts at 1.5 Mbit/s, 6 Mbit/s, and 45 Mbit/s.

The general way that any of these PDH rate circuits are transported by SONET or SDH is to map circuits into a *synchronous container* (see Figure 10.2). From here on, we adopt SDH to explain the principles of synchronous networking to avoid duplicating the terminology differences just described.

A synchronous container can be viewed as a subdivision of the basic SDH frame structure, and it consists of a predefined number of 64-kbit/s channels. The entire family of synchronous containers comprises only a few different types, each of which has been sized to accommodate one or more of the common plesiochronous transmission rates, without wasting too much bandwidth. When plesiochronous circuits are mapped into a synchronous container, the plesiochronous channel is synchronized to the frequency of the synchronous container, which is, in turn, synchronous to the basic SDH frame structure (see Figure 10.3).

Before examining the question of exactly what the frame structure is synchronous with, it is worth digressing to discuss a further operation. This is the attachment of an overhead, known as the Path Overhead (POH), to each synchronous container (see Figure 10.2). The idea is that when a plesiochronous circuit has been loaded into a synchronous container, this container has a defined set of POH bytes appended to it, which remain completely unchanged until the synchronous container arrives at its destination. The combination of synchronous container plus POH is known as a VC. The VC POH bytes allow a PTO to

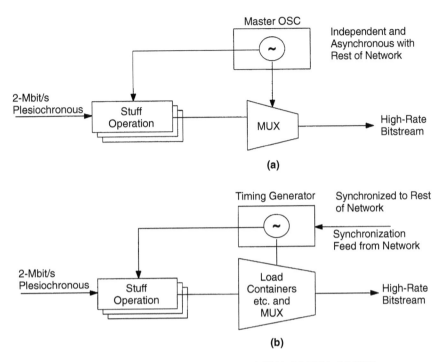

Fig. 10.3 Difference in synchronization between PDH and SDH: (a) PDH; (b) SDH.

monitor several parameters, the most important of which is the error rate of the VC between the points at which it was loaded and unloaded with its plesiochronous payload. This provides a PTO with a capability to monitor end-to-end performance.

Most plesiochronous channels are bidirectional, hence there are usually two continuous streams of VCs traveling in opposite directions between the two endpoints at which the plesiochronous channel enters and leaves the SDH portion of a network. As a result, the job of an SDH-based network can be considered to be to loading its VCs with (usually) conventional PDH channels and then transporting these to their various destinations together with an accurate indication of the quality of the delivered VC payload.

This process of loading containers and then attaching POHs is repeated at several levels in SDH, resulting in the nesting of smaller VCs within larger ones (see Figure 10.4). The nesting hierarchy stops when a VC of the largest defined level is loaded into the payload area of a Synchronous Transport Module (STM). These logical STM signals are what is seen at the interface between any two pieces of SDH equipment, and they can be presented either electrically or, more usually, optically (see Figure 10.5).

Such an interface is referred to as a *Network Node Interface* (NNI) because it is usually confined to the internal interfaces within the network rather than any interface presented to a network user. A *User Network Interface* (UNI) has also been defined. Finally, the reason for the name "virtual container" is that unlike the STM signals that appear at the NNI, VC signals are never presented to the outside world. They exist only within pieces of SDH equipment or within STM signals; hence an SDH network element can have NNI and PDH interfaces but never VC interfaces.

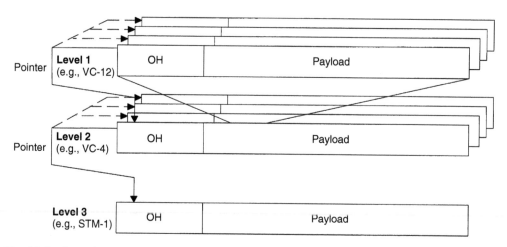

Fig. 10.4 General representation of a three-layer synchronous multiplexer structure.

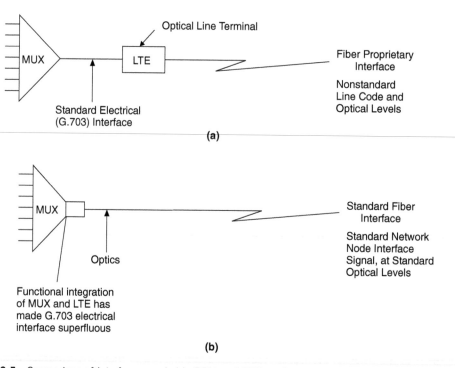

Fig. 10.5 Comparison of interfaces needed in PDH and SDH environments: (a) current PDH; (b) SDH.

Returning to the question of exactly what is synchronized to what, the basic problem is that it is very difficult to maintain a complete transmission network in rigid synchronization for all time. Even if we could tolerate the delays introduced by the addition of the numerous "wander buffers" necessary to accommodate the slow changes in transmission medium delay, there is no guarantee that different PTOs' networks would all be synchronized to the same master clock. For a network based on the SDH, this problem translates to that of how to synchronously multiplex and de-multiplex many individual VCs, which, because they have been created in disparate parts of the same, or even different, SDH networks, might have slightly different short-term bitrates.

The Concept of Pointers

The solution adopted by SDH is to associate a "pointer" with each VC so that when the VC is multiplexed, along with others, into a larger VC, its phase offset in bytes can be identified relative to some reference point in this larger VC (see Figure 10.6).

Furthermore, there is also a mechanism for allowing the value of this pointer to change if, for some reason, there is a loss of synchronization and the smaller-capacity VC is running either slightly slower or slightly faster than the larger VC. In fact, each of the smaller-capacity VCs has its own pointer, which can change independently of any of the others. Although the use of these pointers still entails some input buffers, these are much smaller than would be required if there were no mechanism for changing the phase of a small-capacity VC within a larger one; hence the problem of excessive delays can be contained. We can now picture an SDH network where the majority of VCs, both large and small, are well synchronized to each other, but, at the same time, there are a few that are not so well synchronized, and every so often the increasing strain of their asynchronism has to be relieved by a byte-sized slip relative to the majority of the other VCs in the network.

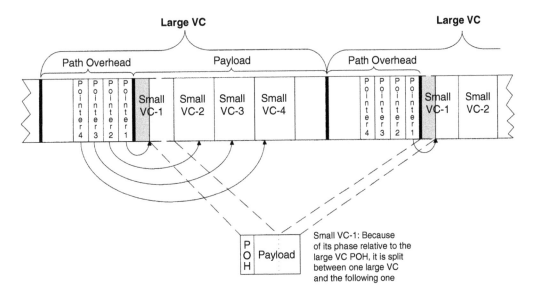

Fig. 10.6 Pointers allow a small VC to have arbitrary phase with respect to a large VC.

The crucial point is that we retain our ability to locate the management and control information in the VC's POH bytes because the pointer value associated with the VC is recalculated whenever a slip occurs.

The pointer mechanism previously described is at the very heart of the SDH standard. It is this mechanism that enables us to construct networks that are nearly, but not completely, synchronous, and yet still allows us to easily locate each traffic channel (VC) together with its associated management and control information (i.e., POH), but without incurring large penalties in transmission delay. It could be argued that SDH networks are not really synchronous at all but are actually very tightly controlled asynchronous networks. However, the fact that we have quantized the slips due to this asynchronism means that it is now possible, at any time, to locate and route any of the traffic paths within an SDH network. This, together with network management software, gives us the traffic-routing flexibility that was very difficult to achieve using PDH-based equipment. In terms of actual network hardware, it opens the way to the production of economically viable drop-and-insert multiplexers and cross-connects.

SDH Standards

There are now many ITU-T recommendations describing various aspects of SDH, but the centerpiece of the whole group is undoubtedly ITU-T G.707. This standard defines the basic SDH multiplexing structure and the ways that non-SDH traffic channels can be mapped into the SDH virtual containers. Other standards in the group deal with such things as the functionality of multiplexers (G.783), the management requirements of such equipment (G.784), and the specific recommendations for line systems (G.958). Optical interfaces for all types of SDH equipment are covered in G.957. Finally, there are two more standards, G.803 and G.805, which address the way that entire SDH-based networks should be constructed so they can interwork successfully with other such networks and, even more important, so that the management of these networks can be brought under software control. Nevertheless, because of its central role in SDH, we will concentrate mainly on explaining G.707.

Bearing in mind the nesting of smaller VCs within larger ones, and thence into STMs, the best way to appreciate the details of the SDH multiplexing standard, G.707, is to follow the progress of a bidirectional 2-Mbit/s plesiochronous circuit, which, for part of its journey, is transported across an SDH-based network (see Figure 10.7).

Such a 2-Mbit/s circuit could be a channel between two PSTN switches, or it could be a private leased line that is connecting two PBXs. Although the 2-Mbit/s circuit is bidirectional, the SDH operations are identical in both directions. Thus we will concentrate on just one direction of transmission, that from A to D.

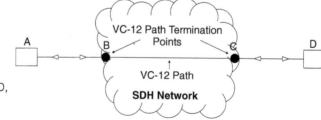

Fig. 10.7 Two-Mbit/s plesiochronous circuit from A to D, which is transported by an SDH network for part of its journey.

At point B, where the 2-Mbit/s circuit meets the SDH network, the first operation is to take the incoming plesiochronous bitstream and selectively add "stuffing" bits to "pad out" this bitstream to the exact rate required to fill the appropriate synchronous container. In this case, the synchronous container size would be a C12, which is sufficiently large to accommodate a 2-Mbit/s plesiochronous bitstream at the limits of its 10-ppm tolerance, together with some additional "fixed-stuffing" bytes. The stuffing of the plesiochronous bitstream should ideally be done relative to the clock to which the whole SDH network is beating; however, there is a chance that the SDH network element (e.g., multiplexer) that is performing the stuffing operation is not quite synchronous with the rest of the network. In this case C12, which it creates, is similarly asynchronous.

Path Overhead Information

Having mapped the 2-Mbit/s bitstream into the C12, the next operation is to generate and attach the Path Overhead (POH) byte, which enables this C12 to be identified, monitored for errors, and routed through the SDH network. The addition of this POH byte to the C12 creates a virtual container 12 (VC-12). As mentioned in The Basis of SONET and SDH section, the idea is that the POH stays attached to its C12 all the way from the point where it was generated to the point at which the 2-Mbit/s payload exits the SDH network. These two points are the "path termination" points for this VC-12, with the continuous stream of VCs between them being referred to as the "path." Between the two path termination points at B and C, there is no legitimate mechanism for altering any of the information in the POH; hence if the receiving path termination detects any discrepancy between the POH and the content of the VC-12 payload (i.e., the C12), this indicates that the VC-12 payload has somehow become corrupted during its journey across the SDH network.

Although path-level monitoring is sufficient for a PTO to ascertain what error rate is being inflicted by the SDH network on its customers' 2-Mbit/s circuit, it provides no information whatsoever on the source of the errors—that is, which network element has gone faulty. This task is dealt with by the addition of still further overhead information, which will be described shortly. Before this, it is necessary to examine the way in which several VC-12s are multiplexed into a higher-rate signal.

Multiplexing Virtual Containers

The general principle of this multiplexing operation is fairly straightforward. Several VC-12s, which should, hopefully, be synchronous with one another, are loaded into a larger synchronous container, which subsequently has its own POH added, thus creating a larger (i.e., higher-bitrate) VC (see Figure 10.4). In ETSI countries, this operation often, but not always, results in the creation of a VC-4, which has a payload that is large enough to accommodate up to 63 VC-12s.

Unfortunately, as discussed in The Basis of SONET and SDH section, complications arise in this operation because the network element that is performing this task might not itself have created all the VC-12s (see Figure 10.8). This leads to the possibility that not all the VC-12s are completely synchronous with one another or, more important, with the VC-4 into which they are being loaded.

The solution to this problem comes in two parts. First, the internal structure of the VC-4 has been purposely designed to allow each VC-12 to run slightly faster, or slower, than the VC-4 rate. This is done by designating certain bytes in the C4 as overflow bytes (one per

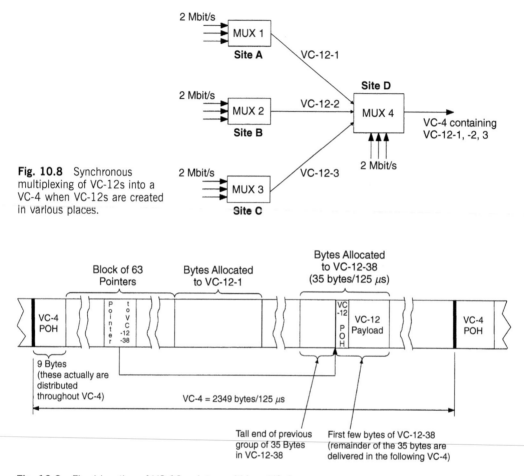

Fig. 10.8 Synchronous multiplexing of VC-12s into a VC-4 when VC-12s are created in various places.

Fig. 10.9 Fixed location of VC-12 pointers within a VC-4.

VC-12) to cope with a VC-12 that is running too fast. On the other hand, when a VC-12 is running slow, occasionally a VC-12 byte can be repeated. Second, this (hopefully) infrequent change of phase of a VC-12 relative to its VC-4 is recorded by means of a pointer. A VC-4 maintains one pointer for each of the VC-12s within its payload, and each pointer registers the offset in bytes between the first byte of the VC-4 and the POH byte of a particular VC-12. Each pointer is located in a predefined position within the VC-4 so that, when the VC-4 POH has been located, it is a simple matter of counting bytes to locate each of the 63 pointers to the VC-12s (see Figure 10.9).

The trigger for a rephasing of a VC-12 relative to its VC-4 is when the fill of the VC-12 input buffer (into which the incoming VC-12 is written prior to loading into the VC-4) exceeds a predetermined threshold. At this point, a fairly conventional justification operation occurs, with either an extra VC-12 byte being loaded into the VC-4 (into the overflow position mentioned earlier) or, conversely, one byte being repeated (see Figure 10.10). Either way, the resulting change of phase of VC-12 relative to VC-4 is tracked by a corresponding change in pointer value.

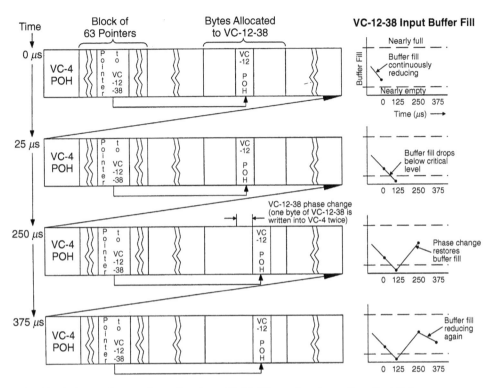

Fig. 10.10 Pointer adjustment resulting from VC-12-38 running slower than the VC-4.

Because of the importance of pointers for locating low-order VCs (e.g., VC-12s) within high-order VCs (e.g., VC-4s), the combination of a low-order VC plus its pointer is referred to as a *Tributary Unit* (TU). In this case, the combination of a VC-12 plus its pointer constitutes a TU-12.

Channels and Tributary Unit Groups

Having discussed the way in which up to 63 individual TU-12s are multiplexed into a VC-4, it is now necessary to examine more closely the internal structure of the VC-4. For this purpose, it is helpful to consider a VC-4 not as a linear string of bytes but instead as a two-dimensional block of bytes that is arranged as nine rows, each of 261 bytes.

For transmission purposes, this block is serialized by scanning left to right, top to bottom (see Figure 10.11). With this structure, a TU-12 can be seen to occupy four widely separated, nine byte columns, rather than the contiguous block of bytes, as suggested by Figures 10.9 and 10.10 (see Figure 10.12).

Because a VC-4 repeats every 125 μs, this implies that a TU-12 contains 36 bytes per 125 μs—that is, rather more than the 32 bytes nominally required by a plesiochronous 2.048-Mbit/s signal. This, however, is consistent with the need for a TU-12 to include a pointer, overflow byte positions, VC-12 overhead byte, plesiochronous stuffing, and so forth. Each such group of four columns represents a separate channel within the VC-4, and when a

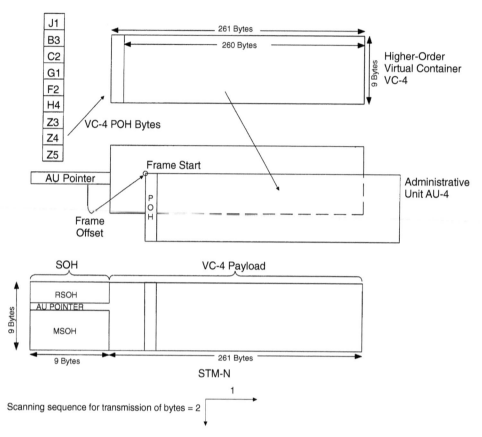

Fig. 10.11 A VC-4 as a block of 261 × 9 bytes together with its incorporation within an STM signal.

VC-12 slips phase relative to the VC-4, it slips within its own channel; that is, it does not overflow into another channel because this would obviously corrupt the data in the other channel. As mentioned earlier, the position of the pointer within this channel is constant; only the VC-12 part of the TU-12 is allowed to wander in phase.

A group of three such TU-12 channels within a VC-4 is known as a *Tributary Unit Group* (TUG). As with the concept of a channel, a TUG is not a multiplexing level (such as a VC-12) but a group of defined byte positions within a VC-4. Some of these positions are reserved for TU pointers, while the others are for the rest of the TUs (i.e., the VCs). The reason for introducing the concept of a TUG, instead of staying with that of a channel, is that, as mentioned in The Basis of SONET and SDH section, a C12 and hence a VC-12 is not the only size of synchronous container that has been defined (see Figure 10.13).

For example, there exists a C2 that is designed to accommodate the North American 6.3-Mbit/s plesiochronous rate. In the event of a PTO needing to transport both 2-Mbit/s and 6.3-Mbit/s circuits, they can both be accommodated within the same VC-4 by assigning some TUGs to carry groups of three TU-12s, while other TUGs are assigned to each carry a single TU-2.

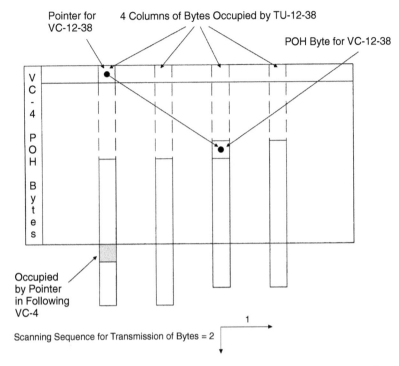

Fig. 10.12 Internal structure of VC-4 showing fixed pointer locations and distribution of a single TU-12 over four separate columns.

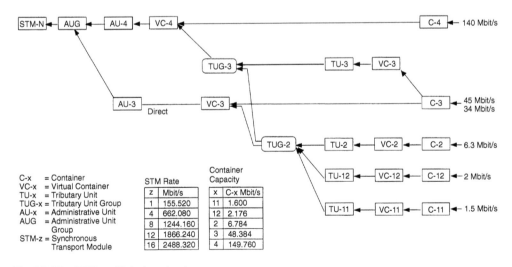

Fig. 10.13 SDH multiplexing structure.

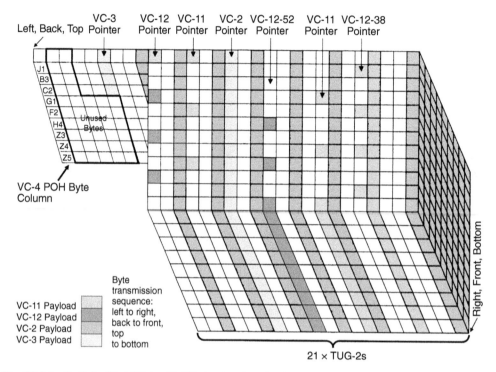

Fig. 10.14   Partially filled VC-4 with TU structured payload.

The TUG in this example is known as a TUG-2 because at 12 columns of 9 bytes it is large enough to accommodate a single TU-2. Beyond this, a TUG-3 has been defined. This is slightly larger than seven TUG-2s and is designed to accommodate a single 45-Mbit/s circuit after it has been mapped into its appropriate synchronous container (in this case, a VC-3). A 34-Mbit/s signal can also be mapped into a VC-3 in an operation that appears somewhat wasteful of bandwidth because the C3 has been sized for a 45-Mbit/s signal. It is interesting to note that, rather than use a more aptly sized container, this bandwidth sacrifice was agreed to by European PTOs to reduce the potential network control problem posed by a larger variety of VCs.

A TUG-3 is one-third of the payload capacity of a VC-4, and together with the TUG-2, it constitutes an extremely flexible mechanism for partitioning the payload bandwidth of a VC-4. The potentially complicated nature of this partitioning can be better understood by replacing the two-dimensional representation of the VC-4 payload structure by a three-dimensional one (see Figure 10.14).

For this representation, the order of transmission is left to right, back to front, top to bottom. Figure 10.14 shows not only TU-12s and TU-3s, as we have already discussed, but also examples of other types of TUs, notably the TU-11, which is the TU used to accommodate a 1.5-Mbit/s (DS1) signal in North America. This payload flexibility extends to alternative mechanisms for constructing a VC-4 payload in which, for instance, an unstructured 140-Mbit/s signal is mapped directly into the C4, without any need for recourse to the notion of TUGs. Because of the variety of possible VC-4 payloads, part of the VC-4 overhead (i.e., the H4 byte) is reserved for indicating the exact structure of this payload. This ability of a single VC-4 to carry a mixture of smaller VCs within its payload is considered to be especially useful in the access portion of the network, where a PTO usually will not have the freedom to dedicate particular VC-4s to carrying a single type of lower order VC.

Fitting a VC-4 into a Synchronous Transport Module

Although a VC-4 can have its payload constructed in a variety of ways, its POH conforms to the same principles as those of a VC-12; that is, it is generated and attached at the point where the C4 is loaded and remains unchanged until the C4 is unloaded at its destination. As with the VC-12, the VC-4 POH is capable of indicating that errors have been introduced into the VC-4 payload during its journey, but it is not capable of identifying which network element was responsible. This problem is solved by the addition of yet another set of overhead bytes to the VC-4, known as Section Overhead (SOH) bytes. The combination of the SOH plus the VC-4 is termed a Synchronous Transport Module (STM), but another way of looking at this structure is to regard the VC-4 as fitting neatly into the payload area of the STM (see Figure 10.11).

Like the VC-4, the STM repeats every 125 µs. To appreciate the structure of an STM, and particularly the SOH, it is best to revert to using the two-dimensional representation of a block of bytes. Figure 10.11 shows the standard representation of the smallest STM, known as an STM-1, which has nine rows of 270 bytes each. This produces a transmission rate of 155.520 Mbit/s.

The significance of the STM SOH is that, unlike the VC-4 POH, it is generated afresh by every network element that handles the VC-4. This handling of the VC-4 includes the operations of creating, multiplexing, and routing of VC-4s within the network, even if such multiplexing or routing happens to be completely inflexible (e.g., hardwired). When a network element (i.e., line terminal, multiplexer, or cross-connect) receives an STM, it immediately examines the relevant bytes of the SOH to determine whether any errors have been introduced into the payload (i.e., the VC-4). Unless this particular network element happens to contain the VC-4 path termination point, it subsequently calculates a new, replacement set of SOH bytes that are then attached to the VC-4 for onward transmission to the next network element (see Figure 10.15).

This section-by-section monitoring gives the PTO a powerful tool for locating the source of any poor performance within the network and complements the capabilities of the VC POHs, which are solely concerned with end to end, rather than section-by-section (i.e., network element–to–network element), issues. It could be argued that the POH monitoring is redundant because the end-to-end path performance could be synthesized from the individual section indications. However, not only would this be difficult to do, especially if the VC-4 journey happened to traverse the networks of more than one PTO, but it would not necessarily detect all the errors in the VC-4 because it is quite possible for errors to be generated within the confines of a transited network element—that is, within that portion of a network element between where the old SOH has been removed and a new one has been added (e.g., between X and Y of the MUX at site C in Figure 10.15).

Both the VC POHs and STM SOHs have other duties in addition to the performance monitoring described above, but the SOHs shoulder by far the larger part of this extra burden—hence the reason for the much larger number of bytes in the SOH than, for example, in a VC-4 POH (see Figure 10.11). These additional duties include the STM alignment function, the carriage of the network management channels, Engineer Orderwire

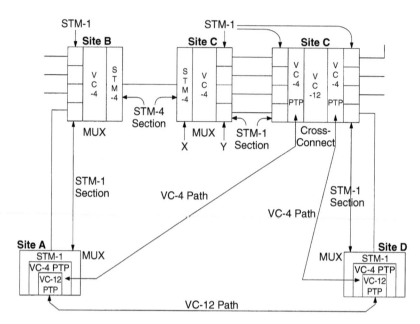

Fig. 10.15 VC-12 paths between sites A and D, showing the VC-4 and STM sections involved.

channels, data channels reserved for the PTO, and synchronization signaling channels. Even when these have been accommodated, there is substantial unallocated capacity that is being kept in reserve to service future network control requirements that have not yet been identified.

Further Use of Pointers

The use of STMs and their associated SOHs entails a few additional complications beyond those previously discussed. An STM might well be generated by a network element that did not have the privilege of also generating the particular VC-4 that it is attempting to load. This immediately introduces the possibility of a slight asynchronism between STM and VC-4, and, as with the VC-12, this problem is also solved by a slip mechanism plus pointer. The pointer bytes occupy defined positions within the SOH and indicate the offset, in bytes, between themselves and the first byte of the VC-4 POH. The main difference between this and the VC-12 pointer is that when a VC-4 slips its phase relative to the STM SOH, it does so three bytes at a time, rather than the single-byte phase change experienced by the VC-12.

A second difference between this case and that of a VC-12 is that the combination of a VC-4 plus its pointer is known as an *Administrative Unit 4* (AU4), rather than a TU-4, when the pointer is located in an STM. The AU4 is used in all ETSI countries as the size of traffic block on which networks are planned and operated. There is also an AU3, which is used mainly in North America. This refers to an alternative construction of an STM, whereby the payload consists not of a single VC-4 but instead of a group of three VC-3s, together with their associated pointers. A further difference between the STM SOH and a VC POH is that the SOH can be divided into two parts, known as the *Multiplexer Section OverHead* (MSOH) and the *Regenerator Section OverHead* (RSOH) (see Figure 10.11). The reason for this is that

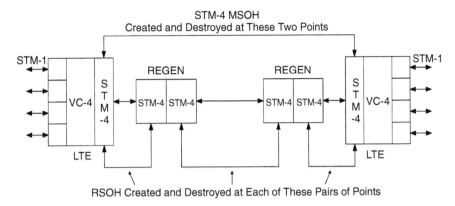

Fig. 10.16 Line transmission system showing the MSOH operating between LTEs, while only the RSOH is recalculated between each pair of regenerators.

on long line transmission systems, the regenerators do not need to perform the rather costly, and in this case unnecessary, operation of generating and destroying the complete SOH. Instead, only a subset of an SOH is processed, leading to a reduction in gate count, power consumption, and so on, but still preserving the ability to detect traffic errors and access some management channels. On the other hand, the line terminal equipment processes both the RSOH and MSOH (see Figure 10.16).

Finally, as with VCs, STMs come in various sizes. As mentioned earlier, the smallest size is termed an STM-1, and it can accommodate a single VC-4. However, larger sizes exist whose bitrates are integer multiples of the basic STM-1 rate. ITU-T G.707 currently recognizes the STM-4, STM-16, STM-64 and STM-256, and it is likely that in the future STM-1024 will be used. For all these higher-rate STMs, the construction mechanism is the same: The payload is produced by straight byte interleaving of the tributary VC-4s, while the SOH is constructed in a more complicated way, particularly in relation to the way the error-checking bytes are calculated.

Other Sizes of VCs and Payloads

The preceding description of the loading of a 2-Mbit/s circuit into a VC-12, a VC-4, and thence an STM-1 mentioned the existence of other sizes of VC. The complete family of VCs, together with their allowed multiplexing routes up to STM-1, are shown in Figure 10.13. So far mappings into these synchronous containers have been defined for all the common plesiochronous bitrates, together with a few others, notably the 125-Mbit/s FDDI signal. This latter mapping is somewhat wasteful of transmission bandwidth because it loads the 125-Mbit/s FDDI signal into a VC-4, which has a payload capacity of around 149 Mbit/s. It might be thought that this degree of inefficiency is more or less inevitable for any further type of signal whose bitrate does not correspond roughly with that of one of the existing synchronous containers. In fact, this is not necessarily so because ITU-T G.707 contains a provision for the concatenation of TU-2s, AU3s, and AU4s.

As a hypothetical example, to illustrate the use of concatenated TU-2s, an incoming service signal at, say, 16 Mbit/s is mapped into a group of three VC-2s, known collectively as a VC-2-3c. (see Figure 10.17).

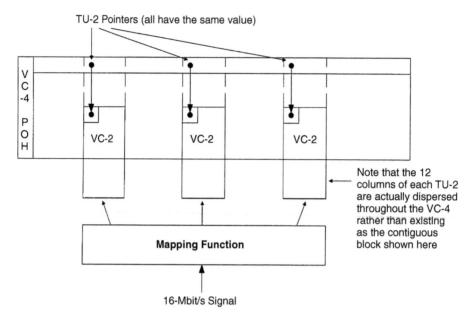

Fig. 10.17   Loading of a hypothetical 16-Mbit/s signal into three concatenated VC-2s (VC-2-3c) within a VC-4.

These three VC-2s are loaded into the VC-4 with identical pointer values and are subsequently transported, as a group, across the entire SDH network. The best current example of the use of this technique is in the area of video transmission, where a TV signal is digitally encoded at around 32 Mbit/s and subsequently loaded into a concatenated group of five VC-2s (VC-2-5c). This mapping allows up to four such video signals to be transported in a single VC-4. If, instead, the normal mapping into a VC-3 had been used, then only three video signals could have been accommodated within one VC-4—hence a useful increase in efficiency by using concatenation. Despite this, the concept of concatenating VCs as a mechanism for increasing SDH's flexibility has never really caught on, and it is likely that all such nonstandard bitrates will be handled by other means (e.g., ATM or IP).

So far in this discussion, all the examples of a service rate signal being mapped into a synchronous container have assumed that the SDH network element makes no use of any structure that might be present in the service rate signal. In fact, this assumption is always true when the service rate signal is plesiochronous. However, because of the way the PSTN service is usually operated (with all of the switches running synchronously with one another), it was decided to endow SDH with a second class of mapping, known as *byte-synchronous mapping*. Currently, the only byte-synchronous mapping defined is for a 2-Mbit/s signal, which has a G.704 frame structure (i.e., is byte oriented) and is synchronous to the SDH network or, more precisely, synchronous with the network element that is mapping it into a synchronous container. In the course of a byte-synchronous mapping, the SDH network element locks onto the incoming G.704 frame alignment word and subsequently proceeds to load this, and every other byte in the G.704 frame, into predefined positions in the C12 (see Figure 10.18).

This results in a situation in which any channel of a 30-channel group can be easily located when the location of the VC-12 POH has been established. The advantage of this

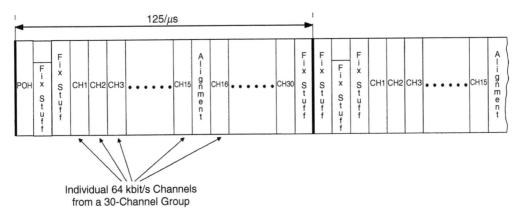

Individual 64 kbit/s Channels
from a 30-Channel Group

Fig. 10.18 Byte-synchronous mapping of a G.704 structured 2.048-Mbit/s signal into a VC-12, showing the fixed location of each of the 30 × 64-kbit/s channels.

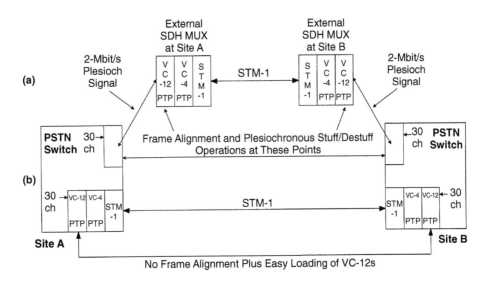

Fig. 10.19 Comparison of plesiochronous and byte-synchronous mappings of 30 channel groups between two PSTN switches.

mapping becomes apparent when considering two PSTN switches that are interconnected by several groups of 30 channels that are transported over an SDH network (see Figure 10.19).

It is easy for such a PSTN switch to operate as part of the SDH network by generating and byte-synchronously loading its own VC-12s, because it already knows the location of all 30 channels. It obviously does not need to search for frame alignment because this too is already known. At the receiving PSTN switch, because the VC-12s were mapped byte synchronously, there is also no need for a frame alignment operation because this follows automatically when the phase of the VC-12 is known. The removal of the G.704 frame

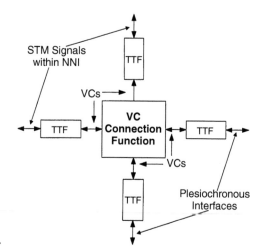

Fig. 10.20 Generalized representation of the traffic paths through an SDH network element.

alignment process not only leads to a reduction in gate count (and hence cost) but also reduces the delay associated with this operation. These advantages apply not just to the PSTN example previously given, but, in fact, to any network where visibility and routing of individual 64-kbit/s channels is required; hence they could also apply to networks of PBXs or 64-kbit/s cross-connects.

SDH Network Elements

Because traffic-routing flexibility is one of the reasons for the existence of SDH, the simplest way of considering any SDH network element is as a group of transport termination functions (TTFs) surrounding a traffic path connectivity function (see Figure 10.20).

The TTFs are essentially the conversion functions from the VC level to either the STM or plesiochronous signal levels, while the connectivity function allows VCs to be routed between the various TTFs. More than anything else, it is the size and flexibility of the connectivity function that distinguishes one type of network element from another. For example, a cross-connect might have a large connectivity function, capable of simultaneously routing a large number of VCs of a given type between a large number of TTFs, with very few restrictions on this connectivity. This, of course, is consistent with the normal role of a cross-connect, as a major traffic-routing node in a transport network, where electronically controllable traffic-routing flexibility is its *raison d'être*.

At the other end of the scale comes *Line Terminating Equipment* (LTE), which normally incorporates a multiplexer as well. The connectivity between tributaries and aggregate ports of such a multiplexer is normally hardwired, so it has no flexibility whatsoever. Between these two extremes come drop-and-insert (add-drop) multiplexers, which attempt to strike a balance that leads to adequate, rather than comprehensive, routing flexibility, with an attendant reduction in equipment costs.

All of the previously mentioned types of network element normally assume that any interconnection will be based on optical fibers. It is, however, also possible to use other interconnection media, in particular radio transmission. The problem in using radio inter-

Transport Termination Function

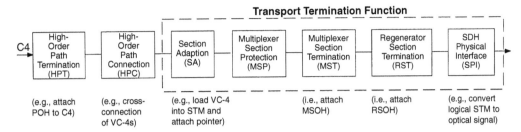

C4 →	High-Order Path Termination (HPT)	High-Order Path Connection (HPC)	Section Adaption (SA)	Multiplexer Section Protection (MSP)	Multiplexer Section Termination (MST)	Regenerator Section Termination (RST)	SDH Physical Interface (SPI)
	(e.g., attach POH to C4)	(e.g., cross-connection of VC-4s)	(e.g., load VC-4 into STM and attach pointer)		(i.e., attach MSOH)	(i.e., attach RSOH)	(e.g., convert logical STM to optical signal)

Fig. 10.21 Fragment of an SDH network element showing how it is broken down into atomic functions according to ITU-T G.783.

connection is that, unlike optical fibers, the available spectrum is finite and in short supply. The liberal use of overhead capacity in SDH only exacerbates the already difficult problem of squeezing the traffic information into the existing arrangement of radio channel bandwidths. These problems are by no means insuperable, but they do tend to restrict radio interconnection of NNI signals to the lower end of the SDH range.

The reduction in cost that accompanies a network element with restricted connectivity results not only from simpler hardware but also from simpler control software. This is somewhat surprising at first sight because it has often been said that SDH does not really make economic sense without a large measure of software control over all network element functionality and, in particular, routing flexibility. However, the complexity of even relatively simple SDH network elements leads to a situation where control of complete networks rapidly becomes unmanageable unless some restrictions are put on the traffic-routing complexity. In short, every additional piece of traffic-routing flexibility is a potential network control headache.

The general problem of control of an entire SDH network, notwithstanding the traffic-routing problem, is almost impossible if it is composed of network elements from different manufacturers that do not have some form of common control interface. The network element functional standard ITU-T G.783 standardizes the functions that each of the network elements actually performs. This recommendation describes how the functionality of a network element can be decomposed into the basic "atomic" elements of functionality, together with how such atomic functions can be combined and thereafter exchange both traffic and control information (see Figure 10.21).

The rules for combination of these atomic functions enable a variety of network elements to be synthesized, with differing traffic capacities and routing capabilities. Indeed, one of the problems is the potentially large number of different ways a conformant network element can be constructed. It is rather difficult for software-based control systems to cope with this variety, and hence a more restricted range of network element functionalities has been stardardized. G.783 was created primarily to describe different types of drop-and-insert multiplexers; however, the general principles and, indeed, much of the detail is just as applicable to LTEs and large cross-connects. Moreover, the formalism required to describe complete SDH networks was produced very much with this same functional decomposition in mind. We shall see later how these atomic functions can be paired between network elements on opposite sides of a network to produce a complete traffic path that is sufficiently well defined for a computer to recognize.

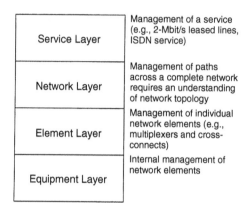

Service Layer	Management of a service (e.g., 2-Mbit/s leased lines, ISDN service)
Network Layer	Management of paths across a complete network requires an understanding of network topology
Element Layer	Management of individual network elements (e.g., multiplexers and cross-connects)
Equipment Layer	Internal management of network elements

Fig. 10.22 Network management layering for the control of telecom networks and services.

Control and Management

Control and management within any type of telecommunications network (not just an SDH network) can best be viewed in terms of a series of layers (see Figure 10.22).

At the lowest level, there is internal control of an individual network element, which performs internal housekeeping functions and deals in alarm and control primitives. ITU-T G.783 has rigorously defined a minimum set of control and alarm primitives for each of the SDH atomic functions, and these are the basis of all SDH management. However, beyond this, ITU-T G.784 describes how such primitive information should be processed and ordered to produce derived information, such as error rates, which is stored in logs of defined duration and reported at set intervals, and so on. At the lowest level, some of this information might look rather different from that specified in G.783/4, and the like; however, the internal network element control system takes these and from them synthesizes the information that is required by the next level in the management hierarchy (i.e., the element manager).

The element manager is a piece of software that can control many individual network elements (usually in the range 10–1000), but it can only control them as individual elements and does not have any view of the traffic relationships between them. Normally it is located remotely from the elements it is controlling, and more often than not, it runs on some form of workstation. The interface between the element manager and the network elements is an obvious area for standardization because without this there is little chance of a single element manager controlling network elements from more than one manufacturer. Despite the progress that has been made in rigorously describing the network element atomic functions, it has still proved very difficult to agree on a software description of them. The approach adopted by ITU-T, ANSI, and ETSI has been to describe complete network elements as collections of "objects" in line with the rules of "object-oriented" programming.

The idea is that an object is a software entity that has both attributes and behavior. External stimuli (i.e., information, commands, etc.) trigger an object to behave in a certain way—for example, change its attributes, transmit information, obey commands, and so on. The claim is that a standardized object could be looked on as the software equivalent of a hardware integrated circuit. One of the biggest areas of disagreement in generating an agreed on standard set of SDH objects is the question of whether an object should represent a piece of functionality or a (small) piece of hardware. Although ITU-T G.783 functionally decomposes SDH equipment, its says almost nothing about the way such atomic functions are split or combined in any real hardware implementation. The current view within both ITU-T and ETSI is that the set of objects (collectively known as the *information model*) should present both a functional and a physical view of the network element that they represent. This view is given in ITU-T G.774, a very large recommendation that describes not only the information model but also its application to functions such as performance monitoring, multiplexer section protection, and so on.

Not only is it necessary to have a standardized information model so that the network element and element manager can understand one another; it is also necessary to have an agreed on message set to go with it. Fortunately, this flows reasonably easily from the definition of the objects themselves. However, the existence of such a message set then leads to the requirement for an agreed on information protocol stack with which to transport it. For information transferred from a network element direct to its element manager, ITU-T G.773 details several allowed protocol stacks that split into two groups: those that have a full seven-layer structure and those that have a "short stack," where layers 4, 5, and 6 are absent. Most observers favor the use of the heavier, but more flexible, seven-layer protocol stacks for SDH networks.

Not only are there defined protocol stacks for information transfer direct from element manager to network element, but there is also a defined protocol stack for information transfer between individual network elements. In general, the majority of information transfer between network elements is actually information arising from an element manager, which is being relayed by an intermediate network element to a more remote element (see Figure 10.23).

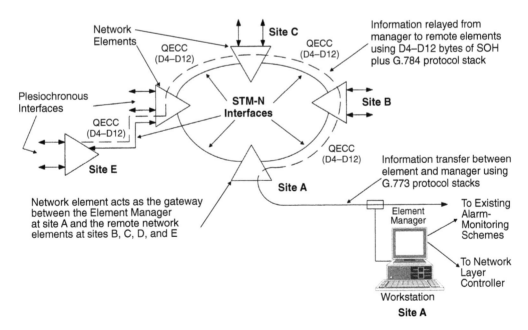

Fig. 10.23 Control channels and protocols within an SDH network.

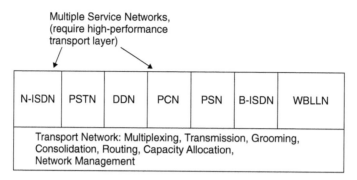

Fig. 10.24   Managed transmission network; SDH network as a bearer for other services.

This flow of management information among network elements, element managers, and yet more managers at the network and service levels gives rise to the concept of a *Telecommunications Management Network* (TMN).

The main recommendation concerning the TMN is ITU-T M.3010, which attempts to define a series of interfaces between different management entities in such a network. It is not confined purely to SDH networks, but SDH networks were the first to implement an M.3010-style TMN.

SDH-Based Networks

Until a few years ago, SDH-based networks were viewed as the transmission bedrock supporting all other terrestrial telecommunications services (see Figure 10.24).

As already mentioned, the main advantages of such a network are the ease and precision with which the available network bandwidth can be partitioned among the higher-layer services, together with accurate monitoring of the quality of the transmission links. Despite this, the control and management of complete national networks is a difficult problem that requires some degree of standardization in the way such networks are functionally decomposed, in much the same way as the individual network elements have already been functionally decomposed into their atomic elements by ITU-T G.783.

The basic idea behind the functional decomposition described in ITU-T G.803 and G.805 is that a transport network can be stratified into a number of layers. Each layer provides a service to the layer above it, and is in turn a client of the layer below it, in much the same way the ISO seven-layer information transfer model consists of layers that participate in client–server relationships with their vertical nearest neighbors (see Figure 10.25).

This layering within an SDH network could be viewed as a subdivision of ISO layer 1. As an example of a non-SDH client–server relationship in a transport network, consider Figure 10.26, which shows a 64-kbit/s circuit as a client of the 2-Mbit/service layer (i.e., the 2-Mbit/s layer), which can be viewed as a 2-Mbit/s network, transporting the 64-kbit/s circuit between its desired end points. To do this, the 2-Mbit/s layer will probably call on the services of the 8-Mbit/s layer, and so on up to the 140-Mbit/s layer. The SDH counterpart of this simple PDH example is slightly more complicated in that the transport layers are divided between those concerned with end-to-end networking (i.e., the path layers) and

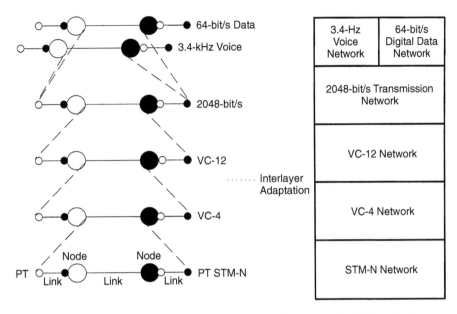

Fig. 10.25   Layering and client–server relationships between the layers of an SDH network.

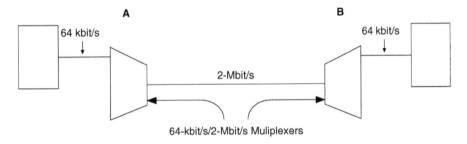

Fig. 10.26   A 64-kbit/s circuit making use of a 2-Mbit/s service to get from A to B.

those concerned with transport between each pair of SDH network elements along the route (i.e., the STM section layer).

As an additional complication, there are two path layers: the lower-order paths consisting of VC-1s, VC-2s, or VC-3s, and the higher-order paths, which are VC-4s in ETSI countries, but could also be VC-3s (see Figure 10.27). Usually when a client layer makes use of a server layer, it is necessary to adapt the client signal to a format that can be carried by the server layer (see Figure 10.28).

Plesiochronous "bit stuffing" and SDH pointer adjustments are examples of adaptation functions. The adaptation function is only one of the network atomic functions that have been described in G.803, and it is indeed fortunate that G.803 has been developed in full recognition of the contents of G.783, because many of the equipment and network atomic functions are identical.

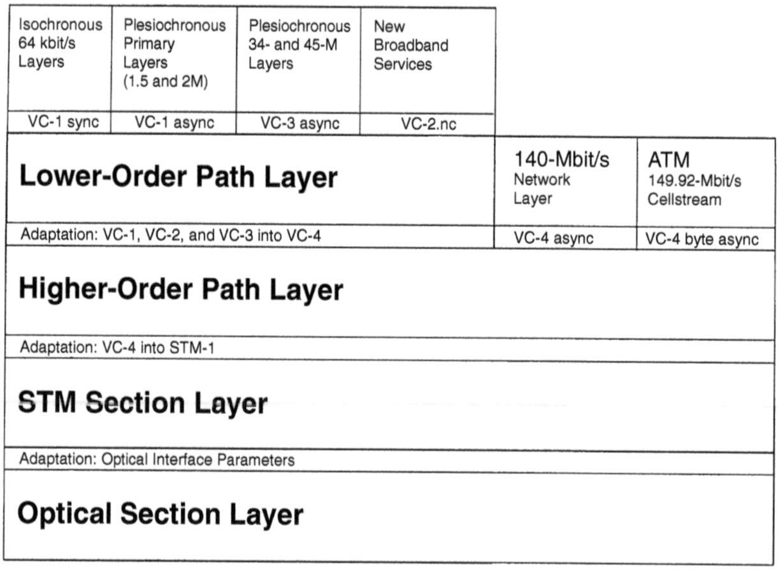

Isochronous 64 kbit/s Layers
Plesiochronous Primary Layers (1.5 and 2M)
Plesiochronous 34- and 45-M Layers
New Broadband Services
VC-1 sync
VC-1 async
VC-3 async
VC-2.nc
Lower-Order Path Layer
140-Mbit/s Network Layer
ATM 149.92-Mbit/s Cellstream
Adaptation: VC-1, VC-2, and VC-3 into VC-4
VC-4 async
VC-4 byte async
Higher-Order Path Layer
Adaptation: VC-4 into STM-1
STM Section Layer
Adaptation: Optical Interface Parameters
Optical Section Layer

Fig. 10.27 Layering in an SDH transport network.

Link Connection in Client Layer
Client Layer
Server Layer
Adaptation Function (i.e., server layer adapts client-layer information to server-layer format)
Path Terminations (i.e., server layer adds overheads to adapted information)
Path in Server Layer

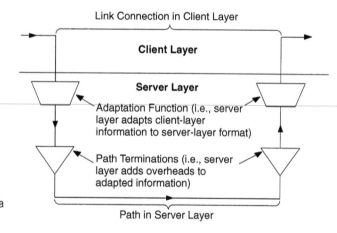

Fig. 10.28 Handling of a link connection in a client layer by a path in a server layer.

G.805 not only describes a series of vertical client–server relationships for an SDH transport network; it also gives a structure to each of the individual layers. Each layer can be viewed as a network in its own right, which can be partitioned into a series of subnetworks (see Figure 10.29).

These subnetworks are connected together by "link connections" and can, if necessary, be further subdivided into yet smaller subnetworks. The logical place to stop this process is where a whole subnetwork is completely contained within one network element. As an example, consider a VC-12 cross-connect. The switching matrix (connectivity function) within this element can be considered as a subnetwork, which is capable of making VC-12–layer subnetwork connections from one port to another (see Figure 10.30).

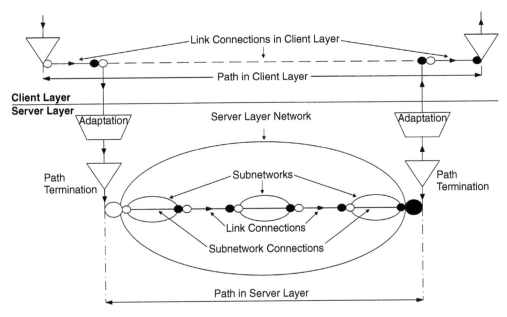

Fig. 10.29 Partitioning of a transport-layer network into a series of subnetworks.

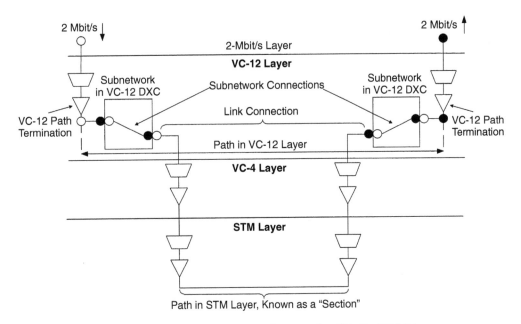

Fig. 10.30 Use of two VC-12 cross-connects to produce a path within the VC-12 layer.

To connect one of these VC-12s to another cross-connect, a link connection is now required, which will need to make use of the VC-4 and STM layers. This concatenation of subnetwork and link connections across the VC-12 path layer network begins and ends at the points where the VC-12 is created and destroyed—namely, the path termination points. It is at these points that the VC-12 POH is added or removed, exposing only the C12 synchronous container. This same path termination function is recognized in both G.783 and G.803, and it is located immediately next to the adaptation function previously described. On the assumption that the C12 in question is carrying a 2-Mbit/s circuit, the combination of adaptation functions, termination functions, and the chain of subnetwork and link connections in the VC-12 layer has succeeded in transporting this 2-Mbit/s circuit between the two end points of a single 2-Mbit/s link connection (see Figure 10.30). From this point, the preceding analysis can now be repeated in the 2-Mbit/s layer, where the particular 2-Mbit/s circuit previously cited will probably be found serving as part of a link connection for a number of 64-kbit/s circuits (see Figure 10.26).

This formal description of an SDH network opens up the prospect of real control and management of large networks consisting of network elements from several manufacturers. The pairing of path termination functions on opposite sides of a network, together with a similar operation on the same functions in the STM layer, allows a PTO to accurately monitor the service that the SDH network is delivering and to pinpoint those network elements responsible for any poor performance. Beyond that, the SDH network model greatly facilitates management of the all-important traffic flexibility points; that is, those subnetworks that consist of an electronically controllable connectivity function.

SDH Network Topologies

Once again, traffic-routing flexibility, its control and physical distribution within an SDH network, is one of the most important influences on the topologies in which SDH equipment is deployed. The most obvious manifestation of this is the drop-and-insert ring topology that is finding favor in the PTO access networks (see Figure 10.31).

The idea behind a drop-and-insert ring is that the ring structure can give a high degree of protection against cable cuts, and so on, because of its potential for routing traffic either way around the ring. In fact, one of the biggest advantages of this topology, in control terms, is the limitation on the rerouting possibilities for any traffic affected by a cable break. Anything more complicated than a simple clockwise/counterclockwise routing decision requires up-to-date knowledge of a rather more extensive portion of an SDH network than just a simple ring.

The nodes of such a ring are populated by drop-and-insert multiplexers that have a restricted traffic-routing flexibility that is tailored to the requirements of a ring. This gives a relatively low-cost ring implementation, which nevertheless, when viewed as a single entity, appears as a restricted form of cross-connect. The versatility of drop-and-insert rings also extends to drop-and-insert chains, which can be considered as a "flattened" version of a conventional ring (see Figure 10.32).

Beyond this, a ring can also be made from a chain of drop-and-insert multiplexers, whose ends have been joined by an SDH line system. The main problem with implementing this type of ring, at least in the access area, is that the existing layout of cables and ducts usually takes the form of a star, which formerly linked the old analogue local exchanges to their district switching center. In many cases, a partial physical ring can be created by a small extension of the existing cable pattern. This can be supplemented by creating a logical

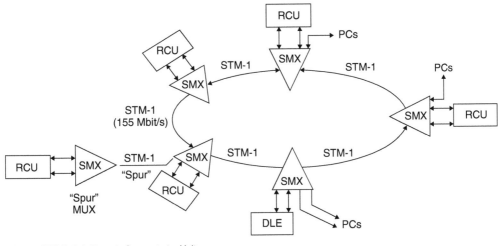

RCU = PSTN Switch Remote Concentrator Unit
PC = Private Circuits (2 Mbit/s)
DLE = Digital Local Exchange
SMX = SDH Multiplexer

Fig. 10.31 SDH multiplexers deployed in a drop-and-insert ring.

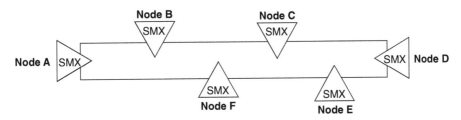

Fig. 10.32 Drop-and-insert chain produced by "flattening" a conventional drop-and-insert ring.

ring when existing cables are laid in a star arrangement, although this obviously affords less protection against cable breaks (see Figure 10.33).

Finally, it is sometimes possible to produce a physical ring by linking some of the ring nodes with microwave radio rather than optical fiber. Outside of the access areas (i.e., within the transmission core of the average PTO network), the normal topology advocated is that of a mesh of cross-connects interconnected by point-to-point transmission systems (see Figure 10.34).

For this application, the SDH line transmission systems, like their PDH counterparts, require almost no routing flexibility. This is more than compensated for by the cross-connects, which provide complete traffic-routing flexibility at a variety of VC rates. Used in this way, cross-connects can be viewed as electronic replacements for the present-day digital distribution frames. The benefits of this deployment of flexibility are alleged to be those of easier traffic path provisioning and a fast, efficient scheme for restoring failed paths that requires the absolute minimum of standby transmission capacity. This last benefit is

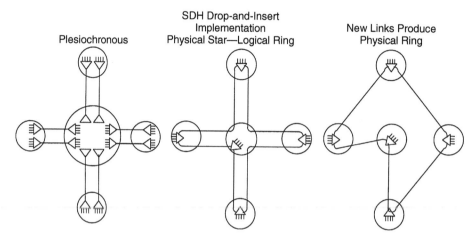

Fig. 10.33 Comparison of topologies possible with PDH and SDH multiplexers.

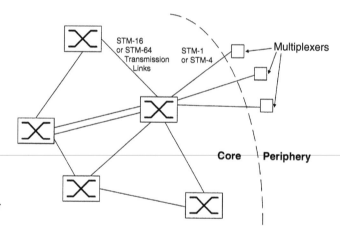

Fig. 10.34 Example of a core transmission network topology that relies on cross-connects for flexibility.

heavily dependent on the control that is exercized over the cross-connects, and here there are some potential problems.

The main control problem is that of database integrity. With a network of meshed cross-connects, when a transmission link fails the path restoration action will usually involve rerouting all the affected paths through several alternative cross-connects. To do this efficiently, the network control system must rapidly command simultaneous switching actions in all of these cross-connects, which usually implies that the control system has a predetermined plan of action that is based on the spare bandwidth that is thought to be available on the relevant transmission links and cross-connects. Unfortunately, the integrity of the control system's database might have been compromised because of other recent reconfiguration activities (e.g., new circuits setup), links temporarily taken out of service for maintenance, and so on. This problem escalates rapidly as the number of cross-connects in a network increases.

There are several potential solutions to this problem. One is simply not to use the cross-connects for protection against transmission failures and instead to rely on transmission systems having $1 + 1$ or $1:N$ protection. In this case the cross-connects are used solely for offline management of the network's transmission capacity. This is often called "facilities management" in North America.

An alternative is to deploy the cross-connects in a rather more bounded topology than the completely free mesh topology previously assumed. A limiting case of this idea is to deploy them in a ring. The restriction on routing choices imposed by such a topology greatly eases the control problem, albeit with some significant increase in the standby line transmission capacity that must be available to allow a traffic reroute.

On some occasions, networks based on this topology have been adopted by new operators, who have had more freedom to lay their cables in rings. Usually the network elements themselves are a halfway house between add-drop multiplexers and cross-connects, but what really makes them attractive is the bandwidth saving that can be achieved with the so-called *Shared Protection Ring* mechanism (MS-SPRing). This is usually a ring version of the familiar $1:N$ protection scheme previously used on point-to-point line systems. Compared to the alternative $1 + 1$ protection scheme, it can easily give an improvement of two or three times, especially with the even traffic distributions found in the core of the network.

Conclusion

SDH and SONET became widely established in an era when voice was the dominant traffic type in telecommunications networks. As the demand for data transmission has grown, other protocols better suited to data traffic have emerged, such Next Generation SDH, Multi-protocol Label Switching, and Provider Backbone Bridging with Traffic Engineering. However, the deployment of SDH is so widespread that it will continue to be used for many years to come, particularly in conjunction with Next Generation SDH, which enables much more efficient use of SDH capacity when carrying data.

11 Next Generation Synchronous Digital Hierarchy

Ken Guild
University of Essex

Introduction

In the last 20 years, Synchronous Digital Hierarchy (SDH) has established itself as the predominant optical fiber transmission technique in European operator networks. The SDH format is very similar to the North American SONET format but with some modifications to accommodate the different legacy Plesiochronous Digital Hierarchy (PDH) transmission schemes used in Europe. Although this chapter focuses primarily on SDH, the fundamental principles of Next Generation SDH (NG-SDH) are identical to those of its SONET counterpart, except for the different multiplexing hierarchies between SDH and SONET, which are described in Chapter 10.

Recently there have been a number of discussions about whether SDH/SONET deployment is now in decline and soon to be phased out of operators' networks in favor of packet optical transport systems with Wavelength Division Multiplexed (WDM) interfaces. It is worth spending a few moments reflecting on this possibility by analyzing the market share and trend of packet- and circuit-switched optical equipment to put such comments into perspective. According to Infonetics Research, sales of optical network hardware increased consistently in the four years from 2003 to 2007 to a total worldwide market value of $13.9 billion—a 19 percent jump from 2006.

Although a billion-dollar market, the packet optical transport systems market is growing because of the rapid increase in consumer and business packet-based services transporting voice, video, and data across converged IP infrastructures. The remaining $12.9 billion market share is in WDM and SDH/SONET transport equipment. Although the spending trend now appears to be shifting toward the area of WDM equipment (29 percent growth in 2007) compared with SDH/SONET (9 percent growth in 2007), SDH/SONET accounts for 61 percent of worldwide optical equipment sales. Although it is anticipated that growth in sales of SDH/SONET will gradually decline in the next few years, the huge installed base of SONET/SDH equipment is not likely to be replaced in the foreseeable future.

SDH is favored by network operators for its ability to offer resilient bandwidth and infrastructure scalability both nationally and internationally with proven Operation, Administration, Management, and Provisioning (OAM&P) capabilities. The rise of the Internet and widespread adoption of Internet technologies deployed within corporate intranets has fueled the explosive growth of bursty IP traffic and subsequently the deployment of large IP/MPLS routers with large-capacity Wide Area Network (WAN) interfaces. In practice, WANs typically interface to an SDH transport network. However, because SDH was originally conceived to transport 64-kbit/s voice circuits, it can be very inefficient for transporting data packets.

Subsequently, many equipment vendors at the turn of the century produced proprietary solutions that optimized packet transport over SDH networks. Unfortunately, this led to large-scale incompatibility, and the International Telecommunication Union (ITU) was forced to undertake the process of augmenting the existing SDH recommendations with three internationally agreed on techniques that have made SDH more "data friendly." These new features are often collectively referred to as "next generation" and consist of Virtual Concatenation (VCAT), Link Capacity Adjustment Scheme (LCAS), and Generic Framing Procedure (GFP). The advantage of these features is that they can be implemented as interface cards at the edge of an existing SDH network and therefore offer a piecemeal upgrade process to incumbent operators who are experiencing relentless growth of their packet-based services. In this way, next-generation SDH offers a solution to increase the lifetime of the existing SDH infrastructure to transport packet data more efficiently without radical changes to the underlying transport network and without sacrificing the tried and trusted OAM&P features of SDH.

Concatenation

Concatenation is the process of combining X SDH containers into a larger container that provides a bandwidth X times bigger than an individual container. Both low-order containers (e.g., VC-11, VC-12, and VC-2) and high-order containers (e.g., VC-3, VC-4) can be concatenated in either contiguous or virtual concatenation modes. Although VCAT is one of the main features of NG-SDH, contiguous concatenation in SDH networks will be briefly described for comparative purposes.

Contiguous Concatenation

Contiguous concatenation was conceived in the early days of SDH to accommodate high-speed data applications that use protocols such as Asynchronous Transfer Mode (ATM) and High Level Data Link Control (HDLC). The process of contiguous concatenation allows larger bandwidth containers to be created by simply summing the bandwidth of X smaller

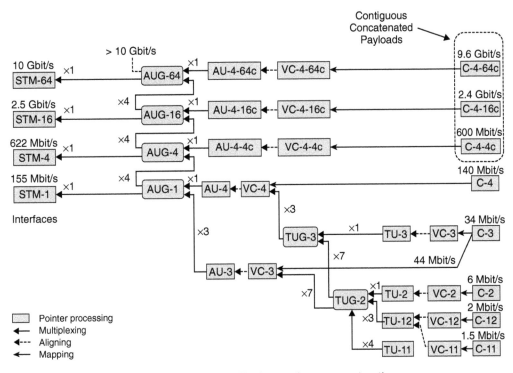

Fig. 11.1 SDH extended multiplex structures allowing contiguous concatenation.

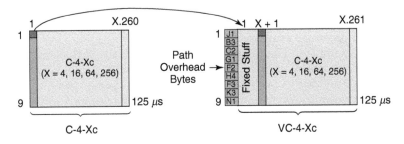

Fig. 11.2 Contiguous-concatenation frame structure for VC-4-Xc.

containers (C-n); the result is a contiguous concatenated payload container referred to as a C-n-Xc. ITU-T recommendation G.707 has defined X to be limited to orders of four (i.e., X = 4, 16, 64, 256) (see Figure 11.1). For the transport of a C-n-Xc between two endpoints, the intermediate network elements in the transport network must not split the concatenated container into smaller pieces. Each network element must therefore have concatenated payload functionality.

Figure 11.2 illustrates the frame structure of a contiguous payload container, VC-4-Xc, in a typical SDH matrix format. The repeated frame every 125 μs consists of nine rows and X × 261 columns in which each cell contains a single byte. The VC-4-Xc is constructed by byte-interleaving (v.) the X individual VC-4s such that the columns containing the

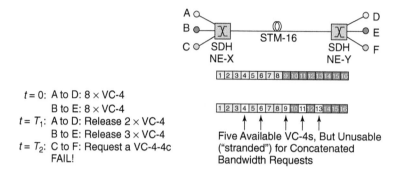

$t = 0$: A to D: $8 \times$ VC-4
           B to E: $8 \times$ VC-4
$t = T_1$: A to D: Release $2 \times$ VC-4
           B to E: Release $3 \times$ VC-4
$t = T_2$: C to F: Request a VC-4-4c
           FAIL!

Five Available VC-4s, But Unusable
("stranded") for Concatenated
Bandwidth Requests

Fig. 11.3   Example of a fragmentation problem with contiguous concatenated payloads.

Operational, Administration, Management, and Prorisioning (OA&M) Path Overhead (POH) of the individual VC-4s are located in the X columns of the VC-4-Xc. The first column is used as the POH for the whole VC4-Xc, and, as with any AU pointer, an AU-4-Xc pointer indicates the position of J1, which is the position of the first byte of the VC-4-Xc container. The remaining X-1 columns contain fixed byte values to indicate concatenation and thus link all the units together. The remaining X × 260 columns provide the payload area of the VC-4-Xc that has exactly the same size as a C-4-Xc. Because the whole contiguous container is transported as a unit across the network, the integrity of the bit sequence is guaranteed.

After the upgrade or installation of equipment supporting contiguous concatenation, an issue regarding bandwidth fragmentation appeared. As a result of the continuous setup and teardown of trails through the network, the total payload area in an STM-N might become fragmented and lead to an inefficient use of network bandwidth. As with the problem experienced with computer hard disks, a virtual container fragmentation problem is exhibited when physically available SDH paths exist but are not usable because of the noncontiguous nature of the available containers.

Consider the example in Figure 11.3 in which two SDH network elements (NE-X and NE-Y) are interconnected by an STM-16 link. Initially, $8 \times$ VC-4s are connected between clients the A-and-D and $8 \times$ VC-4s between B and E. After a period of time ($t = T_1$), the A-to-D connection releases $2 \times$ VC-4s and the B-to-E connection releases $3 \times$ VC-4s; this results in a total of five VC-4s becoming available between nodes X and Y. To illustrate the fragmentation problem, consider the case when a VC-4-4c connection is requested between C and F at time T_2. Even though there is enough bandwidth between the two nodes, the restriction of keeping the four VC-4 containers together as one concatenated container means that the connection cannot be established and the connection request has to be denied.

Over time and for large networks, the unused bandwidth containers accumulate and result in suboptimal use of network bandwidth and a higher number of connection failures for contiguously concatenated payload services. One solution would be to regroom the network to recover this "stranded" bandwidth. However, in some situations regrooming might not be an option because this is akin to "defragging" the network and is potentially risky for a network operator. The alternative is to simply increase the network capacity and tolerate the inefficiencies.

Another disadvantage of contiguous concatenation is the mismatch in bandwidth granularity between connectionless and packet-oriented technologies, such as IP or Ethernet, and the rigid bandwidth granularity provided by contiguous concatenation. For example,

Service	Bitrate	Contiguous Concatenation
Ethernet	10 Mbit/s	VC-3 (20%)
Fast Ethernet	100 Mbit/s	VC-4 (67%)
Gigabit Ethernet	1000 Mbit/s	VC-4-16c (42%)
FICON	1000 Mbit/s	VC-4-16c (35%)
ESCON	200 Mbit/s	VC-4-4c (33%)

Fig. 11.4 Contiguous-concatenation service mappings and efficiencies.

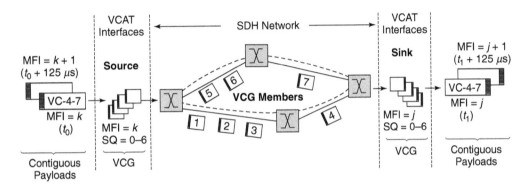

Fig. 11.5 Generic diagram of the principle of operation.

to transport a single Gigabit Ethernet service, it would be necessary to allocate a VC-4-16c container, which has a capacity of 2.4 Gbit/s—more than double the necessary bandwidth. Further examples are given in Figure 11.4.

Virtual Concatenation

Virtual concatenation is a solution, standardized by ITU-T G.707, that allows granular increments of bandwidth in single VC-n units and solves the problems previously highlighted for contiguous concatenation. VCAT also creates a continuous payload equivalent to X times the VC-n units at the source, but each container is individually transported across the SDH network toward the destination (see Figure 11.5). The set of X containers is known as a *Virtual Container Group* (VCG), and each VC is a member of it; the resulting payload is termed a VC-n-Xv, where "v" denotes "virtual." The attraction of such a solution is that only the edge interfaces of an existing SDH network have to be upgraded to perform VCAT, and the underlying SDH infrastructure does not need to be aware that individual VC-ns might be part of a virtually concatenated container. Recall that contiguous concatenation requires that intermediate network elements have concatenation functionality.

Because each member of a VCG can be diversely routed across different paths in the network, inherently the end-to-end resilience of the connection increases because the total client traffic is distributed across multiple members of the VCG. A failure in one or more members results in reduced bandwidth but not loss of connectivity. However, each member of the VCG will experience different end-to-end delay because of differences in physical path lengths. In addition, different routes through the network might also mean that members traverse different numbers of network elements, each contributing an additional switching delay. The difference between the fastest and slowest VC-n in a VCG is known as the differential delay. The receiving interface has to be aware of this so that the delay can be compensated for before reconstituting the original contiguous payload.

In addition, if a protection switch occurs that affects the path of individual members, this delay compensation must be dynamic. The realignment process is achieved by using variable-length buffers for each member of the VCG, and the recommendations state that it should be possible to compensate for a maximum differential delay of 256 ms. Because light propagates through fiber at 5 ms/1000 km, this is equivalent to a differential delay of 50,000 km. The maximum differential delay achievable is vendor specific and typically less than this figure. The Multi-Frame Indicator (MFI) in the path overhead is defined to detect the differential delay and is discussed in detail in the next section. To reassemble the constituent VCG members, which might arrive in a different order from how they were transmitted, a Unique Sequence Number (SQ) is assigned to each VC-n member of the VCG by the Network Management System (NMS).

VCAT has been defined for both low- and high-order SDH. Figure 11.6 serves to demonstrate the flexibility of capacity assignment to SDH clients using VCAT for both low- and high-order containers.

In comparison with contiguous concatenation, virtual concatenation offers an efficient method of transporting packet services, as illustrated in Figure 11.7. Whereas contiguous concatenation offers poor container utilization, virtual concatenation allows containers to closely match the required amount of bandwidth for the client service, with efficiencies typically greater than 90 percent. A new SDH service enabled through VCAT is a fractional Ethernet service in which a customer connects through a Gigabit Ethernet interface but can

	Container	X	Individual Capacity	Virtual Capacity
Lower Order	VC-11-Xv	1 to 64	1600 Kbit/s	1600 to 102,400 Kbit/s
	VC-12-Xv	1 to 64	2176 Kbit/s	2176 to 139,264 Kbit/s
	VC-2-Xv	1 to 64	6784 Kbit/s	6784 to 434,176 Kbit/s
Higher Order	VC-3-Xv	1 to 256	48,384 Kbit/s	48,384 to 12,386 Kbit/s
	VC-4-Xv	1 to 256	149,760 Kbit/s	149,760 to 38,338,560 Kbit/s

Fig. 11.6 Low- and high-order VCAT payload containers and their capacities.

Service	Bitrate	Contiguous Concatenation	Virtual Concatenation
Ethernet	10 Mbit/s	VC-3 (20%)	VC-12-5v (92%)
Fast Ethernet	100 Mbit/s	VC-4 (67%)	VC-3-2v (100%)
Gigabit Ethernet	1000 Mbit/s	VC-4-16c (42%)	VC-4-7v (95%)
FICON	850 Mbit/s	VC-4-16c (35%)	VC-4-6v (94%)
ESCON	200 Mbit/s	VC-4-4c (33%)	VC-3-4v (100%)

Fig. 11.7 Comparison of transport efficiencies for contiguous and virtually concatenated solutions.

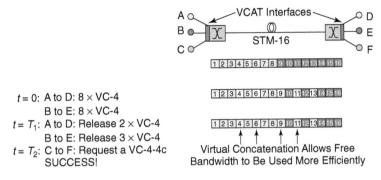

$t = 0$: A to D: $8 \times$ VC-4
      B to E: $8 \times$ VC-4
$t = T_1$: A to D: Release $2 \times$ VC-4
       B to E: Release $3 \times$ VC-4
$t = T_2$: C to F: Request a VC-4-4c
        SUCCESS!

Virtual Concatenation Allows Free
Bandwidth to Be Used More Efficiently

Fig. 11.8 Virtual concatenation as a means to overcome the problem of fragmentation.

be given a fraction of this total bandwidth as transmission capacity in accordance with its usage requirements.

VCAT also eliminates the fragmentation problem associated with contiguous concatenation. Returning to the previous example where a request for a 622-Mbit/s (VC-4-4c) connection from C to F failed using contiguous concatenation, virtual concatenation only requires the interfaces to be upgraded and allows the VC-4s to be routed independently across the SDH network (see Figure 11.8). The existing SDH network elements continue to switch at the VC-4 granularity (in this example) and are transparent to the virtual concatenation process. Now the request to interconnect C and F with a VC-4-4c is successful because VCAT allows the use of the available noncontiguous VC-4s.

High-Order VCAT
Figure 11.9 illustrates the process for a high-order C-n-Xc container that is split across X individual C-n containers of a VC-n-Xv structure. The frame structure of a high-order virtual

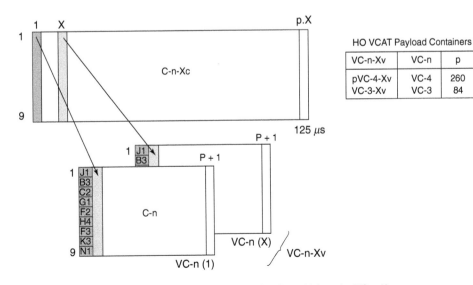

HO VCAT Payload Containers

VC-n-Xv	VC-n	p
pVC-4-Xv	VC-4	260
VC-3-Xv	VC-3	84

Fig. 11.9 Principle of operation of virtual concatenation for a high-order VC-n-Xv.

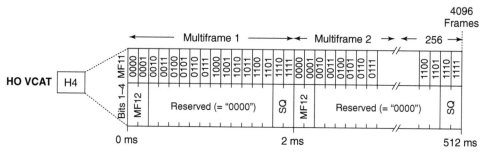

Fig. 11.10 H4 coding for a high-order VCAT OH.

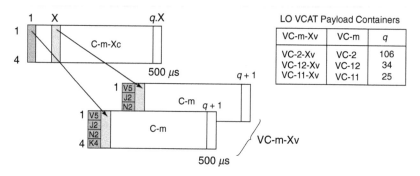

Fig. 11.11 Principle of operation of virtual concatenation for a VC-m-Xv.

container consists of 9 rows by $p + 1$ columns, where p is either 260 or 84 for VC4 and VC3 payloads, respectively. Each VC-n has its own administration and management path overhead, while the remaining columns contain the payload. The H4 byte of each OA&M path overhead contains VCAT path overhead. The signal label in the OA&M path overhead (C2 byte) indicates that the H4 byte is to be used for this purpose.

As previously discussed, the standards require a maximum differential delay tolerance between VCAT members of 256 ms. This has been implemented through the use of a two-stage VCAT multiframe, as illustrated in Figure 11.10. Because each frame is 125 μs, 4096 multiframes provide the necessary 512 ms (±256 ms). The four most significant bits of H4 (MFI1) are used for the first stage, creating 16 frames per multiframe, and the four least significant bits (MFI2) of the first two frames of a multiframe identify each of the 256 multiframes.

To reassemble the members at the receiving interface, the sequence indicator byte is implemented as four bits in frames 15 and 16 of each multiframe.

Low-Order VCAT

Low-order virtual concatenation employs a similar mechanism to route low-order virtual containers over the SDH infrastructure and reassemble the low-order paths at the far end. The frame structure of a VC-m-Xv consists of 4 rows and $(q + 1)$ columns and is repeated every 500 μs. The first 4 columns contain the four POH bytes. The value of q is dependant on the type of container, as shown in the Figure 11.11 inset.

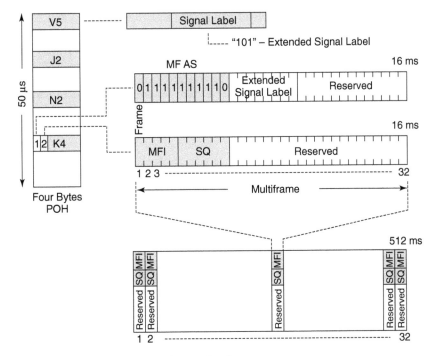

Fig. 11.12   Low-order virtual concatenation overhead.

The frame structure for a low-order VC-m-Xv is shown in Figure 11.12. In the existing POH of VC-ms, there were not enough reserved bits available to be used for the low-order VCAT overhead, and a two-stage process had to be defined to circumvent this problem. This was achieved by defining that when the "signal label bits" of the V5 byte have a value of 101, an extended signal label is to be used. The first bit of K4 is defined as the Multiframe Alignment Signal (MFAS) and is a fixed value of 0111 1111 110 followed by an extended signal label byte, a fixed bit equal to 0, and 12 reserved bits set to 0. The 32-bit multiframe repeats itself every 16 ms (because each bit equals 500 μs). The start of the MFAS is synchronized with the low-order VCAT overhead in bit 2 of the K4 byte, which consists of a 5-bit Multiframe Indicator (MFI), a 6-bit sequence number, and 21 reserved bits set to 0. In this way, a 32-bit control sequence for low-order virtual concatenation is achieved. Because of the multiframe nature of the channel, the granularity of the differential delay measurement achievable using the MFI is 16 ms.

Link Capacity Adjustment Scheme

Virtual concatenation extends the payload transport capability of SDH networks by the ability to match the concatenated container size with the client signal bandwidth. However, if the client signal is packet based, the required transport bandwidth will vary with time, and a mechanism is necessary to adjust the size of the provisioned bandwidth in a manner that does not disrupt the transported service. The link capacity adjustment scheme was standardized (ITU-T G.7042) to provide such a capability in conjunction with virtual concatenation. LCAS relies on a two-way handshake signaling protocol that enables addition and removal of high-order and low-order SDH path(s) into and out of a VCG without interrupting the existing service. LCAS does not provide the setting up and releasing of connections between intermediate nodes because this is the responsibility of Network Management Systems (NMS). Under a network failure scenario, the failed VCAT members are removed from the VCG by LCAS such that a client experiences a reduction in bandwidth rather than a total loss of connectivity. Replacement members can be provisioned through the NMS, and LCAS will hitlessly reintroduce them to the VCG to restore the connection to its original capacity.

LCAS Protocol

LCAS operation requires the VCG transmitter (source) and VCG receiver (sink) to maintain states on each member of the group and to exchange control messages to manage addition and removal of member links within the group. Bandwidth adjustments are initiated by the NMS by forwarding a request to the LCAS controller in the source and destination nodes. The controller then synchronizes the adjustment request using appropriate control packets and monitoring responses from the destination/sink node. Control packets are short-duration frames that are continuously transmitted between nodes. The control packets are used to synchronize the nodes and provide the status of each member and the VCG; they are transported in the H4 and K4 bytes of the path overhead for high-order VCAT and low-order VCAT, respectively (see Figure 11.13). These are reserved bytes that are not used for MFI and SQ in VCAT, as described earlier in the Concatenation section. Even if there is no change in the information that the VCAT contains, control packets are sent continuously, and each control packet describes the state of the member during the next control packet so that all the planned changes of the VCG are hitless.

The state of the member at the source side determines the reconstruction process at the sink, and the control (CTRL) field is used for this purpose. Each member in the LCAS process can be in one of five states. Four bits are defined and the values assigned, as recommended in ITU-T G.7042; they are shown in Figure 11.14. The default mode of VCAT without LCAS

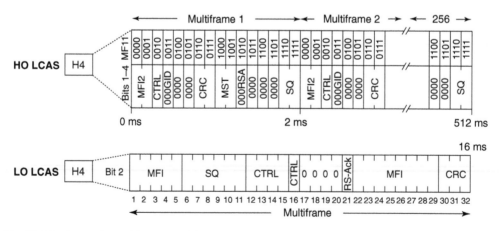

Fig. 11.13   Low-order and high-order LCAS overhead.

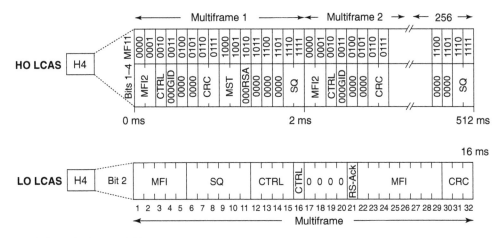

Fig. 11.14 LCAS control bits.

is indicated by all zeros (because the reserved bits of VCAT are all set to zero). Four bits allow for possible future expansions of LCAS with more states.

To check that members are cross-connected to the correct VCG, the Group Identification (GID) bit was introduced. The GID bit of all members of the same VCG has the same value in frames with the same MFI value. In this way, the sink can verify that all arriving members originate from a single VCG source. In SDH, it is common to validate a signal by the correct reception of three to five consecutive identical values. Because such a process would cause a significant delay in LCAS, a Cyclic Redundancy Check (CRC) is performed on every control packet after it has been received, and the contents are rejected if the check fails. The Member Status (MST) field reports the status of all members in a VCG from the sink to the source. At the sink, members belonging to a VCG are either functioning correctly (OK) or not (FAIL); members in the IDLE state are considered to be in the FAIL state. The MST field in each control packet has been limited to eight bits, and the MST of all members is distributed across multiple control packets. In this way, the status of all possible members of a VCG will be refreshed every 64 ms for high-order LCAS and every 128 ms for low-order LCAS. (Recall that each control packet is 2 ms for high-order LCAS, with a maximum of 256 members per VCG; and 16 ms for low-order LCAS, with a maximum of 64 members per VCG.

When the size of a VCG is resized by adding or removing members, the sequence numbers of the group will change. Although the standard defines that new members are added at the end of the VCG, removing one or more members can occur at any point. However, any changes at the source will be detected at the sink, which will consequently change the MST value of the affected members, and this will be relayed back to the source. To acknowledge to the source that a sequence change has been detected at the sink, the Resequence Acknowledgment (RS-Ack) bit is toggled. Before any further changes are committed to the VCG (or before any new MST information is evaluated), the source waits until the RS-Ack is received to acknowledge the change.

As an example of the process between two LCAS interfaces, consider the addition of a member to a VCG with reference to Figure 11.15. When the provisioning of the bandwidth from source to sink through the SDH network is complete, an ADD command from the

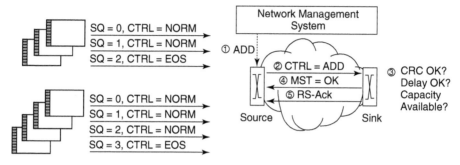

Fig. 11.15   Example of the LCAS process for adding a new member to a VCG.

network management system to the LCAS interface is initiated. The source sends a CTRL = ADD in the control packet to the sink node to notify it of the impending action. The sink node then checks that (a) the CRC of the control packet is correct; (b) the differential delay of the new member relative to existing members is within bounds; and (c) capacity is available to accommodate the new member. If any of these fail, the member is rejected. Otherwise, the sink node sends MST = OK to indicate the success of the end-to-end member connection. The source then makes the new member the last in the sequence using the control packet CTRL = EOS, and the penultimate member sends a CTRL = NORM message. When a confirmation RS-Ack toggle ($0 \rightarrow 1$ or $1 \rightarrow 0$) is received by the source from the sink, the bandwidth of the new member is made available to the VCG.

Removing a member is also initiated by the NMS and achieved by sending a CTRL = IDLE for the member(s) that are to be removed. The sequence number of the remaining members is updated accordingly, and when an RS-Ack is received from the sink, the payload size of the VCG is decreased and distributed over the remaining active members. If required, the NMS can then remove each member's path through the network. In addition to the planned increase or decrease of the payload capacity of a VCG, LCAS offers the ability to automatically decrease the provisioned capacity in the event of network failure. Routing members of a VCG across different paths in the network is a resilience strategy known as diversification. An example of splitting a VCG into two diverse paths is shown in Figure 11.16.

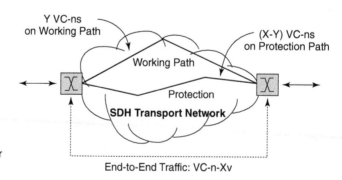

Fig. 11.16   LCAS used for resilience.

In a network failure scenario that affects one or more members of a VCG (but not all members), the contents of the VCG will be in error for a short duration of time because the reassembled payload at the sink will be missing the data being transported by the affected members. LCAS coordination requires the removal of these members from the VCG, and this is achieved by changing the member status to failure mode (MST = FAIL, CTRL = DNU). The VCG payload is therefore reduced and uses only the remaining members, sending the CTRL code NORM or EOS. A message reporting the failed member is sent to the NMS. Because the sequence numbering is not changed, the RS-Ack bit is not toggled. The NMS could reprovision replacement members via an alternative path in the network and reestablish the full VCG payload in a hitless manner as described before.

Clearly, LCAS allows a VCG to continue to carry reduced amounts of traffic even when constituent members fail. This is an important feature if an NG-SDH transport network is used to interconnect IP routers. IP routers detect link failures by the absence of a number of periodic "hello" messages that are sent between adjacent routers; the failed link is advertised to the rest of the IP network, resulting in all routers updating their routing tables. This is an undesirable action, and the graceful reduction in bandwidth offered by LCAS is a desirable feature.

Generic Framing Procedure

Methods of transporting data over SDH networks can be achieved using a variety of methods, and a number of proprietary techniques have appeared on the market. For this reason, the ITU-T defined a Generic Framing Procedure (GFP) to provide a uniform and standardized method of encapsulating multiple client traffic types onto synchronous transmission channels. Compared to existing techniques such as Packet over SDH (POS) or Link Access Protocol SDH (LAPS), GFP supports more client signals, transports them more efficiently, and is capable of scaling to very high bitrates, such as 40 Gbit/s. GFP has seen widespread use also in Resilient Packet Rings (RPRs) and Optical Transport Networks (OTNs) for mapping data onto synchronous optical data units (ODUs), and has even been adapted for use in the Gigabit Passive Optical Network (GPON) standards.

GFP Overview

The GFP, defined in ITU-T G.7041, is a mechanism for mapping constant- and variable-bitrate data into synchronous SDH/SONET envelopes. GFP supports many types of protocols, including those used in Local Area Networks (LANs) and Storage Area Networks (SANs). Functionally, GFP can be divided into common and client-specific aspects, as illustrated in Figure 11.17. Its common aspects include Protocol Data Unit (PDU) delineation, data link synchronization and scrambling, client PDU multiplexing, and client-independent performance monitoring. The mapping of client PDU into the GFP payload, client-specific performance monitoring, and OA&M are the responsibility of the client-specific aspects.

Currently, two modes of client signal adaptation are defined: Framed (GFP-F) and Transparent (GFP-T). GFP-F is a layer-2 encapsulation mode that is optimized for data packet protocols (e.g., Ethernet, IP/PPP, DVB) of variable-size frames. GFP-T is a layer-1 encapsulation or block coded–oriented adaptation mode, optimized for protocols that use 8B/10B line coding (e.g., Fibre Channel, ESCON, 1000Base-T) and that are of constant-size frames.

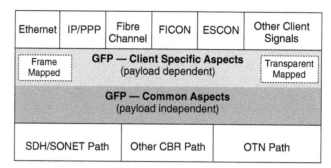

Fig. 11.17 Functional block diagram of GFP.

GFP Frame Structure

A GFP frame consists of a core header and a payload area, as shown in Figure 11.18. The core header supports frame delineation procedures and essential data link operations functions independent of the higher-layer PDUs. Within the header, a 2-byte field indicates the size of the GFP payload in bytes (4–65,535 bytes), and a 2-byte field (cHEC) contains a CRC sequence that protects the integrity of the core header. Rather than relying on a special character for frame delimiting (as is the case for HDLC), the GFP receiver begins the framing process by looking for a 2-byte field that is followed by a correct CRC for that field.

When such a 32-bit pattern is found, it is very likely to be the core header at the beginning of a GFP frame because the probability of such a pattern randomly occurring in the client data is 2^{-32}. The GFP receiver then uses the PLI information to determine where the

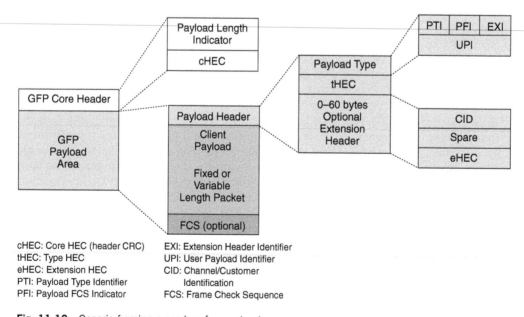

cHEC: Core HEC (header CRC)
tHEC: Type HEC
eHEC: Extension HEC
PTI: Payload Type Identifier
PFI: Payload FCS Indicator

EXI: Extension Header Identifier
UPI: User Payload Identifier
CID: Channel/Customer
    Identification
FCS: Frame Check Sequence

Fig. 11.18 Generic framing procedure frame structure.

end of the GFP frame is. The next core header will occur immediately after the end of this current frame, so if another valid 32-bit pattern is found in that location, the receiver can be sure that it has acquired framing for the GFP stream. The CRC is adequate to provide a simple, single error correction capability for the PLI, which increases the robustness of the GFP framing. In contrast, an error in an HDLC start/end flag has no error correction, which makes HDLC framing more vulnerable to transmission errors.

This low-complexity frame delineation ensures that GFP scales to high bitrates such as 40 Gbit/s. In addition, because GFP supports variable-length PDUs, it does not need complex SAR functions or frame padding to fill unused payload space, resulting in a deterministic transport bandwidth overhead. In contrast, compare HDLC, in which to prevent erroneous packet delineation, to special escape characters that have to be inserted into the datastream if the client data contains the start/end flag; this leads to an unpredictable transport bandwidth because it is influenced by the content of the client data.

The GFP payload area consists of payload header information, a payload field that contains the client data, and an optional Frame Check Sequence (FCS) CRC-32 over the Payload field. There are two types of payload headers. The payload type header is used in all GFP frames except idle frames and consists of a 2-byte Payload Type field (protected by a CRC-16). If additional payload information is required, then a 60-byte extension is available. In these circumstances, the Payload Type field consists of four subfields:

Payload Type Identifier (PTI) consists of three bits and distinguishes between two currently
   defined frame types: client data frames (PTI = 0) and client management frames
   (PTI = 4).
Payload FC Indicator (PFI) indicates whether a payload FCS field is present.
Extension Header Identifier (EXI) is a 4-bit identifier that indicates the type of extension that
   is present. A null header extension implies that an extension header and its associated
   HEC check (eHEC) are not present. A null header extension is generally used in
   situations where all GFP client data frames are mapped into a single transport path
   (SDH container/virtually concatenated container) and carry a single type of payload,
   which is demapped at the receiving end to a single physical port/link. The linear
   extension header signifies multiple types of payload from multiple ports/links
   mapped in a framewise multiplexed manner into a single transport path. In this case,
   the extension header contains two octets. One octet is the Channel ID (CID), which is
   a single byte for a destination port/link ID. The other byte is unused. Two bytes of
   the eHEC are also required for a linear extension header.
User Payload Identifier (UPI) is an octet set according to the type of client that is transported and whether the mapping is frame based or transparent. When PTI = 0, it
   identifies the type of payload and mapping in the data frame. When PTI = 4, the UPI
   defines the management frame type.

Any idle time between GFP frames carrying client data is filled with GFP idle frames. A GFP idle frame consists only of a core header with PLI = 0.

Framed GFP

Framed GFP (GFP-F) is optimized for packet-switching environments and operates on signals that have been packetized or framed at layer 2 or higher by a client service. The diagram in Figure 11.19 illustrates its use in transporting multiple-client IP router traffic

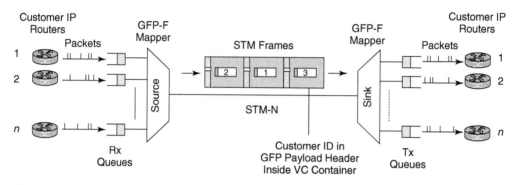

Fig. 11.19 Example of GFP-F usage in transporting multiple-customer IP packets over a single SDH link.

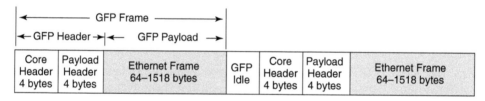

Fig. 11.20 Example of Ethernet frame encapsulation into GFP-F frames.

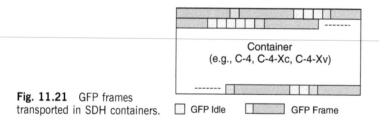

Fig. 11.21 GFP frames transported in SDH containers.

across a common SDH infrastructure. Data is queued before being mapped onto a TDM channel, and the GFP CID field (8 bits) in the payload header is used to de-multiplex at the far end.

Because one client frame is encapsulated into a single GFP frame (see Figure 11.20), there is a delay as the frame is stored and encapsulated prior to transmission. In addition, any preamble or interframe gaps are removed to improve transport efficiency, and these are later reinserted at the receiving equipment. GFP idle frames are inserted to maintain a Constant Bitrate (CBR) at the output stream.

The process of mapping GFP frames into an SDH virtual container is shown in Figure 11.21. The choice of container is determined by the predicted amount of client data. GFP idle frames are inserted to adapt the datarate to the CBR stream of the SDH container and to fill the container in periods of low data activity.

Transparent GFP

Transparent GFP (GFP-T) is a protocol-independent encapsulation method in which all client codewords are decoded and mapped into fixed GFP frames. Unlike GFP-F, the frames are transmitted immediately without waiting for the entire client data packet to be received. For this reason, it is regarded as a layer-1 transport mechanism because it doesn't distinguish between data, headers, or control information and simply transmits the received client bits. The result is a very-low-delay transport mechanism. GFP-T only accepts protocols based on 8B/10B line coding and immediately transcodes to the more efficient 64B/65B superblock version prior to encapsulating into fixed-length GFP-T frames. Several important data protocols use the 8B/10B line code at the physical layer.

Fibre Channel, Fiber Connection (FICON) and Enterprise System Connection (ESCON) are commonly used in modern Storage Area Networks (SANs), and for video distribution, Digital Video Broadcast–Asynchronous Serial Interface (DVB-ASI) is popular. Some of these protocols use line coding to communicate control codes between the end devices for various functions. If the frames from these protocols were carried with GFP-F, the 8B/10B control code information would be lost. Another consideration for the SAN protocols is that they are typically very sensitive to transmission delay (latency) between the two SAN nodes. Framed GFP requires the buffering of an entire client data frame to determine the payload length for the PLI field of the GFP header. For most applications, this is not a problem, but where transparency and low latency are required, transparent GFP should be employed.

Conclusion

As this chapter explained, the early deficiencies of Synchronous Digital Hierarchy have largely been addressed through the introduction of VCAT, LCAS, and GFP into what is now termed "next generation" SDH. The exponential growth in packet-based data and storage services will continue, and there will be competing technologies to NG-SDH in the next decade. However, with such a huge installed base of SDH worldwide, NG-SDH has fueled the continued growth of SDH equipment and is regarded as a trusted and robust technology for providing carrier-class wide-area networking of packet and storage services.

Resources

ANSI (2001) *T1X1.5/2001/062, Revised Draft T105 Sonet Base Standard.*
IEEE Commun. Mag. May 2002,
IEEE (1998) *802.3: CSMA/CD Access Method and Physical Layer Specifications.*
IETF (1990) *RFC 2615, PPP over Sonet/SDH.*
ITU-T (2002) *G.7041/Y.1303, Generic Framing Procedure.*
ITU-T (2001) *G.709, Network Node Interface for Optical Transport Network.*
ITU-T (2001) *Recommendation X.83/Y.1323, Ethernet over LAPS.*
ITU-T (2001) *Recommendation X.85/Y.1321, IP over SDH Using LAPS.*
ITU-T (2000) *G.707/Y.1322, Network Node Interface for the Synchronous Digital Hieracrchy.*

12 Legacy Packet-Switched Data Overlay Networks

Matthew R. Thomas
University of Essex

Paul A. Veitch
BT Design, Adastral Park

David K. Hunter
University of Essex

Introduction to WAN Services

Leased lines have been used extensively to carry corporate data since the 1970s. Mainframe networks of this era used analogue leased lines of a very low datarate (2400 baud) over the Public Switched Telephone Network (PSTN) with modems at either end. During the 1990s, many corporations rented digital leased lines from telecom providers to construct their own data networks. These lines were typically of E0 (64-kbit/s) or T1/E1 (1.544/2.048-Mbit/s) datarates; however, for many customers the cost was prohibitive. A typical company might require data communication over a spread geographical distance (perhaps 1000 km); although it is easy to provision a circuit over such a distance, it would be very expensive. The telecom providers responded by offering *data overlay networks*.

In essence, layer-2 and layer-3 networks were constructed, initially over Plesiochronous Digital Hierarchy (PDH) and later over Synchronous Digital Hierarchy (SDH/SONET). Leased lines were still required for access to the nearest point-of-presence (POP) where the service provider had located layer-2 switches, which themselves were interconnected into Wide-Area Network (WAN) clouds. In their simplest form these overlay networks could be shared by customer traffic over dispersed geographical distances, thus reducing the price point (namely, the price at which demand is relatively high) for the customer. In general, the trade-off for this was lower average bandwidth per customer and differing Service Level Agreements (SLAs).

By the late 1990s and early 2000s, Frame Relay (ITU-T Q.922) had expanded its take-up by corporate customers to become a main player in data overlay networks (ITU-T, 1992). Despite the enormous take-up of Frame Relay, analysts expect the service to give way almost entirely to newer Multiprotocol Label Switching (MPLS) and Ethernet solutions by 2011. In 2003, U.S. customers spent $7.4 billion on Frame Relay services, while in 2005 Frame Relay services in the United States represented a $6.5 billion market. Analysts are expecting a five-year compound annual growth rate (CAGR) of −12.1 percent for the sector (IDC, 2006). In Europe analysts are expecting customer spending to fall at a CAGR of −17 percent from $2.9 billion in 2006 to $1.1 billion in 2011 (IDC, 2007).

Frame Relay differs from earlier offerings such as X.25 (ITU-T, 1996) because it was developed as an OSI (International Standards Organization) layer-2 service designed to relay a frame to its destination (ISO, 1984).

Apart from Frame Relay and X.25, there are two other data services introduced in this chapter that differ significantly from these standard overlay services—namely, ISDN and ATM. Integrated Services Digital Network (ISDN) (ITU-T, 1993a, 1993b, 1993c, 1995, 1997, 1998) is distinctive due to its inclusion within the PSTN. ISDN provides dynamic services, hence, signaling is required, which is provided by Signaling System Seven (SS7). ISDN is a part of the telephone system, and it utilizes bandwidth provided by the telephone trunks.

Asynchronous Transfer Mode (ATM) was designed with a slightly different set of aims in mind. ATM is an extremely efficient way to move data across a network (De Prycker, 1991). From the 1990s onward, ATM was touted by some industry analysts as a possible replacement technology for the underlying SDH/SONET network. The general intention was the convergence of voice and data over a combined infrastructure. The idea was the replacement of the SS7 signaling system with IP voice/data conversion gateways, resulting in the phasing out and removal of the SS7 telephone signaling system with its associated circuit bearers.

It was envisaged that if voice traffic were running over ATM, there would be no need for SDH/SONET to provide bandwidth for both voice and data networks, and so the underlying SDH/SONET functionality could be replaced by part of a single converged network. ATM was being positioned by some vendors to implement such a replacement. During the 1990s, service providers exploited ATM by using converged WAN services, and although the technology did not achieve the critical mass necessary to be a competitor of lower-level SDH/SONET services, it was leveraged effectively both for revenue generation and for early convergence and migration of legacy services.

As such providers rolled out ATM as a high-speed overlay network, it was successfully adopted by many corporations for their wide-area data networks. For example, in the United Kingdom service providers—including BT—have used the ATM network infrastructure to backhaul customer Internet traffic from the Digital Subscriber Line (DSL) local loop via the DSL Access Multiplexer (DSLAM) to the core IP access servers. It would appear that ATM is firmly established; however, vendors of DSLAM technology are now aiming at the IP DSLAM market. Industry analysts are expecting a global migration of services to IP/MPLS and that the associated converged IP technologies will replace ATM.

For example, in 2000 Switzerland's incumbent telecom operator purchased an IP/MPLS router network from Cisco Systems (Rastislav, 2001). This was one of the first major signs of a shift away from ATM solutions, although the network purchase included a large ATM feeder network. The general goal of this project was to converge the Swiss network around an IP and MPLS routing platform. Voice/IP gateways were deployed. Consisting of 35 points of presence located throughout Switzerland, the national Swiss platform that was created ran directly over dark fiber and, unlike many other operators, even bypassed the SDH network.

This IP/MPLS network set the global trend during 2000. The use of dark fiber and the bypassing of the SDH network were pivotal in changing perceptions of MPLS capabilities within the industry. It would be a fair comment that globally there has been a slight slowing of such MPLS adoption, but early adopters have reaped huge financial benefits from the IP/MPLS networks as the corporate data market has significantly expanded.

Today the market for differing low-level technologies has become more crowded. Provider Backbone Bridging with Traffic Engineering (PBB-TE), a redefined Ethernet technology, is one of the latest prospects for low-level adoption (see Chapter 14). This is also known as Provider Backbone Transport (PBT) (Allan et al., 2006). Although PBB-TE is a packet-driven service and vastly different from SDH/SONET circuit-oriented services, it can operate as a replacement for SDH/SONET equipment by replacing and upgrading switches and line cards, while leaving intact the underlying requirement for a low-level multiplexing technology. As such, this technology can be more palatable to telecom operators than complete replacement by ATM or perhaps IP/MPLS. There is currently debate within the industry as to whether or not this technology will be used to update SDH/SONET equipment; it even seems possible that a hybrid GMPLS-signaled PBB-TE solution might be deployed in access networks.

Overlay Network Concepts

Fiber optic and radio transmission media form the lowest level of a country's telecommunications infrastructure. The organization of these media into a coherent architecture is achieved by multiple layers of network infrastructure. The lowest of these layers is most commonly SDH/SONET. The PSTN, Internet, and data connection services offered by service providers can all be categorized as *overlay networks*.

SDH/SONET Networks and Leased Lines

The underlying fiber, submarine cable, and microwave transmission media lie at the heart of any telecommunications network. Almost universally the media and the multiplexers are now standardized as SDH or SONET, which has now almost completely replaced the older Plesiochronous Digital Hierarchy.

SDH/SONET allows circuits of T1/E1 (1.544/2.048 Mbit/s) and higher rates to be configured statically between different geographical locations. These point-to-point services are routed over ring and mesh networks constructed from physical fibers, interconnected via Add-Drop Multiplexers (ADMs) and Digital Cross-Connects (DXCs) of various sizes (Chapter 10). In order to realize greater capacity over existing fibers, Dense Wave Division Multiplexing (DWDM) may be employed. Although capable of forming a multiplex layer itself, with the potential for wide use of optical add-drop multiplexers and optical cross-connect switching, DWDM is normally implemented as a point-to-point technology between SDH/SONET switches to increase the fiber's data-carrying capacity.

The SDH/SONET transmission network provides the raw bandwidth used by all data and telecommunications circuits. Bandwidth is statically provisioned over the SDH/SONET

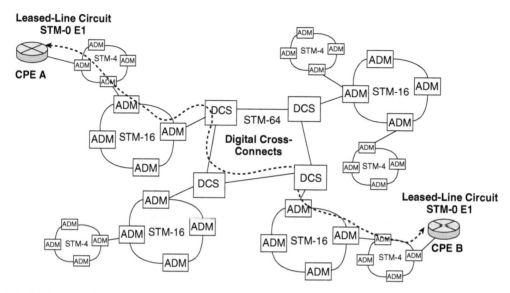

Fig. 12.1   Leased-line circuits over SDH.

network between two geographical locations. SDH/SONET cross-cornect manufacturers provide management software to enable this static provisioning to be carried out by operators on proprietary management platforms. Figure 12.1 shows a simplified SDH/SONET network to illustrate this concept. The SDH/SONET ADMs and the larger Digital Cross-connect Systems (DCSs) provide transparent service at rates of T1/E1 and above.

Figure 12.1 shows that a leased-line network operating at T1/E1 rates and above can be provisioned directly on the underlying SDH/SONET network. For lower-rate lines, additional multiplexers are used to subdivide the bandwidth down to E0 rates. The older PDH network utilized these lower rates directly.

Generic Wide-Area Networks

Figures 12.2 and 12.3 illustrate multiplexing of different customer services over a shared network; Figure 12.2 shows the underlying SDH/SONET network with the Data Switching Equipment (DSE) shaded grey and overlaid in bold (in all figures), while Figure 12.3 shows the type of packet-switched WAN that is implemented. The WAN switches are located at Local Exchanges (LEs) and can multiplex together signals from different sites and customers to provide a lower price point. They are interconnected via leased circuits, themselves provisioned over an SDH/SONET core transmission network.

Ideally connections to the WAN switches are provisioned directly over the local loop at both local and remote customer locations, terminating at the local exchange, where they are fed into the WAN overlay switch. The switch in the LE multiplexes data from several customers, and the data is then carried across the overlay network from one WAN switch to another. Often these networks were sold at a loss, while customers were being added during platform rollout, until enough data was being multiplexed and "oversold" to provide profit to the telecom operator.

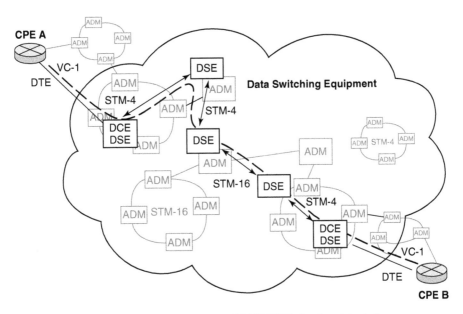

Fig. 12.2 WAN network (bold) with its underlying SDH/SONET structure (shaded).

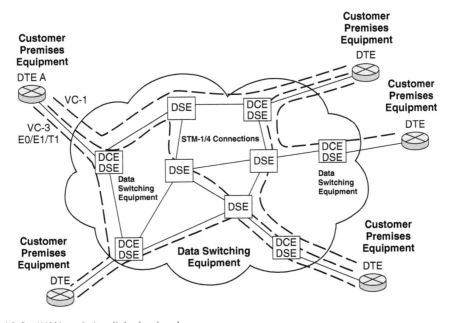

Fig. 12.3 WAN packet-switched network.

The first stage in the connection is from the Customer Premises Equipment (CPE) to the LE. The interface from the CPE to the WAN switch is often designated a User Network Interface (UNI). The CPE end is the Data Terminal Equipment (DTE), and the LE end is the Data Communication Equipment (DCE).

If there is a WAN switch present in the LE, local customers can be connected by an existing copper pair. For higher rates, fiber terminating at the LE might have to be run up to the CPE from a junction box in the street. However, not all LEs have a WAN switch.

Tail Circuits

If there is no WAN switch in the nearest LE, then a leased-line "tail circuit" can be provisioned to another LE with a WAN switch to connect it to the CPE (see Figure 12.4). This bandwidth is normally provisioned over the SDH/SONET network in the same way as a standard leased line.

Avoiding the costs associated with long-haul dedicated leased lines is a major business driver for this type of WAN cloud. Often the costs of provisioning the access SDH/SONET circuits are not passed directly on to the customer and are considered to be part of the development costs of the WAN service. As more customers join the network from remote locations, the number and density of the WAN switches increase, leading to fewer and shorter SDH/SONET tail circuits. Today most national carriers have WAN services with extensive geographic coverage.

Depending on the WAN service involved, the customer is then provisioned some type of virtual circuit or path through the "overlay" WAN cloud between the ingress and the egress WAN switches. This is the service for which the telecom operator bills the customer.

The interfaces between the WAN switches are designated Network Node Interfaces (NNIs). Some technologies, such as Frame Relay, do not have a standard NNI at all but

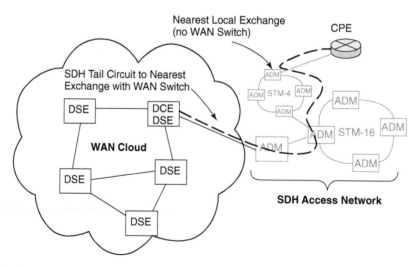

Fig. 12.4   SDH tail circuit.

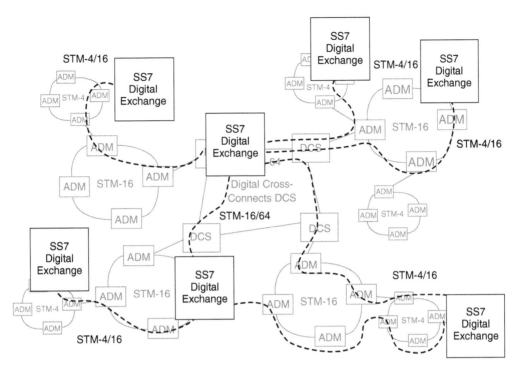

Fig. 12.5 Example PSTN bearer bandwidth.

merely define the UNI portion of the connection, leaving the equipment vendor free to design the NNI portion of the circuit in whatever way is preferred.

The virtual circuits over the WAN cloud can be provisioned in a number of ways. Some WAN technologies such as X.25 include signaling protocols, which allow the dynamic establishment of connections, referred to as Switched Virtual Circuits (SVCs). Other WAN services, such as Frame Relay, do not include an NNI standard or signaling services for circuit establishment. These connections are normally manually provisioned between the various WAN switches and are hence not dynamic, being referred to as Permanent Virtual Circuits (PVCs).

The PSTN as an Overlay Network

The SDH/SONET network also provides the PSTN (SS7) bearer bandwidth (Figure 12.5). Although often considered as a whole, the SDH/SONET and SS7 Digital Exchange PSTN switches are separate; hence the telephone network could itself be characterized as an "overlay network."

The Local Exchange

A key component in any local exchange is the SDH/SONET equipment, which terminates the physical fibers and constitutes the lowest layer of transmission (Figure 12.6). On a fiber-by-fiber basis, more capacity can be gained by using DWDM. Although Optical Cross-

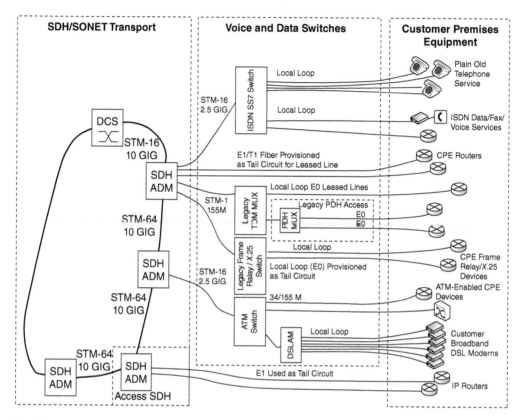

Fig. 12.6   Example local exchange.

Connects (OXCs) using DWDM and optical switching technology have been researched, these have not as yet proved to be cost effective, and presently DWDM is always employed merely as a point-to-point capacity multiplier.

The SDH/SONET network provides all of the bearer bandwidth utilized by both the SS7 digital exchange–switched network and all of the other overlay networks. Access from customer premises to the LE takes a number of forms. Existing copper pair wiring, originally running to the SS7 digital exchange for POTS services, can be reused and provisioned as leased lines, DSL, Frame Relay tail circuits, and so on. The SDH/SONET network allows circuits down to T1/E1 rates to be provisioned end-to-end across the cloud. Lower-bandwidth circuits are provisioned over SDH/SONET by means of a further set of multiplexers and demultiplexers.

Frame Relay Services

Frame Relay is standardized in ITU-T Q.922 (ITU-T, 1992) and was originally conceived as part of the ISDN protocol standards (Figure 12.7). It uses the notion of a Data Link Connection Identifier (DLCI) to identify the individual PVCs established manually across the cloud. The UNI is referred to as the Local Management Interface (LMI). This establishes a PVC to

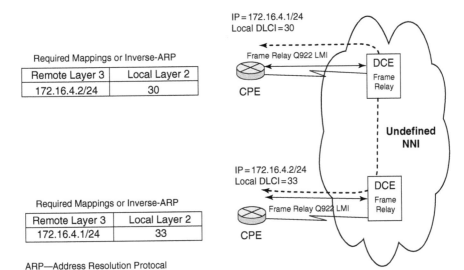

Required Mappings or Inverse-ARP

Remote Layer 3	Local Layer 2
172.16.4.2/24	30

Required Mappings or Inverse-ARP

Remote Layer 3	Local Layer 2
172.16.4.1/24	33

ARP—Address Resolution Protocol

Fig. 12.7   Basic IP–over–Frame Relay services.

the CPE for limited signaling. LMI uses a well-known DLCI number depending on the LMI standard being used. There are three LMI standards: ANSI's T1.617 Annex D (American National Standards Institute, 1991), ITU-T's Q.933A (ITU-T, 2003), and the Cisco, DEC, Strata COM and Nortel (Gang of four) de facto standard.

When this LMI DLCI is established, the switch informs the CPE of the DLCI numbers that the CPE is connected to. The CPE devices are defined as Frame Relay Access Devices (FRAD) and are commonly IP routers with Frame Relay functionality.

Frame Relay defines a service in terms of an access datarate, a Committed Information Rate (CIR), and an Excess Information Rate (EIR). Frame Relay is an alternative to costly higher-rate leased-line connections. Frame Relay services are offered at a maximum port speed of DS-3 (45 Mbit/s), although much lower speeds, with a common access rate of 64 kbit/s, are the norm.

Within this value a customer will pay for a CIR value. This is the rate that is guaranteed for delivery by the network. Packets exceeding this rate are allowed up to the negotiated EIR but are sent marked Discard Eligible (DE). The CIR and the EIR are normally added together to make the access rate. Service Level Agreements (SLAs) often include formulas for allowable short bursts and timing of the CIR and EIR values.

X.25 Services

X.25 was one of the earliest data networks (ITU-T, 1996); it is a true OSI layer-3 protocol and as such was implemented long before the requirements for carrying IP data. Most incumbent network operators had extensive X.25 networks; see for example, Atkins, Cooper, and Gleen (2007). In the 1990s most of the major banks' teller machines ran over such networks, running X.25 software directly. Ironically, today X.25 is more likely to be run over TCP as "X.25 over TCP" (XOT) (Forster, Satz, & Glick, 1994), although X.25 services are still used and offered.

Asynchronous Transfer Mode

ATM as a technology was designed to be an extremely effective way to move data across a transport medium (De Prycker, 1991); it was designed to be flexible, to use bandwidth efficiently through statistical multiplexing, and to be able to accommodate any future service. It was originally standardized for line rates of 155 Mbit/s and 622 Mbit/s, although lower rates were later standardized by the ATM Forum. Unlike IP (Internet Protocol), ATM's packets (or "cells") are of equal length, simplifying ATM switch and cross-connect design and making it easier to design special-purpose hardware for their implementation.

Cells are 53 bytes in length, consisting of a 5-byte header followed by 48 bytes of data (see Figure 12.8). Of particular interest and importance are the Virtual Circuit Indicator (VCI) and Virtual Path Indicator (VPI) fields. Unlike IP, the destination address of a cell is not held inside its header. Instead, the VCI indicates the virtual circuit number on which the cell is being transported.

Often virtual circuits are set up in response to user requests; to do this, signaling packets set up tables in ATM switches that implement "label swapping." The label (i.e., the VCI) on an incoming cell indicates the virtual circuit number—this is looked up in the switch's table to find out the outgoing VCI number and the output port on which the cell should be forwarded. Virtual paths are similar but are indicated by the VPI and are always set up by the network management system over longer periods. Many virtual circuits can be transported within one virtual path (see Figure 12.9).

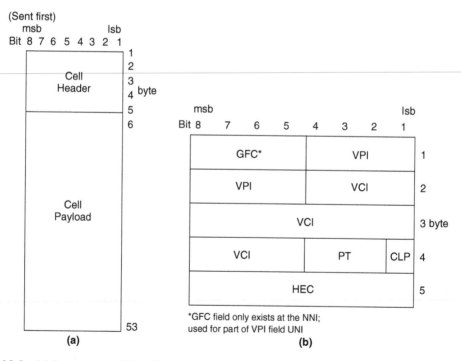

Fig. 12.8   (a) Format of an ATM cell; (b) format of ATM header, including VCI and VPI fields.

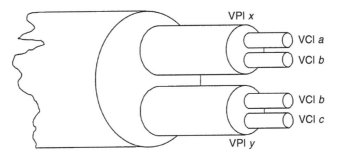

Fig. 12.9 Relationship between virtual circuits (defined by the VCI field) and virtual paths (defined by the VPI field).

There are two types of ATM switching node—ATM cross-connects and ATM exchanges (or service switches). The former are generally smaller, and set up virtual paths through the network under network management control, based on the VPI field. This allows the creation of private virtual networks for business users. Exchanges are larger and perform routing under user-signaling control based on both the VCI and VPI fields.

With the ascent of IP networks in the 1990s, several solutions were proposed to transport IP over ATM. Unfortunately, this resulted in solutions that were complex, difficult to manage, and in some scenarios resulted in excessive computational complexity (Davie & Rekhter, 2001). Many of the lessons learned from this resulted in the creation of Multiprotocol Label Switching (MPLS) (Rosen, Viswanathan, & Callon, 2001) in the late 1990s, which also employs the "label-swapping" scheme, originally implemented in ATM.

Commercial ATM/Frame Relay Services Example

Initially, services such as ATM and Frame Relay were made up of dedicated WAN switches. The original expectation was that ATM switches would eventually dominate data. networking architectures, and many customers decided to install ATM-compatible interfaces to their corporate networks to prepare for such a trend. In terms of ATM adoption, providers limited themselves to overlay networks. Furthermore, toward the end of the 1990s vendors started to offer multiple services on a single-chassis platform.

Alcatel and Cisco Systems were two vendors leading the way in converging features on a single platform. Alcatel's 7670 RSP and MSP (*www.alcatel.com*) were designed specifically to be a single chassis capable of carrying ATM, X.25, Frame Relay, SMDS, and circuit emulation services. Cisco Systems offered the Strata COM BPX and MGX ATM carrier platforms (*www.cisco.com*).

Service providers in the late 1990s started to converge their disparate networks onto similar unified platforms. In the United Kingdom, British Telecom (BT) rolled out a multiple-node Multi-Service Intranet Platform (MSIP) (see Figure 12.10). This state-of-the-art WAN service incorporated BT's commercial corporate data offerings of Framestream, Cellstream, and MegaStream Ethernet onto a single network and greatly improved service delivery time and failure response.

The network was implemented as a three-tier system and incorporated a range of equipment from more than one switch vendor. Trunks between the DSE overlay switches were

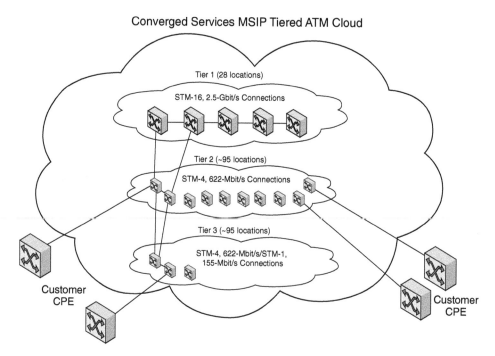

Fig. 12.10 MSIP platform.

implemented in SDH/SONET and included STM-16 rates in the core. As of late 2007, the network encompassed a geographic footprint of approximately 200 distinct switching locations.

The current WAN network has two different tasks. The first is that of a corporate data service offering designed to connect customer locations from geographically dispersed regions without the high cost of leased-line access. In this respect traffic enters and exits the WAN cloud from customer CPE to customer CPE. The second purpose of the converged ATM network is as a feeder network to BT's IP Internet backbone. The Internet backbone in the United Kingdom is dominated by the Internet exchanges in Manchester, London, and Scotland.

Connected to the Internet exchange is a purpose-built high-speed router backbone, incorporated into which are nine major POPs that host the BT access servers. Internet traffic travels over this backbone from other ISPs via the Internet exchanges to the BT IP access servers located at the points of presence.

A home Internet user can attach to the LE using a digital subscriber line. This line is terminated over copper pairs into the LE. The voice and data signals are separated at the exchange, with the voice component patched onto the SS7 digital exchange. The termination of the data part of the signal is then performed on a DSLAM.

Currently the DSLAMs are ATM enabled, and the data signal is carried over an ATM virtual path using the MSIP data overlay network to BT's IP points of presence. The BT access servers can authenticate the customer and then forward authorization requests to another ISP's access servers if the user is not a BT customer. This can be performed over layer-2 tunneling between the BT and ISP access servers.

Currently all of the United Kingdom's home Internet DSL access is configured in this way. The MSIP ATM network carries the DSL home user Internet traffic over the PVCs to the BT access servers at the POPs. The loading of new DSL broadband services over ATM or IP/MPLS cores became a hallmark of the generic model used by early adopters.

ISDN Basic Rate and Primary Rate

ISDN provides leased-line services dynamically "on demand." However, it has pricing-point issues and was widely deployed as a backup service for leased lines. For a short time in the 1990s, ISDN service was also marketed as a "high-bandwidth" Bandwidth-on-Demand (BOD) solution, but the commercial datarates of networks have changed with the introduction of DSL.

ISDN services resemble a data overlay network. As shown in Figure 12.11, CPEs are connected into a shared cloud. The interface from the CPE to the WAN switch is a defined UNI. In the case of ISDN, though, services are offered within the digital exchange switch. This means that "NNI" signaling is in fact integrated with SS7, and the network nodes are the digital telephone exchanges themselves.

Two different types of channel are used from the CPE to the ISDN switch. These are termed bearer channels (B-channels) and data channels (D-channels). Both are time division

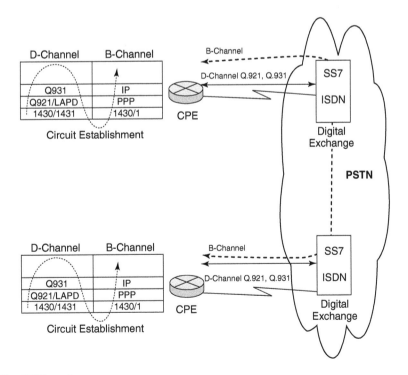

Fig. 12.11 ISDN services.

multiplexed. In the case of narrowband ISDN service, two B-channels are provided for customer data and one D-channel (ITU-T, 1993b). The D-channel is defined between the CPE and the ISDN switch.

ISDN can establish dynamic B-channels across the PSTN. Each CPE must have the D-channel to the local switch established prior to communication. There are slight (national) differences between ISDN switch types, but in most cases the signaling D-channel stays connected between the switch and the CPE. Q.921 (ITU-T, 1997) runs over this channel, and a Terminal Endpoint Identifier (TEI) is assigned to the CPE by the switch. Q.931 (ITU-T, 1998) operates over the layer-2 service established between the switch and the CPE and is used by the CPE to initiate a call to the remote desired endpoint. The connection request is carried across the PSTN, and the call request is forwarded to the digital exchange at the remote end. Q.931 running on the D-channel at the remote end forwards the call request to the TEI of the remote CPE, and the CPE responds if the call is allowed.

Upon the correct response, the two digital exchanges open up a B-channel for the CPE data to be transmitted. The B-channel can be configured to run standard layer-2 services such as Point-to-Point Protocol (PPP) (Simpson 1994). Troubleshooting the ISDN stack can be followed as in Figure 12.11.

ISDN comes in two size bundles. Basic Rate Interface (BRI) is established with a single 16-kbit/s D-channel and two 56/64-kbit/s B-channels referred to as b0 and b1 (ITU-T, 1995). Primary Rate Interface (PRI) is established with a 56- or 64-kbit/s D-channel (ITU-T, 1993c) and 23 or 30 × 64-kbit/s B-channels that correspond to T1/E1 access rates.

Future Trends

The current trend is for many services to converge on a single platform. In fact, convergence is probably the largest goal in telecommunications operator spending and network design today. As the newer IP/MPLS platforms have been deployed, much of the corporate data has moved over from Frame Relay and legacy services to IP/MPLS VPNs.

Clearly data services are moving rapidly to fully converged networks. Newer networks are able to emulate circuits and a host of legacy services, bluring the distinction between them. An X.25 connection, for example, might ride over IP, which in turn rides over an MPLS-derived pseudowire. It is certain that the architecture of the modern telecom provider will continue to evolve toward technologies that achieve critical mass and as such IP and mobile services will play an ever increasing role. The business model adopted is one of converged networks with customer-driven differentiated products and services.

References

Allan, D., et al. (2006) Ethernet as Carrier Transport Infrastructure, *IEEE Commun. Mag.* (2):134–140.
American National Standards Institute (1991) *ANSI T1.617, Annex D.*
Atkins, J. W., Cooper, N.J.P., and Gleen, K. E. (2007) The Evolving Art of Packet Switching, *BT Technol. J.* 25(3/4):222–228.
Davie, B., and Rekhter, Y. (2001) *MPLS: Technology and Applications*, Morgan Kaufmann.
De Prycker, M. (1991) *Asynchronous Transfer Mode: Solution for Broadband ISDN*, Ellis Horwood.
Forster, J., Satz, G., and Glick, G. (1994, May) *X.25 over TCP (XOT)*, IETF RFC 1613.
IDC (2006) *Analyst Report, U.S. Frame Relay Services 2006–2009 Forecast*, Publication ID: IDC1266863.
IDC (2007) *Analyst Report, Western Europe Frame Relay/ATM Services Market Analysis 2007–2011*, Publication ID: IDC1573601.

ISO (International Standards Organization) (1984) *Standard ISO 7498: Open Systems Interconnection—Basic Reference Model.*

ITU-T (1993a) *Recommendation Q.930/I.450: ISDN User-Network Interface Layer 3—General Aspects.*

ITU-T (1993b) *Recommendation Q.920/I.440: ISDN User-Network Interface Data Link Layer—General Aspects.*

ITU-T (1993c) *Recommendation I.431: Primary Rate User-Network Interface—Layer 1 Specification.*

ITU-T (1995) *Recommendation I.430: Basic User-Network Interface—Layer 1 Specification.*

ITU-T (1996) *Recommendation X.25: Interface between Data Terminal Equipment (DTE) and Data Circuit-Terminating Equipment (DCE) for Terminals Operating in the Packet Mode and Connected to Public Data Networks by Dedicated Circuit.*

ITU-T (1997) *Recommendation Q.921/I.441: ISDN User-Network Interface—Data Link Layer Specification.*

ITU-T (1998) *Recommendation Q.931: ISDN User-Network Interface—Layer 3 Specification for Basic Call Control.*

Rosen, E., Viswanathan, A., and Callon, R. (2001) *RFC 3031*, IETF.

Rastislav, S. (2001) Technical Overview of IPSS and BBCS, *Proceedings Swisscom SwiNOG 3.*

Simpson, W. (1994) *The Point-to-Point Protocol (PPP)*, IETF, RFC1661 STD 51.

13 Multiprotocol Label Switching

Martin Reed
University of Essex

Introduction

Multiprotocol Label Switching (MPLS) is a methodology of controlling switched networks that has been designed to be compatible with IP networks. This is unlike many other switched-network solutions that have been developed separately from IP and that inter-operate with variable degrees of success. This chapter starts by introducing the MPLS architecture, which includes the components and the methodology of switching to create a Label Switch Path (LSP). The architecture is supported by signaling that sets up the LSP. The signaling is fairly complex because there are a number of LSP signaling protocols; therefore, only an overview of the main protocols is presented and the motivation for their existence is given. The final parts of this chapter describe the two main applications for MPLS: traffic engineering (including quality of service) and providing carrier-grade Virtual Private Networks (VPNs). These two applications are considered separately, although of course it is possible (and likely) that they will be used together. Before considering the MPLS architecture in detail, the basic principle and history behind its evolution are described.

When considering MPLS in detail it becomes clear that it is a complex architecture and set of protocols. The reason for this is partly that, rather than being created as a new protocol, MPLS has evolved from a set of competing technologies. Another reason for the complexity is that MPLS has been designed to be truly "multiprotocol" in that it can, in theory, support many higher network layer protocols over a variety of link layer protocols. In this chapter the only network layer protocol considered is IP. However, it is possible to encapsulate many higher-layer protocols. There are also MPLS implementations that allow lower-layer protocols to be encapsulated over MPLS, and this is often termed layer-2 encapsulation, pseudowire, or "Any Transport over MPLS" (AToM); see De Ghein (2006) for more information on the latter. Those familiar with the OSI layered network models might notice that this description does not fit the standard model because MPLS often sits in a nonexistent OSI layer between layers 2 and 3. By claiming that MPLS supports higher network layers over many link layers, some say that it is a layer-"2.5" protocol. However, it might be better to think of MPLS as breaking the OSI model rather than trying to make it fit.

The Principle of Label Switching

MPLS switches data using label swapping. This is not a new idea; Asynchronous Transfer Mode (ATM), as shown in Figure 13.1, is a technology that uses label swapping and that preceded MPLS. The basic principle in label swapping is that data is carried in packets with a header that is changed at each switch. The header contains a single identifier (the label) that is only relevant to each pair of neighboring switches. The label tells the switch how to forward the packet data and how to treat the packet in terms of Quality of Service (QoS). Note that this is quite different from IP, which uses an identifier (IP address) that has global significance and is not changed at each "switch" (IP router). While IP can support QoS, it should be noted that IP QoS mechanisms have had mixed reactions from operators and have a number of difficulties.

In the case of ATM, the "packet" is a fixed-length 53-byte cell with a 5-byte header. The header contains two fields: the Virtual Path Identifier (VPI) and the Virtual Circuit Identifier (VCI). In Figure 13.1 it can be seen that the VPI field is swapped at each switch but that the VCI field remains the same throughout the circuit. This is a feature of ATM that allows *trunks* of circuits to be transported through a core part of the network; outside of this core part, VPI switching might not be used, and the VCI identifiers can be swapped. In each switch a table is maintained that maps an incoming label to an outgoing port and new label value. This table provides a simple relationship between incoming and outgoing cells. This can be contrasted with the more complex routing table lookup required by IP routers, where the destination IP address of an incoming packet has to be matched against the longest prefix in the routing table, which can contain more than 150,000 entries for a core Internet router.

Having introduced the label-switching table (VPI/VCI table in ATM), we should note that there needs to be a mechanism to set the values in the table. We will see later that MPLS can use a variety of protocols to support this. ATM has its own set of protocols: a User Network Interface (UNI) between an end device and the first switch and a Private/Public

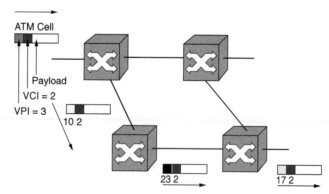

Fig. 13.1 Label swapping as used by ATM.

Network–Network Interface (PNNI) to coordinate path configuration through an ATM network.

The Evolution of MPLS

MPLS is the result of a merging of a number of competing technologies that were put in place to solve the problems of transporting IP over circuit-switched networks, which were widely used in the 1990s. At that time ATM was widely deployed, and some believed that it would be a candidate for an end-to-end communication technology. However, it quickly became clear that IP was to become the dominant end protocol because of its widespread deployment in operating systems. Additionally, Ethernet was substantially cheaper to deploy in the Local Area Network (LAN) than ATM. Consequently, ATM was relegated mostly to a carrier technology for providing *overlay* networks to support connectivity between IP networks. An overlay provides a virtually transparent pipe to a higher-layer protocol such that the two do not interact except for encapsulation of one over the other.

The premise at the time (the 1990s) was that switching IP packets was computationally expensive, whereas switching ATM cells was a relatively simple operation. Thus, it made sense to have a relatively simple ATM switch fabric at the core of a network, where datarates were high. The IP routers were placed nearer the edge, where IP integration was needed and where the forwarding speeds were not as high. An additional factor is that it is difficult for an operator to provide virtual private networks using a pure IP model (De Ghein, 2006). Consequently, it was advantageous for an operator to use ATM, with its comprehensive management support, to provide an overlay to customers that wanted a leased-line service to connect their remote IP networks.

Figure 13.2 shows how IP routers can be connected over an overlay network. In this model the IP layer is completely unaware of the circuit-switched layer, which means that

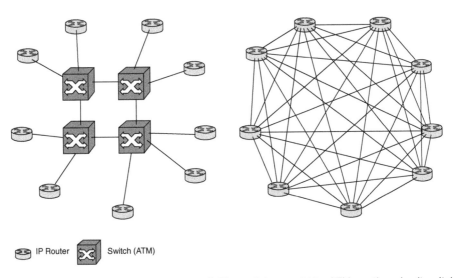

IP Router   Switch (ATM)

Fig. 13.2   IP connected over an overlay network (the switches could be ATM or other circuit-switching technology).

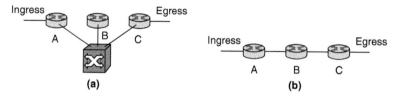

Fig. 13.3 Simplified diagram showing an ATM overlay supplying IP connectivity on two subnets, AB (a) and BC (b). Traffic between A and C has to pass through B even though it seems unnecessary given the physical connectivity.

for N IP routers the number of IP peers is $N(N - 1)$ or approximately N^2. This can cause some problems for IP routing protocols when there is a change in topology (Davie, 2000) and can create additional configuration effort, particularly for VPNs (De Ghein, 2006).

Although there are the aforementioned problems with overlay networks, in the 1990s it was very common to use this type of solution to interconnect IP networks, and it was also used in enterprise IP networks that utilized ATM backbones. However, it soon became clear that ATM and IP did not interoperate very well. ATM and IP have evolved separately and have very different control and management strategies, such that an ATM system cannot natively support IP forwarding. An example of the problem is shown in the highly simplified scenario of Figure 13.3. In this example three routers are connected to an ATM network, which provides the transport infrastructure. B might have one physical connection, but over this connection two separate virtual circuits provide connectivity to A and C, respectively. (In a realistic network, router B would probably have Ethernet ports to stub networks not shown in the diagram.)

In this simple scenario it is assumed that the ATM connections have been manually provisioned to provide the connectivity. Because the ATM and IP *control planes* are unaware of each other, it is not possible for A to know that it has a direct connection over the ATM network to C; instead, it sees B as its neighbor. The effect of this is that traffic passing from A to C has to pass through B, obviating the advantage of the fast ATM switching core. There have been many attempts to make ATM aware of the IP layer using architectures such as Classical IP over ATM (RFC 2225), LAN Emulation (LANE), and Multiprotocol over ATM (MPOA). The three stated techniques are in order of date of introduction and increasing complexity. In particular, it should be mentioned that MPOA helps to solve the problem shown in Figure 13.3. Using these systems requires an ATM management/control plane, an IP management/control plane, and an IP/ATM integration management system. Consequently, in practice it was found that these approaches were too complex and thus unpopular.

Following the difficulties with ATM/IP integration, a number of vendors started work on solutions that provided either improved ATM/IP integration or switching systems for IP that replaced the need for ATM. Examples of the systems proposed are a cell-switching router from Toshiba, an IP switching system from Ipsilon, aggregate route-based IP switching from IBM, and tag switching from Cisco. At the risk of oversimplification, one common theme among the solutions was that the ATM switch fabrics (and sometimes cell formats) were maintained while the fully featured (but IP-incompatible) ATM signaling system was replaced by a new system that integrated with IP. It soon became clear that rather than having competing solutions, a better approach would be to merge the work into a new standard and terminology. Thus, the term *MPLS* was decided on and the IETF started the MPLS working group in 1997 to coordinate development.

As MPLS evolved it moved far beyond a better approach to transporting IP over ATM. In fact, it is more common to use Ethernet as the underlying layer-2 protocol for newly deployed networks. At the embryonic start of MPLS, one of the key reasons for using switching to support IP connectivity was that it was cheaper to perform label switching than the more complex IP forwarding function that had to be implemented in software. However, with the use of improved hardware (such as ASICs and CAMs) the IP forwarding function can now be provided directly in hardware at a cost that almost matches the simpler switch fabric.

The reader might then ask why MPLS is still needed. The answer is that, because IP has become the dominant protocol, a core network that provides IP functionality is needed. However, the IP control plane has difficulties with providing carrier-grade VPNs suitable for replacing legacy leased-line services, and it cannot provide satisfactory traffic engineering. Both of these applications are well supported in MPLS. However, before these key applications of MPLS can be described, MPLS architecture and signaling need to be introduced.

The MPLS Architecture

The MPLS network architecture is defined by the IETF in RFC 3031. This chapter makes frequent use of the IETF standards called Requests for Comments (RFCs), and these can be obtained from *http://www.ietf.org*. Where possible, the RFCs referred to are chosen for their clear descriptions of the technology, and the reader is encouraged to read portions of them for more information.

A simplified view of an MPLS network is shown in Figure 13.4 and compared with an overlay approach. An MPLS network consists of switches; however, unlike the switches of

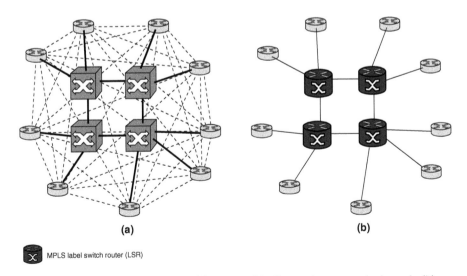

(a)   (b)

⬛X⬛ MPLS label switch router (LSR)

Fig. 13.4 MPLS architecture is shown in (a) compared to the overlay approach shown in (b) as presented in Figure 13.2.

an overlay network, they can be integrated with the IP layer and hence are termed Label Switch Routers (LSRs). Note that the common diagrammatic form of an LSR combines the symbols for a switch and router as shown in Figrue 13.4(a). The figure demonstrates that by making the LSRs aware of the IP layer, the quantity of peering at the IP layer is greatly reduced. This is an immediate benefit for IP routing protocols that have improved performance with a lower number of peers. However, a more important aspect is that MPLS clearly separates the control plane from the forwarding (switching) function. The fact that IP routers closely couple the control of IP paths and the forwarding function is one of the problems with IP, as will be shown later with VPNs and traffic engineering. Now, with MPLS, it is possible to make use of the IP control plane where it is appropriate but to allow the forwarding function to be controlled separately where needed.

Labels in MPLS

The main feature that allows MPLS to function is that the traffic to be transported (normally but not always IP) is encapsulated in a labeled packet. An MPLS label is some form of fixed entity that is associated with each packet and defines how the packet should be switched within an LSR. Furthermore, the labels only have local significance between two neighboring LSRs and are swapped as the packet is forwarded through the LSR. It would be nice to be able to draw a definitive label as used by MPLS. However, if we consider MPLS to be truly "multiprotocol," we see that the label can be a number of different formats depending on the layer-2 technology used for connectivity.

Layer-2 technologies, such as Frame Relay and ATM, already support label swapping. For example, with ATM the VCI and/or VPI fields can be used as labels. However, some layer-2 technologies, such as Ethernet, do not provide a label, and for these types of technologies MPLS defines a *shim* label, as shown in Figure 13.5. In practice, with Ethernet becoming a more common choice for connectivity, the shim label is most widely used. The shim label is between the link layer header and the transported packet. Thus for Ethernet the shim label and IP packet all appear as one data item. Note that the shim label does not contain just the label value; it also contains three other fields:

- *Experimental (Exp):* a 3-bit field originally reserved for experimental use but now used to support IP differentiated services
- *Stack bit:* a 1-bit field that marks the last in a collection of labels termed a *label stack*
- *Time-To-Live (TTL):* an 8-bit field used to provide compatibility with the IP TTL field

While the whole of the shim label is swapped at each LSR, the label field in the shim label is the value used to identify a packet as belonging to a unique Label-Switched Path (LSP).

ATM provides two levels of label (VPI and VCI) to create the notion of traffic trunks that can be switched as a single entity, as shown in Figure 13.1. MPLS can achieve the same

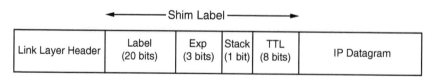

Link Layer Header	Label (20 bits)	Exp (3 bits)	Stack (1 bit)	TTL (8 bits)	IP Datagram

Fig. 13.5 MPLS shim label.

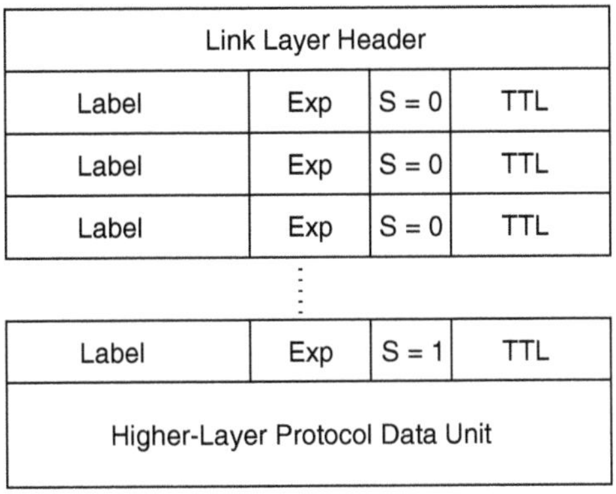

Link Layer Header			
Label	Exp	S = 0	TTL
Label	Exp	S = 0	TTL
Label	Exp	S = 0	TTL

⋮

| Label | Exp | S = 1 | TTL |
| Higher-Layer Protocol Data Unit | | | |

Fig. 13.6 Label stack, allowing multiple labels to be used.

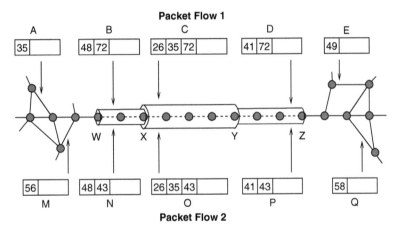

Fig. 13.7 Aggregation of two flows into nested LSP tunnels. Note that the label stack grows by 1 each time the flow enters a new LSP tunnel, and that only the innermost (lowest-level) label needs to be unique. The label values are arbitrary except that labels with the same value intentionally show how the higher-level labels can be the same to create aggregated tunnels.

(and more) through the notion of a label stack. A label stack can consist of multiple shim labels, as shown in Figure 13.6. The stack can be of arbitrary depth, which allows tunnels to be nested to any (practicable) depth required by a network operator. Note that the S-bit is used to identify the last packet in the stack; it is only 1 for the last (lowest) label. Figure 13.7 shows two packet flows in LSPs that pass through two further nested LSP tunnels.

In MPLS terminology this creates a hierarchy of LSP tunnels. As the flow enters a tunnel the label stack has an additional label *pushed* onto it. For example, packet A will progress to W using label swapping of a single label. As the packet leaves W it enters the first LSP tunnel and so here it has an additional label pushed onto the stack. This happens again at X. At Y and Z the packet leaves the tunnel and so the additional label is removed. The scenario in the diagram assumes that the two packet flows can be treated identically as they pass through the tunnels between W and Z, and hence they can be assigned the same topmost label. This can be seen for packets C and O, where the two topmost labels are the same. The use of more than one label in a label stack is used by both traffic engineering and MPLS VPNs, as discussed later. One of the key points in MPLS is that only the topmost label is used for forwarding and neither the lower-level labels nor the IP header are inspected until the end of an LSP tunnel or egress from the MPLS network, respectively.

Forwarding Equivalence Class

The notion of an LSP has been introduced and the fact that packets flow through LSPs. However, there needs to be a mechanism for allocating packets to a particular LSP at the LSP ingress. In MPLS, the term for classifying packets at the ingress is called a Forwarding Equivalence Class (FEC). A FEC is any means of uniquely grouping packets based on the contents of the packet. Examples of FECs include

- Packets with a destination IP address matching an IP prefix in the routing table
- Packets with a certain destination IP address and transport layer port (i.e., a single application flow)
- Packets with the same IP Differentiated Services (DiffServ) code point
- Packets that all leave a core network through the same egress

The first FEC in the preceding list is the most obvious when MPLS is simply used to support conventional IP routing and is often the default FEC if no other is explicitly defined. The last FEC in the list allows specific transit pipes to be constructed in a service provider's network and can be automatically configured by taking information from the BGP protocol.

Processing of Packets in a Label Switch Router

An LSR forwards labeled packets solely based on the information in the top label of the label stack. Part of the design of MPLS is that the LSR does not inspect data after the label. The effect of this is that a labeled packet is transported unchanged through an LSP. This also applies to the lower layers of the stack, which are processed only after the topmost label is removed.

An LSR performs one or more of the following actions on the top label of the label stack:

- *SWAP:* Accept a labeled packet and swap the label for a new label.
- *PUSH* (one of):
  - Accept an unlabeled packet and push a label onto the packet.
  - Accept a labeled packet, SWAP the top-level label, push another level of label onto the label stack.
- POP: Remove a label from top of the label stack.

In addition to these operations, the LSR has to be able to forward IP packets in the normal manner. One reason for this is that the LSR needs to process signaling messages before any LSP has been set up. Another reason is that if the LSR is acting as an egress router, it has to be able to act as a conventional IP router so that it can forward the packets to the correct destination using IP forwarding. It is likely that IP packet forwarding in the conventional

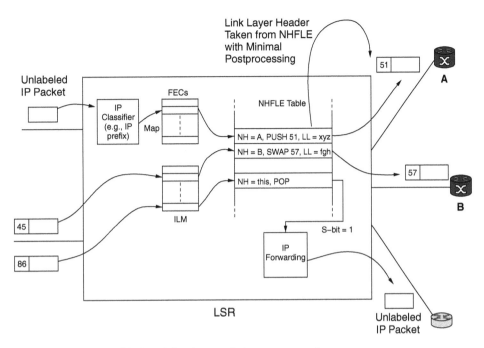

Fig. 13.8 Tables in an LSR that define how packets are processed.

manner will be by *process switching*, which involves passing the packet to a software process rather than directly switching the packet in hardware. Consequently, this type of forwarding is generally limited to network operation and maintenance traffic where possible.

When a labeled packet arrives at an LSR, the label is compared against an Incoming Label Map (ILM), as shown in Figure 13.8. The ILM matches each incoming label value to a Next Hop Label Forwarding Entry (NHLFE), which defines what must be done to the packet and then allows the packet to be passed out of the LSR. The NHLFE contains the following information:

- Next hop for the packet
- One operation from PUSH, SWAP, or POP
- Optionally, link layer encoding for output onto media toward the next hop

The label operations (PUSH, SWAP, POP) have been explained previously. However, it should be noted that in some cases the PUSH operation might put on a number of labels (multiple PUSHes) to create a label stack. This is useful in the ingress LSR for VPN applications. The next hop is needed so that the correct link layer address can be put on the packet. In practice the forwarding process in high-speed LSRs is carried out in an ASIC. In this case it is useful to have the optional link layer encoding (e.g., Ethernet header) available in an (almost) complete form so that it can be taken directly from the NHLFE.

In the case of Ethernet the only additional processing required is to fill in the length field and calculate the Cyclic Redundancy Check (CRC). It is possible that the next hop for the packet might be the same LSR. In some cases this will be because the POP operation has resulted in an unlabeled packet, and the LSR must then forward the packet using conventional IP forwarding. This can be determined from the S-bit of the shim label, which is 1 for the last label in the stack. Hence, a POP operation will result in an unlabeled packet. If the packet is labeled and the next hop is the same LSR, then the packet is fed back into the process described in this paragraph until the next hop is another LSR (or it is forwarded unlabeled).

If an unlabeled packet arrives at the LSR, it will be compared with an entry in the FEC-to-NHLFE Map. This requires the attributes of the packet (e.g., destination IP address) to be compared against the FECs configured in the router, and when a match is found the NHLFE for that FEC is used to encode the packet for forwarding (as described in the previous paragraph). There might be multiple FECs that match the packet; for example, there might be a number of IP address prefixes that the destination IP address matches. In this case there are rules to typically find the lowest-granularity FEC; in the example of the destination IP address prefix, it is typically the longest prefix match that is used, as with conventional IP forwarding rules.

Time-to-Live in MPLS Networks

The TTL field in IP networks is widely used to mitigate IP routing loops and for network trouble shooting. In IP networks the TTL field is typically set to a value in the range of 30 to 255 at the source and decremented by 1 at each router. If there is a routing loop, the TTL field will eventually reach 0, and the packet will be discarded by the router. Routing loops occur either through transient effects of dynamic routing protocols or because of configuration error. Without the TTL-based loop mitigation, a routing loop would cause excessive congestion in the links of the loop.

When a router discards a packet with an expired TTL value, it sends an Internet Control Message Protocol (ICMP) packet to the source with a "time exceeded" message. This latter message is used to good effect by utilities such as *traceroute*, which sends probe UDP packets into the network with TTL values starting at 1 and getting successively bigger. By listening to the ICMP "time exceeded" replies, the traceroute utility can determine the IP address of each hop along a path. This is highly useful for debugging connectivity problems.

Because TTL is so important for both loop mitigation and debugging, it was decided to support its implementation in MPLS. However, because an LSR cannot change the contents of a labeled packet, the TTL field in a labeled IP packet will not decrement as it passes through the LSR. Consequently, the TTL field in the shim label is used and decremented as it passes through each IP router. To make the MPLS TTL field compatible with the IP TTL field, the TTL field from an unlabeled IP packet is copied, decremented by 1, and put into the TTL field of the shim label as the packet enters the LSP. At the end of the tunnel the TTL field from the shim label is copied back into the IP packet.

MPLS designers chose to specify that, when a TTL field expires in a LSR, the packet is dropped and an ICMP packet with a "time exceeded" message is sent. However, as we will see later when discussing VPNs, it is possible that an intermediary LSR will not use the same address space at the two ends of the tunnel. Thus, MPLS defines that the ICMP message is sent to the end of the tunnel and that it is the router at the end that is responsible for sending the ICMP packet back to the source.

Other Issues with Forwarding

The previous section presented an overview of MPLS forwarding architecture, but in practice there are a number of details that require more depth of discussion than possible here. A brief discussion of some of these details with further reading is provided.

The forwarding summary described in this section matches the description in RFC 3031. In practice the breaking up of tasks into various tables (or maps) is somewhat arbitrary, and router vendors often have their own internal architectures that perform tasks by a method that matches their implementation. For example, Cisco has a mechanism termed *Cisco Express Forwarding* (CEF) that is used to perform IP forwarding in an ASIC (De Ghein, 2006). It turns out that CEF is also ideally suited to perform the FEC-to-NHLFE mapping, and hence this latter term is replaced by CEF in most Cisco documentation.

This discussion so far mainly concerns LSRs that use Ethernet for the link layer. It is possible to use the native ATM cell format as the link layer but to use MPLS to control the ATM switch table. It should be noted that some features of ATM require some changes to the way MPLS is implemented compared to Ethernet. For example, ATM only supports a native "label stack" of depth 2 through its VCI and VPI fields so if greater depth is required a shim label has to be carried with the data (the IP datagram). The main MPLS architecture document (RFC 3031) specifies how most of the adaptation to use ATM can be performed; however, RFC 3035 provides more detail.

When ATM was introduced it was mainly designed to create circuits between two endpoints. However, there soon developed an interest in providing one-to-many and many-to-one circuits that in ATM are termed *point-to-multipoint* and *multipoint-to-point*, respectively. MPLS was designed with IP in mind, which is not circuit oriented and effectively allows flows from multiple sources to be treated similarly because they are destined for a common IP prefix. In MPLS this can be achieved through label merging. This is where two incoming labels map to one NHLFE (a single outgoing label). However, this can cause problems with link layers, such as ATM (see RFC 3031 for more detail).

The forwarding architecture presented here shows how unlabeled traffic can enter the MPLS network and be forwarded through it until it leaves unlabeled. To achieve this, the FEC-to-NHLFE map and the ILM-to-NHLFE map have to be populated with coordinated entries. In MPLS it was decided that the main (standardized) mechanism to carry out this task would be signaling. This is the subject of the next section.

Signaling

An MPLS implementation needs a signaling protocol to set up the label switching in the LSRs and thus configure an LSP. There was an early decision in MPLS not to specify one protocol to do this but to allow any number of reasonable protocols to flourish as required. In practice the following four main protocols were standardized:

- *Label Distribution Protocol (LDP)*, specified in RFC 5036
- *BGP-4 extended to carry labels*, specified in RFC 3107
- *RSVP with Traffic Engineering Extensions* (RSVP-TE), specified in RFC 3209
- *Constraint-Routed Ldp* (CR-LDP), specified in RFC 3212

In this section only the first will be considered in detail. The last two are concerned with traffic engineering, which is described in the next section. It should be noted that although the MPLS working group set out to allow a number of protocols to distribute labels, by February 2003 they effectively withdrew CR-LDP (which was a competitor to RSVP-TE), so we are in reality left with three main protocols that have different purposes.

Label Distribution Protocol

LDP is a protocol used to establish label mappings for MPLS networks that do not use traffic engineering. The main FEC used for label mappings is the longest prefix match in the IP routing table of each LSR. An LSR can act in a number of different modes with respect to how labels are determined using LDP (see RFC 3031 and RFC 5036 for more detail). Here the implementation with the most common mode settings is described. LDP defines four types of protocol messages, all of which are sent between immediate peers and use TCP/IP unless stated otherwise:

* Discovery messages are used to announce the presence of an LSR and use UDP/IP.
* Session messages are used to establish communication channels between LSRs.
* Notification messages allow alarms to be sent in case of errors.
* Advertisement messages create, delete, and modify label mappings for a particular FEC.

Only advertisement messages will be considered here.

Advertisement messages allow a variety of modes to obtain (or request) labels. Here we restrict the description to the case where labels are advertised by an LSR to all other connected LSRs as soon as a new prefix is added to the IP routing table of the LSR. For an example, see Figure 13.9 where a new prefix, 10.11.15.0/24, has been advertised by an Interior Gateway Protocol (IGP). As soon as an LSR has this new prefix in its routing table, it sends a *label-mapping* message to its upstream neighbor.

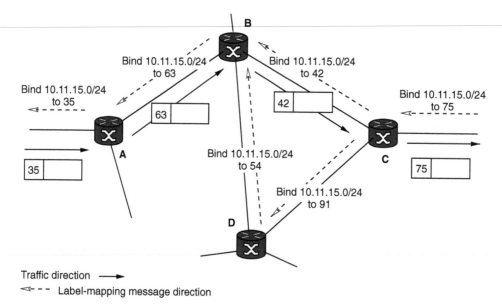

Fig. 13.9 LDP mapping messages for a new FEC (IP address prefix 10.11.15.0/24).

The label-mapping message contains a label value and an encoded form of the FEC that this label applies to. For example, C sends to B the label mapping to *bind* the FEC defined by the IP address prefix 10.11.15.0/24 to the label 42. Then, if B receives unlabeled packets belonging to this FEC, it should label them with 42 before sending them to C. This label will be placed, in the appropriate NHLFE in B. Of course, B will also send a label-mapping message to A; in the example B binds label 63 to this prefix and will put 63 into its ILM and map this to the NHLFE of 42. This process continues until a complete LSP is created. In the mode of operation described here, LSRs will independently send label-mapping messages as soon as the prefix appears in the IP routing table and send the message to all immediate neighbors.

An LSR that receives the label mapping will record this in its NHLFE even if it does not know about the prefix yet (as it might at some future time). In Figure 13.9 the mapping messages are shown only in one direction to emphasize that the mappings are generated by the downstream LSR. However, there is no notion of *split horizon* in LDP, so B also sends a label-mapping message for this FEC to C, even though C does not have immediate use for it, and C keeps this mapping. This is useful if there is a change in topology because label bindings will preexist.

When LDP operates as described here it is said to work as *downstream unsolicited*, with *liberal label retention* and using *independent LSP control mode*. Each of these modes has an alternative, so, for example, rather than mapping messages being sent unsolicited, the messages can be sent only when an upstream router requests a binding *on demand*; rather than liberally retaining a label every time a neighbor sends a binding, the router can *conservatively* use labels only for FECs that it knows about; finally, rather than every LSR independently sending mappings, LSRs can send them only in an *ordered* manner from egress to ingress. It will be up to the network operator to determine which of these modes is most appropriate. Here the most common configuration is described and the reader should refer to RFC 3031, RFC 5036, RFC 5037, or De Ghein (2006) for a more detailed description and arguments about which modes should be used. Later, when RSVP-TE is considered, it will be seen that it naturally uses *downstream on-demand* label distribution mode with *ordered LSP control*.

Traffic Engineering

The MPLS architecture and its native method for distributing labels (LDP) have been introduced. Before considering traffic engineering, it is worthwhile to think about what the technology explained so far achieves. The architecture defines how unlabeled traffic can be labeled and how labeled traffic can be processed in each router. LDP defines how address prefixes obtained from IP routing tables can be used to set up label bindings in each LSR to create LSPs that follow the same paths that would be followed by conventional IP forwarding.

In other words, all of the complex MPLS architecture and signaling described so far achieve nothing except for the fact that packets are switched using label swapping as opposed to IP forwarding using the longest prefix match. In the 1990s this would have been a significant advantage, but modern IP routers can achieve full line rate IP forwarding. Consequently, there has to be an additional need for MPLS for it to be deployed, and the first of these needs described here is traffic engineering. The architecture for traffic engineering using MPLS is described in RFC 2702.

Traffic Engineering (TE) can be broken up into two tasks:

- Traffic-oriented TE forwards packets such that the QoS required by the end application can be controlled.
- Resource-oriented TE manages the resources in the network by sending traffic over appropriate paths such that the load is evenly spread through the network.

Conventional IP has not been very successful at performing either of these tasks because of its inherent connectionless nature and the fact that the forwarding and control functions are closely bound together.

There have been attempts at providing IP QoS mechanisms for traffic-oriented TE in the form of IP Integrated Services (IntServ) and IP Differentiated Services (DiffServ). While these (in particular DiffServ) have had some limited success, they have not achieved widespread adoption in the manner that was originally envisaged and have many problems with deployment. MPLS does not provide an alternative QoS model; instead, it can be said to support IntServ and DiffServ so that they can be deployed in a more straightforward manner.

Integrated Services

IntServ is an IP QoS architecture described in RFC 1633 whereby application flows of real-time services (e.g., video and voice) can have state allocated in routers such that the committed datarate and/or the delay through the routers is controlled within certain bounds. IntServ is a fine-grained approach to IP QoS because each individual flow has resources allocated on every router along the path from sender to receiver. To reserve the resources allocated in each router requires a signaling protocol, and for IntServ this is the Resource Reservation Protocol (RSVP), as defined in RFC 2205.

RSVP operates as shown in Figure 13.10, which shows that there are two signaling messages, termed PATH and RESV. Before a reservation of resources is performed, the traffic source sends out a PATH message, which defines the parameters of the traffic flow in terms of its mean datarate and peak datarate (specified using a token bucket). This PATH message passes from source to sink, and each router along the path notes it but does not reserve router resources at this stage. When the traffic sink receives the PATH message it knows the parameters of the traffic and can then determine what resources should be reserved. The reservation in terms of delay and traffic rate requirements is specified in a RESV message, which originates at the traffic sink and passes back toward the traffic source.

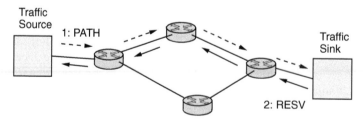

Fig. 13.10   RSVP signaling protocol for IntServ.

The RESV message is processed by each router along the path, and the routers reserve buffer resources to meet the demands of the service (assuming that the traffic sink is authorized to do so). Both the PATH and RESV messages use datagram transport (rather than TCP) and so cannot be transmitted reliably. Consequently, RSVP uses a *soft-state* mechanism whereby the PATH and RESV messages are repeated periodically (30 seconds by default). If a set number of messages are missed (typically three), then it is assumed either that there is a problem or that the reservation is no longer needed; in either case the routers will remove the resource allocations.

IntServ with RSVP was seen as a highly positive contribution to IP QoS in the 1990s when it was introduced. However, there are difficulties with practical deployment. The main difficulty is that IntServ is not scalable for use in core networks and thus, because resources cannot be allocated from end to end, it has limited practical use. The scalability problem can be seen if it is imagined that IntServ was used to control the quality of voice over IP (VoIP). If a 10-Gbit/s backbone is used to carry predominantly VoIP for a national voice carrier network, it is likely that there could be on the order of 100,000 voice calls passing through a router at any one time. If each of these has to have state allocated individually for each VoIP call (as is required by IntServ), then the routers are likely to be severely overloaded with the work needed to classify each packet and check that it has the necessary resources allocated on each output buffer to handle the packet flow.

RSVP-TE

Although IntServ itself is not a practical solution for IP QoS in a core network, the RSVP protocol was identified as suitable for distributing labels in an MPLS network that requires traffic engineering. To support MPLS, RSVP was extended and the scope of operation was changed to form RSVP-TE, as defined in RFC 3209. The main extension incorporated in RSVP-TE is that a label request object was added to the PATH message and a label-mapping message was added to the RESV message. Because LSRs using RSVP-TE request labels to be sent, and the label bindings are sent from downstream node to upstream node, RSVP-TE is said to use *downstream on-demand* label distribution mode with *ordered LSP control*.

The other significant change in RSVP-TE, compared to RSVP for IntServ, is that the scope of the messages is no longer from application sender to application receiver; instead, it is from LSP ingress to LSP egress, as shown in Figure 13.11, where the LSP is set up by RSVP-TE. By specifying that the RSVP-TE signaling only apply to the LSP and not to every

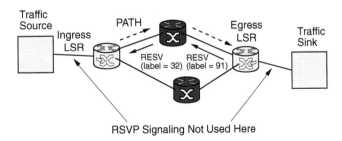

Fig. 13.11 PATH and RESV messages in RSVP-TE.

application flow, the problem of scalability found with RSVP used in IntServ is removed. This can be demonstrated by repeating the example of 100,000 VoIP calls through a path in the network discussed earlier with IntServ. With RSVP-TE, instead of allocating resources for each voice call, a single LSP can be set up that is dimensioned to carry the required number of voice calls. This LSP is often termed a *traffic trunk*. By using RSVP-TE, the LSP can have QoS parameters that are an aggregate for the traffic trunk, and thus LSRs along the path only need to maintain state for one resource reservation, which is directly inferred from the packet label rather than from a complex lookup on the packet header, which was the case with IntServ.

It should be pointed out that deploying RSVP-TE to set up traffic-engineered trunks is not the same as deploying IntServ; it is just that the RSVP protocol has been used for a different application. Of course, using RSVP-TE does not preclude the use of IntServ (using standard RSVP) in the networks attached to the MPLS core (or in the core itself); in fact, it helps to support it. For example, there might be many RSVP sessions in a customer's edge network that need to pass through an operator's MPLS core to get to another of the customer's networks.

Using IntServ, the customer needs to negotiate every RSVP session with the operator's network. This is clearly problematic and makes it hard for the operator to audit requests for resource allocation; in fact, it makes it so hard that this is never used in practice. However, if the customer requests just one traffic trunk from the operator, then this is a scalable solution and it is then up to the customer to decide how many RSVP sessions it wishes to pass through this traffic trunk. This effectively forms an overlay for the RSVP sessions. However, we will see when we consider MPLS VPNs that it is possible to still maintain peering at the IP layer.

Differentiated Services

DiffServ was proposed in RFC 2475 as a scalable alternative to IntServ. Rather than using a fine-grained signaling of resources, as used in IntServ, DiffServ marks packets with a broad class of service. As IP packet definitions did not originally contain a DiffServ field, DiffServ architects chose to redefine the rarely used IP type of service field and use 6 bits for the DiffServ Code Point (DSCP), as specified in RFC 2474. Each DSCP is mapped to a unique class of service that routers should observe through a predefined Per-Hop Behavior (PHB) for each class. The 6-bit DCSP allows for $2^6 = 64$ possible classes. However, in practice there are three commonly defined classes:

- *Default PHB*, which is essentially conventional, best-effort IP forwarding
- *Expedited forwarding* for low-latency and low-loss applications such as VoIP
- *Assured forwarding*, which uses 12 different DSCPs (four subclasses, each with three different priorities) to give a more flexible PHB (see RFC 2597 for a full explanation)

Because DiffServ uses broader classes of service, it is a scalable solution compared to IntServ. In practice, when traffic enters a DiffServ network it is policed (to check that it does not exceed the specified traffic parameters) and is then marked with a DSCP that needs to be supported by routers in the network and by the predefined PHBs. This creates certain problems with the dimensioning of a network to support DiffServ because there is not a reserved amount of resources along a particular path. Instead, it is common to overprovision a DiffServ network such that various possible combinations of edge traffic load will not overburden the core links. In practice it is not always possible to accurately predict which

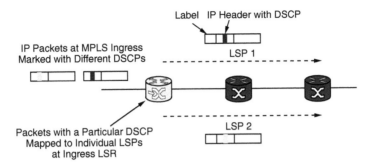

Fig. 13.12 MPLS mapping different DiffServ classes into unique LSPs.

core links will be overburdened because the nature of IP traffic is highly variable, and Diff-Serv does not provide a mechanism to signal traffic demands to the core nodes.

MPLS provides support in two ways: First, it provides a manner of mapping DSCPs to either LSPs or labels; second, it provides a traffic trunk that simplifies the dimensioning problem. MPLS provides two ways of mapping DSCPs into its architecture; the first is shown in Figure 13.12 where two different packet flows that need different PHBs (and thus have different DCSPs) are mapped to two unique LSPs. The LSRs assign the PHBs to the LSP rather than inspecting the DSCP (which is hidden in the IP packet header). A second approach used by MPLS is to assign different DiffServ classes to the same LSP but make use of the three Exp (experimental) bits in the shim label to allocate up to eight PHBs.

While it can be seen that MPLS can support the mapping of DiffServ classes into labels, this in itself does not pose a great advantage. A more significant advantage of MPLS is that traffic-engineered trunks can be used between edge nodes in the MPLS network. In this way it is possible to dimension the network with tighter tolerances than is possible with a conventional IP DiffServ solution. Typically the trunks will be set up using RSVP-TE between customer endpoints and will have QoS parameters that match the customer's Service Level Agreement (SLA). Because the RSVP-TE LSPs allocate resources (typically datarate) it is possible for routers and network management systems to keep track of committed resources in the network. Then, as new traffic trunks are requested, it is possible to plan the network dimensioning appropriately through assigning the trunks to appropriate paths. This requires resource-oriented traffic engineering that IP cannot easily provide but MPLS can.

Resource-Oriented Traffic Engineering

Resource-oriented TE is generally concerned with configuration of paths through the network such that network utilization is maximized. To achieve this usually requires paths to be placed according to resource awareness, not just topology, and the paths must be connection oriented. This is difficult to apply to conventional IP, which uses a shortest-path route and a hop-by-hop approach. As a simple example to demonstrate the problem, consider the scenario shown in Figure 13.13. Here there are two different traffic requests made on the ingress to the network at A and B toward router F. Conventional IP routing will select the shortest path through the network such that the route from C to F will be C, E, F. Clearly

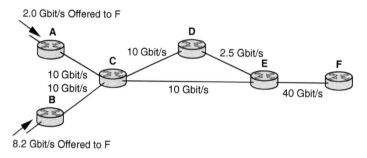

Fig. 13.13   Traffic-engineering problem with conventional IP routing.

the 10.2 Gbit/s of traffic offered will not be accommodated by the 10-Gbit/s link between C and E.

There are some solutions in IP that can help with traffic loading; however, they will have difficulty even with this simple scenario. For example, equal-cost multipath offered by OSPF could split the load evenly between the two paths C, D, E and C, E. However, this would clearly not be a satisfactory solution because, although the 5.2 Gbit/s being offered to C, E is accommodated, the 5.2 Gbit/s offered to C, D, E would overload the D, E link. The problem becomes harder to solve with larger networks. MPLS can solve this type of problem, and this is one of the key applications of MPLS in carrier networks that wish to replace existing circuit-switched systems such as SONET.

MPLS can solve the example problem by the use of *explicit routes*. All of the discussion so far has assumed that LSPs will follow the path that matches the hop-by-hop IP routing tables. However, RSVP-TE supports the inclusion of an Explicit Route Object (ERO). The ERO is a list of next hops that an LSP should take and is sent in the PATH message of RSVP-TE so that the LSP is configured over the desired path. This operation is fairly straightforward: A PATH message is transported through the network; rather than it being forwarded using the IP routing tables, it is transmitted to the next entry in the ERO until the LSP egress is reached. Each router records state (such as the ERO) from the PATH message so that, as the RESV message is sent back from the LSP egress, each LSR can refer to the original ERO to work out the next upstream hop.

Using the mechanism provided by RSVP-TE, it is possible to set up LSPs to meet traffic-engineering requirements that might not match the conventional IP forwarding path. Note that conventional IP forwarding and a dynamic routing protocol are still needed so that the MPLS signaling protocols, and other control and management messages, can be transported using IP.

Consider a solution to the problem set in Figure 13.13 using MPLS with RSVP-TE and using explicit routes. The traffic offered to A can be sent over A, C, D, E, F, while the traffic offered to B can be sent over B, C, E, F. Simple inspection shows that the traffic offered matches network capability. This would not be possible using conventional, destination address–based IP forwarding. The MPLS standards do not state how these explicit routes are planned. While there are schemes that allow a distributed approach to automatic traffic engineering, most network operators prefer to take a centralized, planning-based approach for the configuration of traffic trunks as supported by MPLS. Consequently, it is likely that an MPLS network using traffic engineering to control QoS will have a centralized path allocation system that links with billing and service-level systems. The path allocation system will use SNMP to initiate the RSVP-TE signaling that provisions LSPs, as described in RFC 3812. A discussion of path allocation systems for QoS in MPLS is given by Marzo (2003).

It was earlier stated that MPLS provides the notion of hierarchical LSP tunnels. The main mechanism for provisioning these tunnels is RSVP-TE, and they are often simply termed "TE tunnels." These tunnels can be configured only edge to edge. Alternatively, they can be placed in certain parts of the network to aggregate traffic into trunks that can be used to route traffic around points of congestion. In this latter case, hierarchical LSP tunnels can be created as shown simplistically in Figure 13.7; this figure will be used to explain the signaling requirements for these tunnels.

In the case where an LSP tunnel is used, it should be noted that the LSRs at the ingress and egress of the tunnel (W, X, Y, and Z) are members of both levels of the hierarchy. This means that for the purposes of signaling they each have two neighbors along the direction of the LSP. Consider W, which is sending packet B and has two labels: 48 as the topmost label and 72 as the next (and lowest) on the stack. The topmost label (48) is relevant to the next hop in the tunnel. However, the next label (72) is carried through the tunnel until it reaches Z, where it is used. The consequence of this is that W has to participate in signaling with both its immediate physical neighbor and Z. There is no mechanism specified in MPLS to automatically discover peers at the end of tunnels, so they have to be specified manually (or by some other automatically driven system). In practice an LSP tunnel will have both ends identified on setup through the management function, so this same function can also configure each end LSR with the peering information.

Fast Reroute

MPLS traffic-engineering extensions are aimed mainly at a core network provider whose goal is to replace a traditional circuit switched network with an IP/MPLS core. It should be noted that one of the reasons that a pure IP core is not suitable for the replacement of a circuit-switched network is that its recovery time from failure is counted in seconds, whereas circuit-switched networks, such as SONET, can achieve protection switching on the order of tens of milliseconds (50 ms is the often quoted figure). MPLS can provide protection switching on the order of tens of milliseconds using a number of different techniques, as described by RFC 4090 and Marzo (2003). Here only one *local repair technique* is discussed, and this is termed *facility backup* by RFC 4090.

Facility backup is a local repair technique that provides a backup path to protect a link. The backup path can provide protection for any number of LSPs that pass over the link to be protected; this is attractive to an operator because it does not require individual configuration for the many LSPs that might pass over the link. Figure 13.14 shows an example of MPLS facility backup. It assumes that the paths are configured using RSVP-TE and that the backup path is preconfigured so that there is no delay due to path computation or signaling. Figure 13.14(a) shows one LSP in normal operation over the path A, B, C, D. The local repair path B, E, F, C is used to protect link BC and is set up using RSVP-TE before the failure occurs. Before the point of failure no traffic is passed into the backup LSP (although the links on the backup path can carry other LSPs).

The facility backup method makes elegant use of the label stack to provide the backup path such that switching is only required at the *head end* of the protected link, which is LSR B. Note that in normal operation B puts label 23 onto traffic destined for C on the LSP. When a failure occurs on link BC, B detects this quickly from the physical layer and initiates the

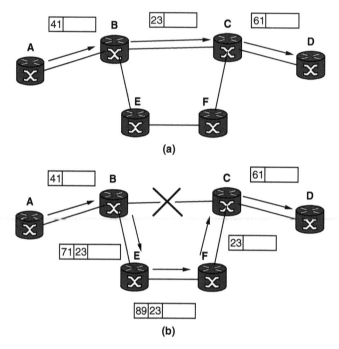

Fig. 13.14   MPLS fast reroute: (a) the LSP in normal operation; (b) the backup LSP in use following a failure on link BC. Note that the backup LSP makes use of a label stack.

head-end protection switching. All this requires is the labeling of packets with 23 (as before) and the tunneling of the packets through the backup path B, E, F, C, as shown in Figure 13.14(b). To pass the packet into the backup path tunnel only requires B to push label 71 onto the packet and output to E. The backup LSP is configured such that when the packet reaches C the top-level label has been popped, revealing label 23, as was used before the protection event.

There is one point shown in Figure 13.14(b) that applies generally to LSP tunnels but has not been described yet; this is the ability of LSRs to perform Penultimate Hop Popping (PHP). This can occur on the LSR immediately before the end of the tunnel. In Figure 13.14, F is the penultimate hop before the end of the backup tunnel at C. Following the earlier discussion of LSP tunnels, it would be expected that as the packet enters C (on the backup LSP) it has two labels on it. The first will be examined and the result will be that a pop is needed and that the destination is C. Then the packet will be examined a second time at C, and the topmost label will be swapped to 61. However, in practice it can be known that C will perform this pop operation and two lookups for every packet entering the backup LSP. Thus, it is possible for the pop operation to be performed at F and the lowest-level label is then exposed directly to C for forwarding onto the original tail end of the LSP. This reduces the workload on the egress LSR and is generally used at an LSP tunnel egress.

Summary of Traffic Engineering

The TE features of MPLS are one of the strongest reasons for deploying it, because it offers facilities that are difficult or impossible to replicate using conventional IP. In particular, MPLS can provide traffic trunks that can replace the overlay systems used with circuit-switched networks. This has advantages because the MPLS architecture is designed to be compatible with IP, whereas an overlay gives rise to the problems discussed earlier.

It should be noted that MPLS TE systems can (and probably will) coexist with LDP within the network. Additionally, while there will be a path configuration system for MPLS TE tunnels, this needs to coexist with a suitable Interior Gateway Protocol (IGP) that will configure the conventional IP routing tables such that IP connectivity is provided for signaling and Operation, Administration, and Maintenance (OAM) requirements.

MPLS Virtual Private Networks

The second major application for MPLS described in this chapter is to provide a layer-3 VPN. This type of VPN provides a replacement for a leased-line service but with some controlled layer-3 interaction between provider and customer networks. This service is the most popular reason that operators have deployed MPLS—because they need a replacement for the ATM and Frame Relay services they were offering to customers. One motivation for MPLS compared to ATM and Frame Relay is that MPLS has been designed to support IP. Some of the arguments supporting this were presented at the start of the chapter; however, there are others that are described in RFC 2547 or by De Ghein (2006) and Davie (2000).

A layer-3 VPN allows a customer's IP network to *peer* with a provider's network at the IP layer. This is difficult to achieve using conventional IP because it typically requires complex configuration of many Generic Routing Encapsulation (GRE) tunnels and very careful control over routes advertised between the provider, the customer, and other customer networks. This section shows how MPLS can be a scalable method of providing VPNs.

MPLS VPN Architecture

The MPLS VPN architecture is shown in Figure 13.15 for one provider and two customers spread over five sites. The goal of the VPN is to provide each customer with connectivity between its sites in such a manner that traffic between that customer's sites cannot (inadvertently) *leak* into another customer's network. In other words, it provides a private network but over a common infrastructure. However, with the layer-3 VPN we note that it is possible to allow some (or as much as needed) IP connectivity between customers, the provider, and possibly the Internet. The scenario in Figure 13.15 assumes that each customer site (A, B, C, D, and E) is running a conventional IP network and that the provider is running an IP/MPLS network (there are other options not discussed here). The routers (or LSRs) have terms applied to them depending on their purpose and locations:

- *Customer Edge (CE):* This is usually owned and controlled by the customer and is the customer site gateway to the VPN. It will be using the IP address space of the customer on all of its interfaces, which could be globally addressable or a private network address space. Each CE is connected to one or more provider edge LSR.
- *Provider Edge (PE):* This is a provider-controlled LSR that supplies IP interfaces to the customer. Each of these interfaces is connected to at most one CE, and each will be using the IP address space of the customer. The PE also has internal interfaces to the provider's core network that will use the IP address space of the provider's network (which is likely to be a private address space).
- *Provider (P):* The P LSRs provide the backbone connectivity of the provider's network and only use the provider's IP address space.

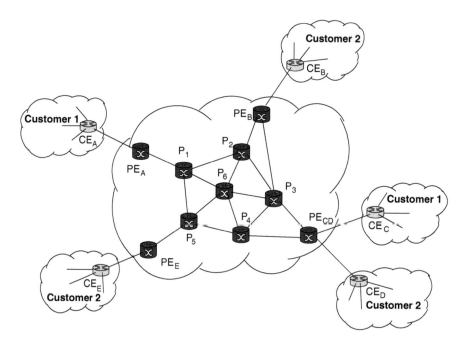

Fig. 13.15   MPLS VPN architecture.

As can be seen from the preceding description, there can be a large number of address spaces to manage in the collection of VPNs because it is likely that each customer is using a private address space range that might overlap with another customer's address space. This is not a problem for the CE and P routers/LSRs because they only use one address space on all interfaces. However, the PE routers have to manage multiple address spaces. Consequently, the PE Label Switch Routers have a routing table for each connected customer and one for the provider's IP space. The main task for the MPLS VPN architecture is to support transport between the PE LSRs without having to run multiple routing tables in them.

For the purpose of routing protocols, the peering is organized such that each CE router only peers with the PE LSR it is immediately adjacent to. Specifically, it does not need to peer with another CE at another of the customer's sites or with the core network. This significantly simplifies the configuration and operating complexity of the CE router.

IP Route Dissemination

The key to the operation of the MPLS VPN is that each PE maintains a *Per-Site Forwarding Table* (PSFT) for each customer site that is directly connected to it. While this is an additional burden, it is limited to the number of customers connected to each PE, which is likely to be small (compared to the total number of sites in the whole network). However, a P LSR has to carry the traffic for potentially all VPNs. If each P LSR had to cope with the address space of each customer, it would require a PSFT for each VPN, which is clearly not scalable. Consequently, the core makes use of MPLS tunnels between each PE such that the customer address space is never seen in the P LSRs because the packets are encapsulated inside an LSP. This means that each P LSR only has to run some form of IGP, such as OSPF, without needing to import any other routes. If the provider network (P and internal PE interfaces) uses an IGP and implements LDP as described earlier, then all the necessary LSPs between the PE routers will be provisioned automatically. However, we will see that this is not the only labeling of traffic needed in the network.

The key protocol that enables the MPLS VPN is the Border Gateway Protocol (BGP). BGP is implemented on the PE LSRs with the external interface (EBGP) toward a CE; the other PEs are Internal BGP (IBGP) interfaces. For this application BGP version 4 is used but with multiprotocol extensions as specified in RFC 2283. The multiprotocol extensions specified for BGP have nothing to do with the term "multiprotocol" as used in MPLS. Instead, the multiprotocol extensions to BGP allow multiple address spaces to be used, and each address space can be given an identifier. This 64-bit identifier is often written ASN:x, where ASN is the Autonomous System Number of the provider (a globally assigned number by IANA) and x is a number assigned by the provider to uniquely identify each customer. This identifier is termed a Routing Designator (RD), and one is configured for each customer, requiring that a corresponding PSFT be enabled for each customer connected to a PE.

As with standard BGP (RFC 17171), each PE has to be configured to peer with every other PE as an IBGP interface. Configuration of which external sites should be "connected" together is implemented by BGP route targets specified in RFC 4364; each route target is a BGP extended community. The route targets are configured for each RD on each PE. BGP will only import or export routes that share a common routing target. As an example of a configuration, consider how PE_A would be set up to allow connectivity between site A and site B, which are Customer-1 sites. First, PE_A knows to *speak* to PE_B through manual BGP configuration. Next, both PE_A and PE_B need to be assigned a common RD for the address space of customer 1; this could be called 1:1. To make PE_A and PE_B share the routes for RD 1:1, they are each configured with the same route target for the RD. If this configuration is followed through for each PE and each customer RD, there will be one VPN for each customer and the VPNs will be disjoint from each other.

However, MPLS VPN offers an additional feature whereby it is possible for routes to be imported and exported between sites of different customers or even the Internet. This is achieved by configuring additional route targets such that where routes are to be shared, the route targets are the same. If customer 1 and customer 2 wish to share connectivity, but only between sites B and C, then a new route target identifier is created and installed for the correct RDs in PE_B and PE_C. If the new route target identifier is 1:300, then in PE_B RD 1:2 (for the customer-2 address space) would have the 1:300 routing target added for importing and exporting routes. In PE_{CD} RD 1:1 (for customer 1) would have the target identifier 1:300 added for importing and exporting the site-B routes into site C. Note that PE_{CD} has two customer sites attached and will have two PSFTs identified by RD 1:1 and RD 1:2. It would be up to the customers to agree between themselves not to have overlapping addresses at their sites, but when this is done they can establish connectivity over the provider VPN infrastructure.

The preceding discussion has shown how the PE LSRs run BGP and the P LSRs run an IGP such as OSPF. There is one further requirement for full routing dissemination: The CE routers and PE routers need to peer for routing information. Note that the CE only peers with its immediate PE neighbor and can do so using an IGP such as OSPF (although other options are possible). It is worth exploring how the routes are disseminated for a path between CE_E and CE_D and where site E is only connected to the VPN for customer 2. First CE_E will learn about routes in site E through the IGP running in site E; these will be exported (and possibly summarized) into the IGP running between CE_E and PE_E such that PE_E can learn them and add them to a PSFT for customer 2, which has RD 1:2. These routes are redistributed into the EBGP interface at PE_E, and BGP will associate these routes with an RD (1:2) and one or more route targets.

BGP passes these routes, with the route target(s), to the other BGP PE LSRs over the IBGP interface. Because site E is only to be "connected" to other customer-2 sites, we will assume a single route target of 1:2. Only the PE LSRs with the same route target specification accept the routes, so for example PE_A ignores these routes because it only uses a route target of 1:1. However, the PSFT for RD 1:2 on PE_{CD} has route target 1:2 specified so it imports the routes; the PSFT for RD 1:1 on PE_{CD} (i.e., customer 1) does not have this route target and so ignores the routes.

Now the routes are added to the PSFT for RD 1:2 at PE_{CD}, and these can be redistributed to the OSPF instance running between PE_{CD} and CE_D. Now CE_D has learned the routes and can redistribute them into the IGP running at site D. This process is carried out in the reverse direction and for all end sites that share common route targets. It should be noted that at no point in this example have the routes in the P LSRs been distributed to the customer sites, nor have the customer site routes been redistributed to the P LSRs. This is by design; the PE and P LSRs will run their own instance of an IGP that simply allows signaling to be transported through the core and with an address space that can be hidden from customers. It will be seen that there is no need to redistribute the BGP routes into the core IGP.

LSP Paths and the Forwarding Model for an MPLS VPN

In a conventional core IP network that uses an IGP in the core and BGP at the edge, there is a requirement for at least aggregated BGP routes to be redistributed into the IGP. The reason for this is that if a packet for an external network is in transit through the core, the core routers need to know the next hop. An alternative solution is to run IBGP on every router in the core so that it can learn about the global routing table. However, neither solution is useful here because we want to support overlapping edge address spaces. Additionally, even if there are not overlapping address spaces, neither solution is popular because either OSPF has to be burdened with all (or many aggregated) external routes, or, with the full BGP solution, there can be complex route-filtering operations required. MPLS VPN provides a more elegant solution.

The route dissemination model has been defined. Note that the PE and P routers (as shown in Figure 13.15) are assumed to run an IGP that only concerns itself with the internal address space of the core network. Each of the IBGP interfaces on the PE routers is advertised in the IGP as a full host route (32-bit mask). Then, if LDP is enabled in the PE and P LSRs, the LSRs will have these host routes in the routing tables and will note each of them as an FEC that needs a label. Consequently, LDP will construct a full mesh of LSPs between each of the internal interfaces of the PE LSRs. Now packets can be forwarded between the PE routers using the LSPs, and because MPLS does not inspect the IP packet, there is no need for the P routers to know the external addresses or routes.

There is a uniqueness problem with the solution just presented because a labeled packet arriving at an egress PE can have packets destined for different customers. Using LDP as

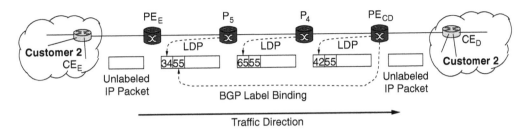

Fig. 13.16 Packet path between site E and site D. The encapsulating LSP is shown highlighting the use of an additional label advertised by BGP.

previously described results in LSPs that transport packets to the egress LSR without needing an IP prefix lookup. However, if packets were simply placed inside these LSPs at the ingress, then at the egress there would be no way to distinguish between packets destined for different customer sites connected to the PE. The IP addresses cannot be used to differentiate site-specific traffic because the different customer sites connected to the PE might use overlapping address spaces. This problem is overcome by the use of a label stack operation at the ingress PE router.

To see the solution (and another explanation of the problem), note the example shown in Figure 13.16, which takes two common customer sites, and the associated path, from Figure 13.15. The packet from site E, destined for site D, is forwarded using conventional IP from CE_E to PE_E. At PE_E it enters the interface belonging to RD 1:2 and so there is a lookup in the associated PSFT. The PSFT is used to configure a unique FEC-to-NHLFE map for each RD. This determines that the packet needs two labels pushed onto it. The first label pushed onto the packet is the label used by the egress PE to distinguish between different RDs.

In the example of Figure 13.16 this is label 55. This label value is passed from PE_{CD} to PE_E using BGP to carry the label binding. Thus BGP becomes a label distribution protocol by including a label-mapping object as an extension to BGP as specified in RFC 3107. This label will only be used at the egress (PE_E) to distinguish packets destined for site D from those destined for site C; it is not a label advertised by the next physical hop, which is P_5. Thus, P_5 will advertise a label using LDP such that it can forward the packet to the internal interface of PE_{CD}. In Figure 13.16, P_5 advertises label binding 34 to PE_E, and this is pushed onto the top of the stack. Now the packet is forwarded along the LDP-provisioned LSP to PE_{CD} using only the top label. When the packet enters PE_{CD}, the top-level label is removed, exposing label 55, which specifies that the label stack should be popped and the (unlabeled) packet should use the PSFT for RD 1:2. In practice the NHLFE table for this label will specify the link layer encapsulation and outgoing port (to CE_D) directly so that an IP table lookup is not required. Finally, the unlabeled packet reaches CE_D and is forwarded to the destination customer site using conventional IP forwarding.

Summary of MPLS VPN

MPLS VPN is one of the biggest applications of MPLS used by core network providers. It is an elegant solution to providing a VPN that allows peering between the customer and the provider IP network. One of the most important features of VPN, from the operator's viewpoint, is that it is a scalable implementation in terms of both configuration management and processor load on the LSRs. The latter is made possible by the fact that the external network address space for each VPN only has to be maintained on each PE router connected to the customer site. No single LSR has to maintain all the routes for all of the VPNs. The key enabling technology for MPLS VPN is the use of BGP at the edge, such that it can maintain routing tables independently for each customer (allowing overlapping address spaces) and can distribute labels to encapsulate traffic for each VPN in its own unique LSP.

Conclusion

MPLS evolved from a methodology to improve switching performance in IP networks, in particular as a replacement for IP over ATM. In practice, the original motivation of forwarding performance was not the reason for its widespread deployment. Instead, it was the fact that MPLS allows IP control and forwarding functions to be separated that has brought new functionality to IP networks. This chapter has shown how MPLS can provide traffic engineering and VPNs in a manner that is difficult to achieve using conventional IP technology. Additionally, MPLS offers significant IP integration that overlay solutions, such as ATM, SONET, and Frame Relay, cannot.

It should be noted that MPLS is a complex technology and that only an overview is presented here. Other points of note are that MPLS provides extensive support to a number of higher-layer protocols and many lower–link layer protocols. This has been extended significantly in generalized MPLS such that MPLS signaling can be used as the control plane for a wide range of circuit-switched technologies.

References

Davie, B. S., and Rekhter, Y. (2000) *MPLS: Multiprotocol Label Switching Technology and Application*, Morgan Kaufmann.

De Ghein, L. (2006) *MPLS Fundamentals*, Cisco Press.

Marzo, J. L., Calle, E., Scoglio, C., and Anjah, T. (2003) QoS online routing and MPLS multilevel protection: A survey, *IEEE Communi. Mag.* 41(10):126–132.

14 Next Generation Ethernet

Alan McGuire
British Telecommunications Plc

Introduction

Ethernet as a technology completely dominates the enterprise environment and is rapidly entrenching itself in the residential market as a result of the widespread uptake of broadband services. For many corporate customers it is the interface of choice because of its ubiquity, simplicity, and cost effectiveness. This is driving the need to develop end-to-end Ethernet service solutions and has seen the introduction of a new industry forum—the Metro Ethernet Forum (MEF)—dedicated to developing such solutions, along with the development of Ethernet technology by standards bodies such as IEEE and ITU-T.

Ethernet services can be delivered over a wide variety of technologies, and Ethernet's popularity as an interface has encouraged network technologies such as SDH/SONET—were not originally optimized to provide packet transport—to successfully evolve to embrace Ethernet services. It has also allowed IP/MPLS to extend its capabilities into new areas.

In this chapter we will begin by asking what is meant by carrier-grade Ethernet before examining the main features of the Ethernet service architecture and the central construct of an Ethernet Virtual Connection (EVC) and its role in describing Ethernet services. It will become clear that this architecture allows Ethernet services to be delivered over many technologies, and although we will indicate some examples, we will focus on how Ethernet as a technology can deliver such services. To meet the growth in demand in the carrier environment, solutions based on Ethernet technology must address the requirements of scalability, performance, resiliency, and manageability. This has resulted in a number of evolutionary changes to Ethernet technology, and we will describe how some of these requirements are being addressed. Nevertheless, it is acknowledged that such solutions will have to compete in the marketplace with the other technologies that can provide the infrastructure to deliver Ethernet services.

403

What Is Carrier-Grade Ethernet?

Those working on Ethernet will find themselves frequently hearing the term "carrier-grade Ethernet." But what does it mean? The reality is that there are many definitions of "carrier-grade," and it can often be assumed that it is whatever the speaker wants it to mean. In some cases it might mean specific technological solutions for the delivery of Ethernet services over a WAN technology such as SDH/SONET or MPLS and how these are constructed to deliver a solution, or it might mean implementations based purely on Ethernet technology, a combination of the two, or simply a service that meets a service level agreement that is more than just best effort.

The view offered here of what constitutes a carrier-grade, or indeed carrier-class, solution is any solution that delivers providers with a set of features that allow the service provider/network provider to offer services that can meet defined performance criteria in the form of a service level agreement and provide operational capabilities to manage both the service and the network. There is a big step change in the requirements from those needed in many enterprise networks to those required in the wide area network. Some of the requirements that often appear in the network provider space include, but are not limited to

- A decoupling of the description of a service from any underlying technology. The service provider wants to avoid the close coupling of service features to a particular technology because this makes any form of future technology migration while retaining that service a very difficult proposition.
- Operational support systems that provide capabilities, such as management of EVCs, inventory management, fault management, configuration management, and tools that allow the Service Level Agreement (SLA) to be managed. For many providers this means element management systems and network management systems that communicate via well-defined interfaces, such as those being developed by the TeleManagement Forum (TMF).
- Operations, Administration, and Maintenance (OAM) tools that provide both always-on monitoring tools and on-demand diagnostics. This has been traditionally viewed as a major weakness for Ethernet in the WAN environment compared to other technologies, and there has been considerable work carried out by standards bodies in recent years to address this, which is described in this chapter.
- The description of SLAs in terms of standardized parameters.
- Survivability mechanisms that enable service level agreements to be met.
- Security features such as isolation of control and management capabilities between the network provider and the customer, access security (e.g., role-driven operator permissions, authentication, audit trails, admission control lists), configuration security (including configuration history and change control), mechanisms to prevent denial of service, protection against MAC address spoofing, port security, and many others.
- In-Service Software Upgrades (ISSUs) that do not impact traffic.
- Redundant and hot-swappable components, such as switch fabrics, power supply units, control cards, and fans. An implementation should avoid single points of failure.

It should be evident that this is not an exhaustive list, but at the same time it allows some variation of how an implementation can be seen as supporting carrier-grade capabilities. In what follows we will consider a service framework that provides a means of

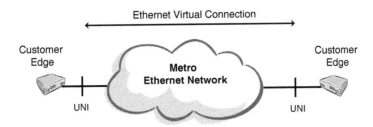

Fig. 14.1 Simplified view of the Ethernet services model.

describing Ethernet service features in a manner that is independent of the underlying delivery mechanism.

Ethernet Service Framework

In its simplest form the reference model of a Metro Ethernet Network (MEN) is as illustrated in Figure 14.1.

A detailed description of the MEF architecture can be found in MEF 4 (2004). It has two major components: the Customer (or subscriber) Edge (CE) equipment and the MEN transport network. The physical demarcation between these components is the User Network Interface (UNI). The service model that uses this architecture is described in terms of what is visible at the CE. The advantage of such an approach is that the internal structure and technologies used inside the MEN are not visible to the customer. As such the delivery of Ethernet services may be provided in the MEN in a number of way, including

- Over Ethernet itself, according to the 802.1 and 802.3 standards of the IEEE.
- Over access technologies that may be fiber or copper based.
- By mapping the Ethernet frames into SDH or PDH using Generic Framing Procedure (GFP) to provide point-to-point private line–type services (Van Helvoort, 2005; ITU-T G.7041, 2005b).
- By transporting Ethernet across an IP/MPLS network (Hussain, 2006). This can be used to provide a point-to-point Virtual Private Wire Service (VPWS) or LAN emulation in the form of a Virtual Private LAN Service (VPLS). In both cases the customers' Ethernet frames are encapsulated with one or more MPLS headers at a provider edge device that faces the CE; they are then transported to another provider edge device that de-encapsulates the frames and passes them onto another CE.

The set of UNI reference points that form an association for the purpose of supporting a service is referred to as an Ethernet Virtual Connection. An EVC can be thought of as a layer-2 virtual private network. Some examples of EVCs are illustrated in Figure 14.2.

The association formed by an EVC may be a simple point-to-point relationship consisting of two UNIs where traffic from the ingress UNI is delivered to the egress UNI. Alternatively, if more than two UNIs are associated they can form a multipoint-to-multipoint EVC, and in this form of EVC it is possible to add or remove UNIs as the service evolves. The behavior of a multipoint-to-multipoint EVC depends on the traffic type, which is determined by the Ethernet address in any individual frame. Ethernet frames with the broadcast address or with multicast addresses that are inserted into the EVC at one UNI are delivered

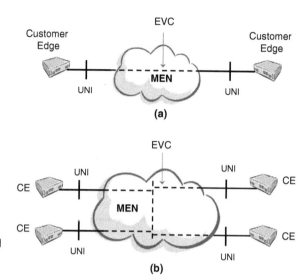

Fig. 14.2 Examples of EVCs: (a) point-to-point line service and (b) E-LAN service (multipoint-to-multipoint).

to all other UNIs. On the other hand, frames with unicast addresses can be delivered to all other UNIs in the EVC only when the address is unknown, or they can be learned, via MAC learning, and when learned are delivered to the required UNI.

The simplest form of service is a dedicated point-to-point Ethernet connection, which is referred to as an E-Line Service by the MEF and as an Ethernet Private Line (EPL) by the ITU-T (MEF 6, 2004a; ITU-T G.8011, 2004a; ITU-T G.8011.1, 2004b). The service is analogous to existing private circuits and to two LANs attached by an EPL: It looks like a wire. An EPL can be subdivided into two classes. In EPL type 1 the information transferred between the UNIs is at the Ethernet MAC frame level. Everything else, such as the preamble and interframe gap, is discarded. The presentation to the EPL is an Ethernet interface, and the Ethernet frames are mapped into the underlying server technology that provides the EPL. For transport over SDH, for example, the MAC frames are mapped into the appropriately sized virtual container via frame-based GFP (GFP-F). In EPL type 2 an 8B/10B-line–coded signal is mapped using transparent GFP (GFP-T) and provides low latency combined with transparency to control-codes. EPL type 2 is only defined for Gigabit Ethernet.

In an EPL a resource is reserved for each service instance and is therefore dedicated to a customer. It is also possible to share network resources between a number of customers, allowing an increase in network efficiency. This is referred to as an Ethernet Virtual Private Line (EVPL) (ITU-T G.8011.2, 2005a). A consequence of sharing is that in the event of congestion, frames might be dropped.

Multipoint EVCs offer an E-LAN service, in MEF terminology, or an Ethernet Private LAN (EPLAN), where a resource is dedicated, and an Ethernet Virtual Private LAN (EVPLAN) for shared resource, in ITU-T terminology.

The main issues that are of concern in providing such services are how to specify the bandwidth of the service (in the worst case all traffic is simultaneously directed to a single UNI, and designing to manage this bandwidth would result in an inefficient solution), how protection will be provided, discard policy under congestion, and the amount of buffering that should be provided.

The EVPLAN is the most complex service to manage, and this has to be balanced against the efficiency gains obtained from sharing transport and switching infrastructure between customers. The main issues in providing such a service are similar to those of an EPLAN but are magnified by the sharing of resources between customers.

For each of the service types described, the service can be described by means of a number of parameters, including

- Connectivity (e.g., point-to-point, point-to-multipoint, multipoint-to-multipoint) and the list of UNIs associated with the EVC
- Transfer characteristics of the EVC, including what addresses can be delivered, drop precedence, and class of service
- Bandwidth profile, including committed information rate and committed burst size
- Connectivity monitoring (e.g., none, on demand, or proactive)
- Whether preservation of a customer's VLAN ID allocation and class of service markings is provided
- Survivability

Operations, Administration, and Maintenance

As part of providing a service level agreement, it is necessary to be able to monitor the service so that it can be verified that the agreement is being met, and to be able to quickly identify faults and take appropriate actions. Traditionally Ethernet has not provided the tools that allow the service or network provider to remotely manage the network and to achieve these goals quickly and cost effectively. In this environment it is also necessary to have tools that are provided in the Ethernet layer and do not rely on client layer technologies such as IP. For this reason there has been much activity by both the IEEE and the ITU-T to develop these capabilities. Within the IEEE standards, OAM can be subdivided into OAM that operates on a physical link and that does not pass through a bridge and that provides features such as discovery (which is necessary to allow the ends of a link to communicate with each other regarding their capabilities and configuration), performance monitoring, remote loopback (to aid with installation and fault finding), fault detection, and statistics collection. This was described in 802.3ah and is also referred to as Ethernet in the First Mile. It was first published in 2004 and has since been incorporated into the 802.3 standard (IEEE 802.3, 2005a).

The second form of OAM provides capabilities that can be used on a per-service basis and can be passed end to end through bridges and is termed, in the language of the IEEE, *Continuity Fault Management* (CFM). The framework for this allows a customer service to be monitored end to end regardless of any underlying technology and also takes into account that a service may be delivered by more than one provider. Work on this was carried out in parallel between the IEEE (802.1ag) and the ITU-T (Y.1731) (IEEE 802.1ag, 2007b; ITU-T Y1731, 2006c). Although both build on many of the features that are common in point-to-point connection oriented technologies the connectionless nature of Ethernet and the need to support multipoint-to-multipoint services means that the OAM models for Ethernet are slightly different. There are some minor differences between the standards, including terminology regarding common functionality, but the major difference is that the latter also covers aspects related to performance management.

The 802.1ag OAM model provides a hierarchy of up to eight levels (maintenance levels) that allow end users, service providers, and operators to run their own OAM traffic

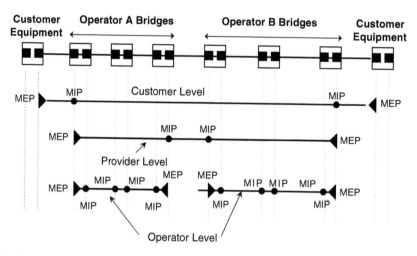

Fig. 14.3   Example of Ethernet OAM entities.

independently of what is run at other levels. The number of levels required depends on the scenario under consideration. OAM traffic at a higher level is transported transparently by equipment at a lower level (e.g., the customer's OAM is transported transparently by the provider's switches).

The entity to be managed is called a Maintenance Entity (ME) in Y.1731 language. An ME can be defined between appropriate points such as administrative domains, between switches, or even a single link. An example of the scope of ME and how they can organized into levels is illustrated in Figure 14.3.

The ME construct can be used in a multipoint environment by grouping MEs into Maintenance Entity Groups (MEGs) so that a multipoint construct with N endpoints has $N(N-1)/2$ MEs associated with it. Each MEG has a unique ID, and this is used in OAM messages to confirm that the messages are for their intended recipients. At the ends of the ME are functions called MEG End Points (MEPs) that generate, receive, and process OAM frames that are used to manage the ME. Within the ME there can also be some intermediate points, MEG Intermediate Points (MIPs), which can respond to OAM messages from a MEP but do not originate them. OAM messages can be sent to a MIP to allow activities such as loopbacks.

The OAM standards define not only the Operations, Administration, and Maintenance architecture but also the OAM messages and the format of OAM frames. The most important OAM message is the Continuity Check (CC) message, which is sent out periodically by a MEP to the other MEPs in a maintenance entity by using a multicast address. The periodicity of these heartbeat messages can be altered over a range from 300 per second to 1 every 10 minutes depending on the application. The CC message provides a number of features:

- Loss of continuity, which occurs when a MEP does not receive CC messages over a configured period of time
- Discovery of missing or unexpected MEPs
- Service cross-connect or service ID mismatch
- Detection of forwarding loops

To verify the bidirectional connectivity between a MIP and a MEP, loopback messages (which are analogous to ping messages) are sent from the MEP to the MIP. This provides a diagnostic tool that can simply be sent once or as often as required. It can be sent to successive MIPs to determine the location of a fault. A link trace message can also be used to identify the path between MIPs and MEPs, providing a sequence from source to sink. Both standards also provide a number of other message types.

At the present time there is little in the way of operator experience with either 802.1ag or Y.1731, but it is expected that they will bring significant improvements to network operations.

The Evolution of Ethernet

As Ethernet has evolved it has faced a number of scaling challenges because of its nature. An excellent overview of many existing capabilities can be found in Clark and Hamilton (2007). Its underlying behavior is derived from its origins as a shared-channel broadcast network technology. As Myers (2004) notes, this model is simple:

> Hosts could be attached and re-attached at any location on the network; no manual configuration was required; and any host could reach all other hosts on the network with a single broadcast message. Over the years, as Ethernet has been almost completely transformed, this service model has remained remarkably unchanged. It is the need to support broadcast as a first-class service in today's switched environment that plagues Ethernet's scalability and reliability.

The reason for this is that broadcast networks are ultimately not scalable. What we require are solutions that retain the fundamental behaviors of Ethernet for customers while overcoming the limitations of broadcast services in the carrier environment.

Attempts to limit the impact of the broadcast service model use mechanisms to partition or segment the network into multiple "loop-free" domains. Why loop free? The broadcast model floods frames throughout the network because end-system locations are not explicitly, known. If there are loops present, a single broadcast frame is duplicated exponentially, resulting in catastrophic behavior. To this end considerable effort has been spent on the development of spanning tree protocols to compute a spanning tree-forwarding topology with no loops. We will return to the subject of spanning tree protocols when we discuss resiliency and protection.

Initially the segmentation into multiple domains to limit the size of broadcast domains was achieved by means of Virtual LANs, or VLANs, and as the name suggests it was primarily intended for the enterprise environment (IEEE 802.1D, 2004; IEEE 802.1Q, 2005b). This was done by adding a short tag to the Ethernet frame to identify VLAN instances. The number of instances that could be supported was limited to 4094, with two other instances reserved for particular applications. To improve scalability, support for a second VLAN tag within the frame was specified as part of the provider bridging activity (IEEE 802.1ad, in progress). This allowed providers to impose their own partitioning without disturbing customer partitioning by allocating the inner tag to the customer and the outer tag to the provider.

The inner tag is referred to as the Customer VLAN (C-VLAN) and the outer tag as the Service VLAN (S-VLAN) tag. However, this was done via stacking identifiers that were designed for single-Enterprise–scale requirements. The S-VLAN limits the number of service instances in a network to 4094. For the C-VLAN, stacking means that the service tag is not only a transport separator function; it is also a customer identification function. An 802.1ad S-VID therefore represents both a community of interest in the form of an individual service instance and a specific topology (a spanning tree instance) within the provider network.

The double-tagging mechanism of 802.1ad only gets us so far. It does not address the fundamental scaling issues associated with broadcast domains and flat addressing (increased MAC table sizes), nor does it offer any real solution to the virtualization of services over a common Ethernet infrastructure.

This has scaling limitations when compared with a hierarchical relationship, where the different levels of hierarchy are functionally decoupled. Nor does simple partitioning provide any real control over how a carrier's facilities are utilized, and the use of multiple spanning trees only increases the degree of indeterminism in network operations.

To operate such networks, the operator has until now had little option but to severely limit the scale and scope of applicability of Ethernet networks in a service context or to rely on it as a means of interconnect for other networks, such as IP. To get to the next step of supporting millions of service instances in Ethernet, a new approach is required.

To enable true hierarchical scaling, virtualization, and full isolation of the provider infrastructure from customer broadcast, domains, the IEEE has embarked on the development of Provider Backbone Bridges (PBBs) (IEEE 802.1ah, in progress). It addresses two primary problems:

- The scalability of MAC address tables. Without PBBs the service provider equipment needs to learn all the customer MAC addresses. Where the customer edge device is a router, the learning is minimized, but if the edge device is a bridge, it is also necessary to learn all of the customer end-station addresses. Limits on the size of bridge MAC tables offer limited scalability as the number of customers increases.
- The scalability of service instances. The limitation of a 12-bit VLAN-ID (VID) is that it only supports 4094 service instances.

The development of 802.1ah represents a significant step toward making Ethernet suitable for carriers. It does this by mapping customer frames into provider frames (hence it is also known as MAC-in-MAC) and in doing so provides the benefits of hierarchy by isolating customer and provider concerns.

Hierarchy is of particular benefit when the customer base consists of a large number of relatively small communities of interest that can be mapped onto a common transport network. It reduces the amount of provisioning and forwarding state in the network core and consequently reduces the load and ongoing cost of performing service assurance and fault management.

To address the scalability of service instances a new 24-bit Service Instance Identifier (I-SID) has been created that provides support for approximately 17 million service instances, thereby removing the limitations imposed by the 12-bit VID. Furthermore, to provide backward compatibility with existing bridges, a 12-bit VID referred to as the Backbone-VID, or B-VID, is used as the outer VID, which is used as part of the forwarding decision.

The evolution of Ethernet as specified by the IEEE, including the 802.1ah frame format, is illustrated in Figure 14.4.

In contrast to 802.1Q and the subsequent 802.1ad, which simply partitioned the forwarding of a single-layer network, 802.1ah implies complete recursion such that customer domains are completely encapsulated and isolated from the provider Ethernet domain. Yet 802.1ah still uses the same original service model, with spanning tree protocols used to prevent loops. As such it still inherits the limitations of the existing service model. What is needed is a change in the service model inside the carrier network. In what follows we describe a method for how this can be achieved (Allan, 2006).

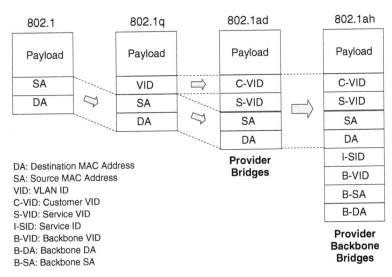

| 802.1 | 802.1q | 802.1ad | 802.1ah |

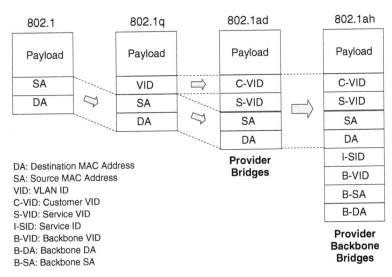

DA: Destination MAC Address
SA: Source MAC Address
VID: VLAN ID
C-VID: Customer VID
S-VID: Service VID
I-SID: Service ID
B-VID: Backbone VID
B-DA: Backbone DA
B-SA: Backbone SA

Fig. 14.4 Evolution of Ethernet (the frame formats have been simplified).

The introduction of hierarchy isolates customers from the operations of the network provider and allows the provider to implement different forwarding modes with the objective of engineering the network by providing determinism and predictability. This is a significant benefit to the provider because it allows for deterministic traffic, fault management, and performance management of customer services. New forwarding modes are required to overcome the limitations of Ethernet networks, which are limited to trivial physically loop-free topologies or constraints imposed by spanning trees. Fortunately there are options based on alternative provisioning and control/management planes that can be employed to permit the Ethernet infrastructure to be engineered. The isolation provided by hierarchy means that a provider now has complete control of its MAC address and VLAN space. This separation allows new forwarding modes to be exploited.

One feature that is of interest and not supported by Ethernet is the capability to set up and pin connection-oriented paths across an Ethernet network in a scalable manner. One approach is to create cross-connect entries using VLAN tags, where forwarding is based purely on the outer VLAN tag regardless of the MAC addresses associated with the incoming frames. If the tags are used in a global context, then, as described earlier, the number of connections in the network is limited to 4094. To overcome this it is possible to carry out per-port translation (as described in 802.1ad) of the VLAN tag in a manner similar to the use of the Virtual Path Identifier (VPI) and Virtual Channel Identifier (VCI) in ATM, thereby allowing the VID to have local rather than global significance. This is illustrated in Figure 14.5.

A variation on this theme that permits further scaling is to use both the inner and outer VLAN tags to create an extended 24-bit cross-connect identifier, allowing approximately 16 million VLANs per port. This extended mode is, however, not consistent with IEEE standards, where only the outer tag is involved in the forwarding process.

A further approach is to minimize changes to the Ethernet forwarding model (and hardware) by recognizing that VLANs are ultimately a scarce resource and that it is therefore

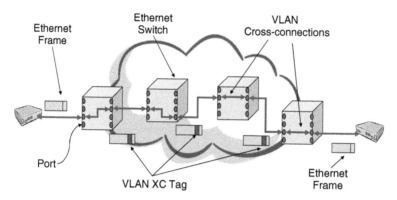

Fig. 14.5 VLAN cross-connection.

of interest to examine alternative mechanisms to scale the connection space. Ethernet switches that employ Independent VLAN Learning (IVL) forward frames on the basis of a full 60-bit lookup of both the VLAN tag and the destination MAC address in each packet. It is clear that simply examining the entries of a forwarding table is not sufficient to describe the form of the forwarding—in Ethernet it is the MAC learning mechanism that results in connectionless behavior. Consequently, it is possible to provide a more deterministic form of forwarding with the same hardware simply by turning off some of the bridging functions.

Bridging employs MAC learning and flooding to populate the forwarding tables of Ethernet switches, but these forwarding tables can also be configured by means of a control plane or via management configuration (e.g., via the command line interface). This allows for the VLAN tag and destination MAC address combination to be exploited in a different manner. There is a benefit to maintaining one of the primary features of Ethernet—namely, globally unique MAC addresses—but there is, as we have seen for VLAN cross-connects, no need to maintain the global uniqueness of the entire VLAN identifier range.

The requirement for a global VLAN identifier comes not from any fundamental requirement of the unicast forwarding behavior of an IVL switch but from the need to constrain broadcast behavior as a result of flooding. It is therefore possible to assign a range of VLAN identifiers containing V identifiers as only locally significant to a given MAC address. It is then easy to envisage a VLAN identifier in that range as an individual instance identifier for one of a maximum of V connections terminating at the given MAC address. The combination of the VLAN identifier and the MAC address is globally unique, with the MAC address identifying the logical administrative owner of a specific VLAN identifier value.

A control or management plane can now be used to populate the forwarding tables for the designated VLAN identifier range. To create a unidirectional connection, all that is required is to populate the required VLAN/MAC-tuple into each of the network elements along the path. A mirror image operation can be used to create a path in the opposite direction to create a bidirectional path. This gives us the ability to create engineered paths for the designated VLAN range and Ethernet bridged behavior alongside it for the remaining VLANs. This ability to create traffic-engineered paths is imaginatively called *Provider Backbone Bridging with Traffic Engineering* (PBB-TE) and is currently in development in the IEEE

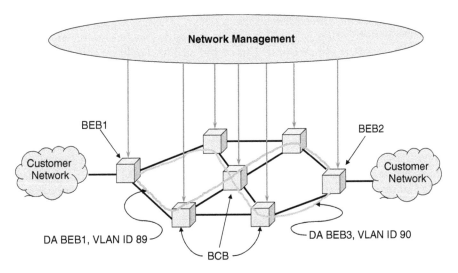

Fig. 14.6 Traffic-engineered paths.

802.1 committee in the form of 802.1Qay (IEEE 802.1Qay, in progress). The application of this technique is illustrated in Figure 14.6.

In the figure, backbone edge bridges, BEB1 and BEB2, correspond to the edge of the provider backbone bridged network and are 802.1ah-compliant devices that offer customer-facing ports. The backbone edge bridges map the customer traffic onto configured Ethernet Switched Paths (ESPs). In this example backbone VLAN identifiers 89 and 90 are within the range set aside for configured behavior. Two ESPs have been configured from BEB1 to BEB2, and the MAC address of BEB2 corresponds to the MAC address component of the forwarding-table entry for each path.

The first ESP path is computed, the B-VLAN identifier (B-VID) 89 is assigned, and the forwarding tables in the intervening Backbone Core Bridges (BCBs) (802.1ad bridges) are configured with the mapping {B-VID = 89/MAC = BEB2}. The same process can be employed to create a corouted return path from BEB2 to BEB1. Note that there is no fundamental restriction from the standpoint of configuration that forces the return path to be corouted. However, asymmetric failures in Ethernet are catastrophic for spanning tree algorithms to the point where significant effort has been expended by standards delegates to ensure that failures are bidirectional, and although a configured ESP does not use a spanning tree, its 802.1ah client frequently does (recall that it is MAC-in-MAC). However, the allocation of bandwidth for each direction need not be the same.

For the example illustrated, the paths can be seen to cross at a BCB, indicating that it is the combination of both the B-VID and MAC that determines a unique forwarding entry. Collisions in either space, such as B-VID 89 or 90 used in combination with another MAC address or as in the example where paths 89/BEB2 and 90/BEB2 cross, are still uniquely resolved to a single egress port.

This technique combines both destination-based forwarding and multiplexing, which is sometimes referred to as shared forwarding. When multiple sources are configured to send traffic to the same VLAN/destination MAC-tuple, knowledge of the source is preserved in the form of the source MAC address, and this is present for both bearer and data plane OAM frames. This differs from other "lossy" multipoint-to-point approaches such as merging, where forwarding is based on a fixed length locally significant label and, on merging, visibility of the source is lost. This can be retrieved only by going to a higher layer in the protocol stack.

Preserving information regarding the source of a frame has significant implications for fault and performance management. When considering the quality of information available on the data plane, a forwarding entry is a multiplex of point-to-point connections that scales as $O(N)$ at intermediate switches while preserving the useful OAM properties of a full mesh at the terminating backbone edge bridge. A path-terminating interface maintains fault and performance management state in proportion to the number of connected peers and is able to identify the source and path of both bearer traffic and OAM traffic on the basis of the embedded source MAC address and VLAN identifier. This is a distinct improvement on the merge example previously described, where additional information is added to the OAM frames to identify the source and actual bearer traffic is not associated with a particular source device. This implies that traffic counts and availability state can be maintained on a per-path pairwise basis and real-time correlation of performance counts is possible.

The replacement of conventional flooding and MAC learning with configuration of the forwarding tables is augmented by the following additional steps to ensure that the method is robust:

- Any discontinuities in the forwarding tables of the switches along the intended path for the connection would normally result in packets being flooded as "unknown." In a mesh topology such behavior would almost certainly be fatal, and as such unknown flooding must be disabled for the designated VLAN identifier range. It is also disabled for both broadcast and multicast traffic.
- MAC learning is not required and might interfere with any form of management/ control population of the forwarding tables. Consequently, MAC learning is also disabled for the designated VLAN identifiers.
- There is no longer any need for a loop-free topology for the delegated VLAN identifiers, so the spanning tree is disabled for the delegated range. This is achieved by assigning the identifiers to a *Multiple Spanning Tree Instance Identifier* to indicate a null spanning tree instance.
- The application of Connection Admission Control (CAC) and a means of accounting for bandwidth assigned in the control/management plane.
- Data plane OAM provides both fault and performance management. As the actual frame transfer function across an Ethernet switch is unmodified (learned or configured), existing OAM can be largely reused. One change is that OAM uses the same MAC address as the user traffic.

The delegation of a small portion of the VLAN identifier range to configured as opposed to learned behavior has a trivial impact on the scaling properties of VLAN partitioned bridging yet simultaneously permits a large number of point-to-point or multipoint-to-point ESPs to be supported. Setting aside as few as 16 VLAN identifiers allows 16 uniquely routed multipoint-to-point multiplexed paths to each of the set of BEBs in the network, while allowing 4078 VLAN identifiers for bridged behavior. This gives a theoretical maximum of some 2^{52} ESPs fully meshing (including resiliency) some 2^{48} devices. It is obvious that the theoretical limit outstrips the capacity of existing implementations.

Unlike the case where flooding and learning are employed, the configured option only requires knowledge of peers on a need-to-know basis.

A configured ESP has complete route freedom (with the exception of corouting for bidirectional connections), and the ability to define multiple paths to any BEB combined with route freedom has a number of implications. When there are a number of paths between any two points in the network, a spanning tree with bridging and auto-learning allows only one of these to be used. This is in sharp contrast to configured behavior, which allows more than one path between any two points and permits criteria beyond simply the shortest path to be used in selecting the routing of any individual path. A wide variety of metrics can be used along with appropriate algorithms and offline computation to select optimal sets of paths with specific attributes and without common points of failure. This means that

- Paths can be engineered.
- Capacity mismatches between physical network build and offered load can be compensated for.
- First-line resiliency in the form of protection can be delegated to the data plane. This is discussed further in the subsection on protection.

When considering a network that uses both learning/bridging and configured behavior, the traffic associated with either forwarding behavior must be distinguishable such that best-effort traffic does not degrade engineered traffic, particularly as the traffic matrix can change dynamically in response to failures and/or maintenance activities. This can be achieved either by instantiating connection state into the switches or via use of class-based queueing and edge policing of class markings. The latter approach can be implemented via the 802.1p class-of-service feature, which has been extended in 802.1ad to include discard eligibility. This is similar to layer-3 QoS mechanisms that are designed around a LAN-style environment where a small amount of QoS-sensitive traffic requires priority over a large bulk of best-effort traffic.

The combination of traffic-engineering–class queueing, CAC, and ESPs results in an environment where there is a correspondence between offered load and committed resources, thereby fostering the ability to offer real and measurable guarantees of performance. Put simply, traffic that is engineered and associated with CAC is given a higher priority than bridged traffic. In particular, the introduction of CAC connections now permits paths to be securely dimensioned so that there is no congestive frame loss. This then provides Ethernet, rather than relying on another technology, with a mechanism to offer EPLs. In addition, it can be envisaged that schemes that allow for some overbooking of capacity can be implemented. We should of course note that many of the capabilities previously described can also be provided in other technologies that can provide engineered paths, but there are also advantages to minimizing the number of technologies that need to be managed in the network.

Protection Mechanisms

The mechanism employed in Ethernet networks to ensure that loops are removed when one or more paths exist between endpoints is the construction of a spanning tree using a spanning tree protocol. Even a small loop can cause significant disruption, and considerable work has been carried out by the IEEE to avoid loops. In many packet technologies the effect of loops is mitigated by means of a Time-to-Live (TTL) field associated with each packet, but Ethernet has no such mechanism and as such it is necessary to create a loop-free topology—in the form of a tree—using a spanning tree protocol.

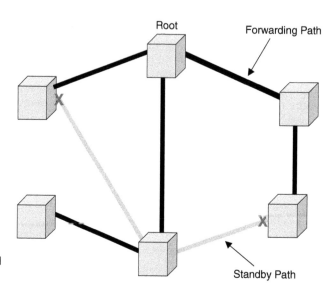

Fig. 14.7 Spanning tree and topology.

To establish a spanning tree, the bridges in the network distribute messages between themselves in the form of Bridge Protocol Data Units (BPDUs). During this exchange of messages, a single bridge is elected as the root bridge and then the spanning tree algorithm calculates the shortest path from each of the other bridges back to the root bridge. When this tree structure has been created only those paths associated with the tree are active and support traffic; all other paths between the bridges are blocked (their associated ports are not part of the spanning tree). This is illustrated in Figure 14.7. A good description of how the spanning tree protocol works can be found in Perlman (2000).

It can be seen from the figure that two paths have been blocked to prevent the creation of forwarding loops and are therefore not used in normal operation. Clearly, if a link in the spanning tree should fail, it is necessary to provide some form of resiliency by recalculating the spanning tree. This is achieved by an aging mechanism, which is triggered when no new BPDUs are received on a port within a fixed period of time, at which point a recalculation occurs.

Although the active topology is recalculated, it is evident that the alternative paths are only usable under failure conditions and hence no load sharing is possible. This has led to alternative solutions for resiliency being considered. A number of spanning tree solutions have been developed:

802.1d—Spanning Tree Protocol (STP) is the original version of a spanning tree. It has a significant limitation for larger networks in that it is slow to converge onto a new tree (sometimes 1 to 2 minutes). This results in a significant disruption to customer services. Although a common spanning tree for the entire network can be deployed regardless of the number of VLANs, this is not the most efficient use of topology and lowers overall network utilization. The reconfiguration time of STP is simply too slow for modern applications, and as a result 802.1w as subsequently described was developed.

802.1s (now integrated into 802.1Q)—The standard describes the Multiple Spanning Tree (MST), which provides a means by which a number of paths can be used across an

Ethernet network, thereby increasing network utilization and load sharing. This is achieved by running multiple instances of the spanning tree over the same network and configuring different bridges as the root for each instance. Network management configuration gives the provider control over the number of spanning tree instances and the assignment of ranges of VLANs to each instance, and in doing so allows creation of a number of logical topologies.

802.1w (now integrated into 802.1D)—The Rapid Spanning Tree Protocol (RSTP) can be seen as a refinement of STP that uses many of the original capabilities but makes a number of changes that enhance performance. The limitation of STP where only the root bridge generates BPDUs and processes topology changes is overcome in RSTP by allowing all bridges to generate BPDUs and respond to topology changes. This and other changes mean that RSTP reduces the response time from a minute or so down to around a second (or less), resulting in an improved customer experience.

It is likely that customers will elect to run a spanning tree on their Ethernet LANs to prevent loops for the same reasons that service providers will choose to implement it. Should the customer's STP interact with the service provider's STP? The simple answer is no. Any changes to a customer's spanning tree could impact the spanning tree of the service provider and consequently impact the performance and availability of the service provider's network.

A service provider's network does, however, need to transport customer spanning tree frames because the blocking of spanning tree control frames between customer sites can lead to undesirable effects, such as loops within the customer's network. The introduction of hierarchy in 802.1ah allows, as described earlier, the complete separation of customer and provider concerns.

When there are multiple Ethernet links between two switches, they can be aggregated together into a Link Aggregation Group (LAG), which can be logically treated as a single link with a higher speed. A Link Aggregation Control Protocol (LACP) is utilized to control the properties of the link by communicating with the switches at either end of the logical link. Traffic is distributed across the individual links by means of a hashing algorithm in a manner that maintains frame ordering for each communication session. A LAG can be used to protect against failure of an individual link in the group by redistributing the traffic between the remaining links. When the failed link has been restored, the LAG distribution function reintroduces it to allow redistribution of traffic.

The ITU-T has defined linear protection-switching schemes for point-to-point VLANs in the form of 1+1 and 1:1 protection-switching architectures in G.8031 (ITU-T, 2006b). In these schemes the route and bandwidth of the protected entity is reserved for a chosen working entity, thereby providing a simple survivability mechanism. This makes for an attractive approach for network providers because it is simpler from an operational perspective to understand the active topology under consideration than it is with mechanisms such as RSTP.

Figure 14.8 illustrates the architecture for 1+1 bidirectional protection switching. Traffic is duplicated on both the working and protection paths, and the appropriate path is selected by the sink. As is common with other transport technologies, the switching architectures provide a number of features that give network providers operational flexibility, including

Nonrevertive switching: The traffic remains on the protection path and does not revert to the working path after the defect has cleared. This is generally employed in 1+1

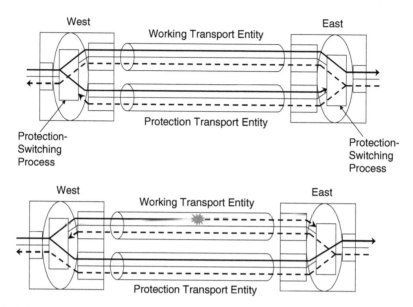

Fig. 14.8 Ethernet 1+1 protection-switching architecture.

schemes and has the advantage that it minimizes the number of times that traffic is impacted by switching events.

Revertive switching: The traffic is returned to the working path from the protection path after the defect has cleared. Generally speaking, the protection path will often have lower performance (e.g., larger delay) than the working path, and for this reason reversion is desirable. If there is an intermittent fault, it is desirable to avoid frequent switching events. This can be achieved by implementing a wait-to-restore timer that waits a period of time before switching can occur. The time can be configured in steps of 1 minute between 5 and 12 minutes.

Hold-off timers: In multilayer networks there are often protection schemes operating at different levels, and it is not desirable to have multiple protection events in different layers occurring simultaneously in response to a single event. It is preferable to give protection schemes in lower layers sufficient time to restore service and only to respond at a higher layer if it is clear that the lower layer has had sufficient time to clear the defect and has failed to do so. To allow this to happen, the defect is not immediately reported in the client layer; instead, the hold-off timer is activated. If at the end of the hold-off time a defect is still present, protection switching will occur in the client layer. The hold-off timer is configurable and generally the range is 0 to 10 seconds in steps of 100 ms.

Forced and manual switch commands: The network provider is allowed to initiate protection switching manually to allow, for example, planned engineering works.

Automatic Protection Switching Protocol (APS): Coordination can take place between both ends and is one of the OAM protocol data units specified in Y.1731.

It is worth noting how protection-switching and STPs should be used. Any ports that are within the protection switching architecture cannot take part in a spanning so as to avoid protection and spanning tree interaction. This can be achieved in two ways:

- Switch the STP off inside the protection domain. Spanning tree protocols can still be used outside of the protection domain. Protection switching can therefore be used internally to a provider's network or as a means of protection between providers, which avoids interprovider spanning tree interactions.
- Assign the working entity and the protection entity to separate spanning trees, where an STP section is contained within the protected domain.

Although the protection schemes previously described are based on VLANs, it is straightforward to employ such schemes in PBB-TE, and they are under consideration as part of 802.1Qay. In addition to the linear protection schemes described in G.8031, there is ongoing activity in the ITU-T to develop Ethernet-based ring protection schemes.

Synchronous Ethernet

As a transport mechanism, it is anticipated in some quarters that Ethernet might displace SDH/SONET for a number of applications, particularly if Ethernet can provide the capabilities required by network providers to manage the network and provide robust and deterministic service offerings to customers. In addition to many of the capabilities described earlier, there would also be a need to support synchronization, both frequency and time. The focus of Synchronous Ethernet is likely to be where the deployment of Ethernet is most prevalent—namely, toward access and aggregation.

The essence of Synchronous Ethernet is that it locks the Ethernet physical layer (in IEEE terms the PHY; in ITU-T terms, the ETY) to a reference clock and distributes it to downstream elements. The introduction of Synchronous Ethernet requires some additional capabilities to be added to Ethernet hardware—namely, the ability to insert and extract the appropriate clock rates into and from the Ethernet PHY and the ability to add appropriate status messaging (containing reference traceability information).

A Synchronous Ethernet solution would be embedded within the transport medium and managed while being secure and free from the fundamental issues that impact synchronization recovery in packet-based solutions. It also permits interworking with existing Ethernet architectures and should not create any significant interworking problems that would affect existing Ethernet standards or interfaces—that is, traffic would continue to run at the MAC frame level and pass from a Synchronous Ethernet to a native Ethernet domain.

There has been significant progress in the development of Synchronous Ethernet standards. The high-level concepts of Synchronous Ethernet and the Synchronous Ethernet Equipment Clock were published in ITU-T recommendations G.8261 (ITU-T, 2006a) and G.8262 (ITU-T, 2007). It is anticipated at the time of writing that G.8261 will be updated in 2008 to include information regarding Synchronous Ethernet messaging.

Next Generation Ethernet Transmission

The transport networks of the telecommunications industry assign bitrates in steps of 4 (e.g., the SDH hierarchy increases from STM-1 through STM-4, STM-16, and STM-64 up to STM-256, covering the range from 155 Mbit/s to 40 Gbit/s). At the same time, the Ethernet community in the IEEE likes to move in steps of 10 (e.g., from 10 Mbit/s to 100 Mbit/s, then 1 GbE followed by 10 GbE). Given the increasing number of 10-GbE interfaces in the enterprise environment and the need for the network operator community to transport these signals, it was only a matter of time before the subject of the next step came to the fore.

Driven by an explosion in anticipated bandwidth for all types of Ethernet networks, the IEEE created a High-Speed Study Group (HSSG) to identify the requirements and applications and decide on the next steps. On the one hand there was a strong requirement from the data center applications side to introduce 40 GbE for high-speed computing/server applications while on the other hand the telecommunications community, driven by network aggregation applications, wanted to develop a 100-GbE solution.

After some considerable debate, the HSSG has decided to combine both 40 GbE and 100 GbE into a single draft standard known as 802.3ba. The group has already set out a number of requirements, including

- Support for the 802.3 frame format. This is hardly surprising given the need for backward compatibility with installed base.
- For 40-GbE solutions, support for at least 100 m on OM3 multimode fiber, at least 10 m over a copper cable assembly, and at least 1 m over a backplane.
- For 100-GbE solutions, support for 40 km on single-mode fiber, 10 km on single-mode fiber, 100 m on OM3 multimode fiber, and at least 10 m over a copper cable assembly.

It should be noted that ITU-T's Study Group 15 has also initiated work to examine how these new Ethernet rates should be supported by the Optical Transport Network (OTN), including extending the multiplexing hierarchy. There is an assumption that the solution developed by the HSSG should be compatible with the OTN hierarchy and vice versa.

The work of these two groups is likely to continue into 2010, and it can be anticipated that there will be some significant engineering challenges along the way. At the present time, a good source of information is the tutorial prepared by the HSSG (IEEE 802.3, 2007a).

Conclusion

In this chapter we have examined a number of aspects related to the development of Ethernet in the wide area network environment. The rate of development of the technology, and indeed technologies that have a strong commercial relationship with Ethernet such as MPLS, OTN, and SDH/SONET, will provide both customers and providers with rich networking capabilities and a plethora of choice. The manner in which Ethernet will be utilized will depend on many factors, including a network provider's current infrastructure, the provider's view of its future network architecture, and the migration path toward it. The rate of technology change means that it is difficult to predict the future of the technology and its relationship to other technologies. Nevertheless, there can be no doubt that Ethernet, in whatever guise, will play a significant role in future networks.

References

Allan, D., et al. (2006) Ethernet as Carrier Transport Infrastructure, *IEEE Commun. Mag.* 2:134–140.

Clark, K., and Hamilton, K. (1999) *Cisco LAN Switching*, Cisco Press.

Hussain, I. (2006) Ethernet Services over MPLS Networks, in Kazi, K. (ed.), *Optical Networking Standards: A Comprehensive Guide for Professionals*, Springer, 425–456.

IEEE (in progress) *802.1ah, IEEE Draft Standard for Local and Metropolitan Area Networks, Virtual Bridged Local Area Networks, Amendment 6: Provider Backbone Bridges.*

IEEE (in progress) *802.1Qay, IEEE Draft Standard for Local and Metropolitan Area Networks, Virtual Bridged Local Area Networks, Amendment 6: Provider Backbone Bridge Traffic Engineering.*

IEEE (in progress) *802.1ad, IEEE Standard for Local and Metropolitan Area Networks, Virtual Bridged Local Area Networks, Amendment 4: Provider Bridges.*

IEEE (2007a) *802.3, High Speed Study Group (HSSG).*

IEEE (2007b) *802.1ag, IEEE Standard for Local and Metropolitan Area Networks, Virtual Bridged Local Area Networks, Amendment 5: Connectivity Fault Management.*

IEEE (2005a) *802.3, IEEE Standard for Information Technology—Telecommunications and Information Exchange between Systems-Specific Requirements, Part 3: Carrier Sense Multiple Access with Collision Detection (CSMA/CD), Access Method and Physical Layer Specifications.*

IEEE (2005b) *802.1Q, IEEE Standard for Local and Metropolitan Area Networks: Virtual Bridged Local Area Networks.*

IEEE (2004) *802.1D, IEEE Standard for Local and Metropolitan Area Networks: MAC Bridges.*

ITU-T (2007) *Recommendation G.8262, Timing Characteristics of Synchronous Ethernet Equipment Slave Clock (EEC).*

ITU-T (2006a) *Recommendation G.8261 Timing and Synchronization Aspects of Packet Networks.*

ITU-T (2006b) *Recommendation G.8031, Ethernet Protection Switching.*

ITU-T (2006c) *Recommendation Y.1731, OAM Functions and Mechanisms for Ethernet-Based Networks.*

ITU-T (2005a) *Recommendation G.8011.2, Ethernet Virtual Private Line Service.*

ITU-T (2005b) *Recommendation G.7041, Generic Framing Procedure (GFP).*

ITU-T (2004a) *Recommendation G.8011, Ethernet over Transport—Ethernet Services Framework.*

ITU-T (2004b) *Recommendation G.8011.1, Ethernet Private Line Service.*

Metro Ethernet Forum (2004a) *Technical Specification MEF 6, Ethernet Services Definitions, Phase 1.*

Metro Ethernet Forum (2004b) *Technical Specification MEF 4, Metro Ethernet Network Architecture Framework, Part 1: Generic Framework.*

Myers, M., Ng, T.S.E., and Zhange, H. (2004) Rethinking the Service Model: Scaling Ethernet to a Million Nodes, *Proceedings ACM SIGCOMM Workshop on Hot Topics in Networking*, November.

Perlman, R. (2000) *Interconnections: Bridges, Routers, Switches, and Internetworking Protocols*, Second Edition, Addison-Wesley.

Van Helvoort, H. (2005) *Next Generation SDH/SONET: Evolution or Revolution?* Wiley.

List of Acronyms

ABS	Analysis by Synthesis	CAC	Connection Admission Control
AC	Alternating Current	CAM	Computer-Aided Manufacturing
ACELP	Algebraic CELP		
ACK	Acknowledgment	CBQ	Class-Based Queuing
ADM	Add-Drop Multiplexer	CBR	Constant Bitrate
ADPCM	Adaptive Differential PCM	CC	Continuity Check
AES	Advanced Encryption Standard	CCS	Common Channel Signaling
AM	Amplitude Modulation	CDMA	Code Division Multiple Access
AMR-WB	Adaptive Multirate Wideband	CE	Customer (or Subscriber) Edge
ANSI	American National Standards Institute	CEF	Cisco Express Forwarding
		CELP	Code Excited Linear Predictive
APD	Avalanche Photodiode	CFM	Continuity Fault Management
APS	Automatic Protection Switching	CID	Channel Identity
ARQ	Automatic Repeat Request	CIF	Common Intermediate Format
ASE	Amplified Spontaneous Emission	CIR	Committed Information Rate
		CMF	Cymomotive Force
ASIC	Application-Specific Integrated Circuit	CML	Chirp-Managed Laser
		CPA	Copolarized Path Attenuation
ASK	Amplitude Shift Keying	CPE	Customer Premises Equipment
ASN	Autonomous System Number	CPM	Cross-Phase Modulation (also XPM)
ATH	Absolute Threshold of Hearing		
ATM	Asynchronous Transfer Mode	CRC	Cyclic Redundancy Check
ATOM	Any Transport over MPLS	CR-LDP	Constraint Routed LDP
AU	Administrative Unit	CSRZ	Carrier-Suppressed Return to Zero (Modulation)
AVC	Advanced Video Coding		
AWG	Arrayed Waveguide Grating	C-VLAN	Customer VLAN
		cw	Continuous Wave
BCB	Backbone Core Bridge	CWDM	Coarse Wavelength Division Multiplexing
BCD	Binary Coded Decimal		
BCH	Bose-Chaudhuri-Hocquenghem (FEC code)		
		DBS	Digital Broadcasting by Satellite
BEB	Backbone Edge Bridge		
BER	Bit Error Rate	DC	Direct Current
BGP	Border Gateway Protocol	DCE	Data Communication Equipment
BOD	Bandwidth on Demand		
BPDU	Bridge Protocol Data Unit	DCF	Dispersion-Compensating Fiber
BRI	Basic Rate Interface	DCM	Dispersion-Compensating Module
B-VID	Backbone VLAN Identifier		

DCS	Digital Cross-Connect System	ESCON	Enterprise System Connection
DCT	Discrete Cosine Transform	ESP	Ethernet Switched Path
DE	Discard Eligible	ETDM	Electrical Time Division Multiplexing
DES	Data Encryption Standard		
DFE	Decision Feedback Equalizer	ETSI	European Telecommunications Standards Institute
DGD	Differential Group Delay		
DiffServ	Differentiated Services	EVC	Ethernet Virtual Connection
DLCI	Data Link Connection Identifier	EVPLAN	Ethernet Virtual Private LAN
DPCM	Differential Pulse Code Modulation	FAS	Frame Alignment Signal
DPSK	Differential Phase Shift Keying	FAW	Frame Alignment Word
DRT	Diagnostic Rhyme Test	FBG	Fiber Bragg Grating
DSBSC	Double-Sideband Suppressed Carrier	FCS	Frame Check Sequence
		FDDI	Fiber Distributed Data Interface
DSCP	DiffServ Code Point		
DSF	Dispersion-Shifted Fiber	FDMA	Frequency Division Multiple Access
DSL	Digital Subscriber Line		
DSLAM	DSL Access Multiplexer	FEC	Forward Error Correction
DSP	Digital Signal Processing (or Processor)	FEC	Forwarding Equivalence Class
		FFE	Feed-Forward Equalizer
DSSS	Direct Sequence Spread Spectrum	FHSS	Frequency Hopping Spread Spectrum
DTE	Data Terminal Equipment	FICON	Fibre Channel Connection
DVB	Digital Video Broadcast	FIFO	First-in-First-out
DVB-ASI	Digital Video Broadcast–Asynchronous Serial Interface	FM	Frequency Modulation
		FOMA	Freedom of Mobile Multimedia Access
DWDM	Dense Wavelength Division Multiplexing	FOT	Frequency of Optimum Transmission
DXC	Digital Cross-Connect		
DXS	Digital Cross-Connect System (also DCS)	FQ	Fair Queuing
		FRA	Fiber Raman Amplifier
		FRAD	Frame Relay Access Device
EBGP	External Border Gateway Protocol	FSK	Frequency Shift Keying
		FSR	Feedback Shift Register
ECC	Error Control Coding	FTM	FEC-to-NHLFE Map
EDC	Electronic Dispersion Compensation	FTP	File Transfer Protocol
		FWM	Four-Wave Mixing
EDFA	Erbium-Doped Fiber Amplifier		
EDWA	Erbium-Doped Waveguide Amplifier	GBE	Gigabit Ethernet
		GEF	Gain-Equalizing Filter
EHI, EXI	Extension Header Identifier	GF	Galois Field
EIR	Excess Information Rate	GFP	Generic Framing Procedure
EIRP	Effective (or Equivalent) Isotropically Radiated Power	GFP-F	Framed GFP
		GID	Group Identification
E-LAN	Ethernet LAN	GMPLS	Generalized Multiprotocol Label Switching
EMP	Electromagnetic Pulse		
EMRP	Effective Monopole Radiated Power	GPDF	Gaussian Probability Density Function
EPL	Ethernet Private Line	GPON	Gigabit PON
EPLAN	Ethernet Private LAN	GPS	Global Positioning by Satellite
ERO	Explicit Route Object	G/T	Gain-to-Noise Temperature
ERP	Effective Radiated Power	GVD	Group Velocity Dispersion

HD	High Definition		MAC	Media Access Control
HDLC	High-Level Data Link Control		MAS	Multiframe Alignment Signal
			ME	Maintenance Entity
HDTV	High Definition TV		MEF	Metro Ethernet Forum
HDV	High Definition Video		MEN	Metro Ethernet Network
HEC	Header Error Control		MEG	Maintenance Entity Group
HF	High Frequency		MEP	MEG End Point
HO	Higher Order		MF	Medium Frequency
HRP	Horizontal Radiation Pattern		MFAS	Multiframe Alignment Signal
HSSG	High Speed Study Group		MFI	Multiframe Indicator
HTTP	Hypertext Transfer Protocol		MIMO	Multiple Input Multiple Output
			MIP	MEG Intermediate Point
IBGP	Internal Border Gateway Protocol		MLSE	Maximum-Likelihood Sequence Estimation
ICMP	Internet Control Message Protocol		MOS	Mean Opinion Score
			MPEG	Moving Picture Experts Group
IDF	Inverse-Dispersion Fiber		MPLS	Multiprotocol Label Switching
IGP	Interior Gateway Protocol		MPOA	Multiprotocol over ATM
ILM	Incoming Label Map		MSE	Mean Squared Error
IntServ	Integrated Services		MSIP	Multi-Service Intranet Platform
IP	Internet Protocol		MSOH	Multiplexer Section Overhead
IPTV	Internet Protocol Television		MSSI	Mid-Span Spectral Inversion
ISDN	Integrated Services Digital Network		MS-SPRing	Multiplex Section Shared Protection Ring
ISI	Intersymbol Interference		MST	Member Status Field
I-SID	Service Instance Identifier		MST	Multiple Spanning Tree
ISSU	In-Service Software Upgrade		MUF	Maximum Usable Frequency
IVL	Independent VLAN			
			NA	Numerical Aperture
LACP	Link Aggregation Control Protocol		NACK	Negative Acknowledgment
			NDF	Negative-Dispersion Fiber
LAG	Link Aggregation Group		NG-SDH	Next Generation SDH
LAN	Local-Area Network		NHLFE	Next-Hop Label-Forwarding Entry
LANE	LAN Emulation			
LAPS	Link Access Protocol SDH		NLSE	Nonlinear Schrödinger Equation
LCAS	Link Capacity Adjustment Scheme		NNI	Network Node Interface
			NMS	Network Management Systems
LDP	Label Distribution Protocol		NRZ	Non-Return to Zero
LDPC	Low-Density Parity Check		NZDSF	Non-Zero Dispersion-Shifted Fiber
LE	Local Exchange			
LEAF	Large Effective Area Fiber			
LED	Light Emitting Diode		OADM	Optical Add-Drop Multiplexer
LFSR	Linear Feedback Shift Register		OAM	Operations, Administration, and Maintenance
LMI	Local Management Interface		OAM&P	Operations, Administration, Management, and Provisioning
LO	Lower Order			
LPF	Linear Predictive Filter			
LSP	Label Switch Path		OC	Optical Carrier
LSR	Label Switch Router		ODU	Optical Data Unit
LTE	Line Terminating Equipment (or Line System Terminal)		OFDM	Orthogonal Frequency Division Multiplex
LUF	Lowest Usable Frequency		OOK	On-Off Keying

OSI	Open Systems Interconnection	QAM	Quadrature Amplitude Modulation
OSPF	Open Shortest Path First		
OTN	Optical Transport Network	QCIF	Quadrature Common Intermediate Format
		QMF	Quadrature Mirror Filter
PE	Provider Edge	QoE	Quality of Experience
PAL	Phase Alternating Line	QoS	Quality of Service
PAM	Pulse Amplitude Modulation	QPSK	Quadrature Phase Shift Keying
PBB	Provider Backbone Bridge		
PBB-TE	Provider Backbone Bridging with Traffic Engineering	RBS	Rayleigh Backscattering
PBT	Provider Backbone Transport	RD	Routing Designator
PCD	Polarization-Dependent Chromatic Dispersion	RDF	Reverse Dispersion Fiber
		RF	Radio Frequency
PCM	Pulse Code Modulation	RGB	Red, Green, Blue
PCT	Paired Comparison Test	RIN	Relative Intensity Noise
PDF	Positive-Dispersion Fiber	RMS	Root Mean Square
PDG	Polarization-Dependent Gain	RPDF	Rectangular Probability Density Function
PDH	Plesiochronous Digital Hierarchy		
		RPR	Resilient Packet Rings
PDL	Polarization-Dependent Loss	RS	Reed-Solomon (Error Correcting Codes)
PDU	Protocol Data Unit		
PE	Provider Edge	RSA	Rivest-Shamir-Adleman (Public Key Encryption Algorithm)
PFI	Payload FC Indicator		
PHB	Per-Hop Behavior	RS-ACK	Resequence Acknowledgement
PHP	Penultimate Hop Popping	RSOH	Regenerator Section Overhead
PHY	Ethernet Physical Layer	RSTP	Rapid Spanning Tree Protocol
p-i-n	Positive-Intrinsic-Negative (semiconductor diode)	RSVP	Reservation Protocol
		RSVP-TE	Reservation Protocol with Traffic Engineering
PLC	Packet Loss Concealment		
PLL	Phase Lock Loop	RTP	Real-Time Transport Protocol
PM	Phase Modulation	RZ	Return to Zero
PMD	Polarization Mode Dispersion		
PN	Pseudonoise	SAN	Storage Area Network
PNNI	Private/Public Network–Network Interface	SBC	Subband Coding
		SBS	Stimulated Brillouin Scattering
POH	Path Overhead	SDH	Synchronous Digital Hierarchy
PON	Passive Optical Network	SECAM	Sequential Color and Memory
POP	Point of Presence	SED	Single Error Detecting
POTS	Plain Old Telephone Service	SID	Service Instance Identifier
PPM	Pulse Position Modulation	SIF	Source Input Format
PPP	Point-to-Point Protocol	SLA	Service Level Agreement
PQ	Priority Queuing	SMF	Single-Mode Fiber
PRBS	Pseudorandom Binary Sequence	SNR	Signal-to-Noise Ratio
		SOA	Semiconductor Optical Amplifier
PRI	Primary Rate Interface		
PSCF	Pure Silica Core Fiber	SOH	Section Overhead
PSFT	Per-Site Forwarding Table	SONET	Synchronous Optical Network
PSK	Phase Shift Keying	SoP	State of Polarization
PSP	Principal State of Polarization	SPE	Synchronous Payload Envelope
PSTN	Public Switched Telephone Network	SPM	Self-Phase Modulation
		SQ	Sequence Number
PTI	Payload Type Identifier	SRS	Stimulated Raman Scattering
PVC	Permanent Virtual Circuit	SS7	Signaling System 7

SSB	Single Sideband (Modulation)		UNI	User Network Interface
SSBSC	Single-Sideband Suppressed Carrier		UPI	User Payload Identifier
STBC	Space-Time Block Codes		VBR	Variable Bit Rate
STM	Synchronous Transport Module		VC	Virtual Container
STP	Spanning Tree Protocol		VCAT	Virtual Concatenation
STS	Synchronous Transport Signal		VCEG	Video Coding Experts Group
STTC	Space–Time Trellis Codes		VCG	Virtual Container Group
SVC	Switched Virtual Circuit		VCI	Virtual Circuit Identifier
S-VLAN	Service VLAN		VCO	Voltage Controlled Oscillator
			VHF	Very High Frequency
TCC	Turbo Convolutional Code		VLAN	Virtual Local Area Network
TCM	Trellis-Coded Modulation		VLC	Variable Length Coding; Variable-Length Coder
TCP	Transmission Control Protocol			
TDFA	Thulium-Doped Fiber Amplifier		VLF	Very Low Frequency
			VoIP	Voice over Internet Protocol
TDM	Time Division Multiplexing		VPI	Virtual Path Identifier
TE	Traffic Engineering		VPLS	Virtual Private LAN Service
TEC	Thermo-Electric Cooler		VPN	Virtual Private Network
TEC	Triple Error Correcting		VPWS	Virtual Private Wire Service
TEI	Terminal Endpoint Identifier		VRML	Virtual Reality Modeling Language
3GPP	Third-Generation Partnership Project			
			VRM-WB	Variable-Rate Multimode Wideband
TMF	TeleManagement Forum			
TMN	Telecommunications Management Network		VRP	Vertical Radiation Pattern
			VSB	Vestigial Sideband Modulation
ToS	Type of Service		VSWR	Voltage Standing Wave Ratio
TPC	Turbo Product Code		VT	Virtual Tributary
TPDF	Triangular Probability Density Function			
			WAN	Wide Area Network
TRDF	Triangular Probability Density Function		W-CDMA	Wideband Code Division Multiple Access
TT&C	Telemetry, Tracking, and Command		WDM	Wavelength Division Multiplexed
TTF	Transport Termination Function		WFQ	Weighted Fair Queuing
TTL	Time-to-Live		WRR	Weighted Round-Robin
TU	Tributary Unit			
TUG	Tributary Unit Group		XOR	Exclusive OR Function
			XOT	X.25 over TCP
UDP	User Datagram Protocol		XPD	Cross-Polarization Discrimination
UHF	Ultra-High Frequency			
UMTS	Universal Mobile Telephone System		XPM	Cross-Phase Modulation (also CPM)

Index

429

Milton Keynes UK
Ingram Content Group UK Ltd.
UKHW051847071024
449327UK00025B/1885